Variability in Human Performance

Variability in Human Performance

Thomas Jay Smith
Robert Henning
Michael G. Wade
Thomas Fisher

CRC Press
Taylor & Francis Group
Boca Raton London New York

CRC Press is an imprint of the
Taylor & Francis Group, an **informa** business

CRC Press
Taylor & Francis Group
6000 Broken Sound Parkway NW, Suite 300
Boca Raton, FL 33487-2742

First issued in paperback 2017

© 2015 by Taylor & Francis Group, LLC
CRC Press is an imprint of Taylor & Francis Group, an Informa business

No claim to original U.S. Government works
Version Date: 20140611

ISBN 13: 978-1-138-07602-0 (pbk)
ISBN 13: 978-1-4665-7971-2 (hbk)

Library of Congress Cataloging-in-Publication Data

Smith, Thomas Jay.
   Variability in human performance / authors, Thomas Jay Smith, Robert A Henning, Michael G. Wade, Thomas Fisher.
      pages cm. -- (Human factors and ergonomics)
   Includes bibliographical references and index.
      ISBN 978-1-4665-7971-2 (hardcover : alk. paper)
   1. Performance technology. 2. Variability (Psychometrics) 3. Human engineering. 4. Evaluation. I. Title.

HF5549.5.P37S654 2015
658.3'14--dc23   2014022615

Visit the Taylor & Francis Web site at
http://www.taylorandfrancis.com

and the CRC Press Web site at
http://www.crcpress.com

Contents

Preface

This book addresses both the scientific and the practical implications of human performance variability by providing a broad perspective on how and why such variability occurs across a number of disciplinary domains.

Variability in performance is ubiquitous to the human condition. As applied to either individuals or to human sociotechnical systems, the term has two basic connotations: (1) differences in individual or system performance at any given point in time; or (2) changes across time in performance within or between individuals or systems. All domains of human science—studies of anthropometry, cognitive or physical proficiency, learning, maturation and aging, health status, disease susceptibility, and language, to name just a few—are grounded in the investigation of performance variability. In practical terms, analysis of performance variability forms the basis of learning and educational assessment, health care, athletic competition, hiring and promotion, organizational effectiveness and productivity, community and/or societal success or failure, indeed entire economic and nation-state systems. Our understanding of how and why evolution occurs dates back to Darwin's original account of the role of performance variability in the emergence of species: descent with modification. With regard to human evolution, phylogenetic origins of human performance variability have been well-documented in the form of evidence for variability in individual and group performance among subhuman species, particularly anthropoid apes.

With its focus on the interaction of human behavior (psychology) with design factors in the performance environment (engineering), the field of human factors/ergonomics (HF/E) is distinctive among human science disciplines in its consideration, in both basic science and practical terms, of the role of design factors in influencing human performance variability. This text will address this role from a number of both individual and human system perspectives.

The approach to human performance variability adopted in the text rests upon the idea of context, or design, specificity in performance; namely, that variability in performance is closely referenced to design factors in the environment in which performance is occurring. Key assumptions of this idea, all of which have strong empirical support, are as follows:

- Behavioral performance is organized as a control systems process, involving closed-loop, reciprocal links between behavioral expression and the environmental context of that expression.
- For a broad variety of modes of motor, cognitive, and learning performance, the preponderance of observed variability in performance is attributable to design factors in the performance environment, a phenomenon termed context specificity, an idea that is entirely consistent with human factors doctrine and systems concepts.

- Context specificity arises as a consequence of three key biological mechanisms: (1) use of behavior to control feedback from design factors in the performance environment, (2) neuronal plasticity, such that as a consequence of such behavioral control, both functional and structural changes in the central nervous system (CNS) occur, and (3) use of behavior to self-regulate energy use in the body.
- Performance variability emerges as an inevitable consequence of context specificity, neuronal plasticity, and biological control systems functioning and limitations.
- Two extensive bodies of empirical evidence, compiled over the past century, support the foregoing ideas: (1) in contrast to innate biological factors or generalized learning factors, task design factors make a much more prominent contribution to observed variability in behavior and performance, and (2) given the first point, the nature and extent of performance variability during interaction with a new design cannot necessarily be predicted in advance.
- The foregoing assumptions can also be readily extended to an analysis of the nature and mechanisms of performance variability in complex human sociotechnical systems.

It is fair to say that the context specificity and control systems perspectives outlined above, in terms of their implications for understanding performance variability, have not received a great deal of research and practical emphasis by the HF/E community. This provides a rationale for the present book. There are few HF/E texts dealing in a dedicated manner with the topic of human performance variability. The neurobiological and biomechanical basis of movement variability (i.e., the kinesiology of performance variability) is better understood than any other mode of performance variability—this topic is addressed in Chapter 2. Empirical evidence for context specificity in cognitive performance and learning as a basis for cognitive performance variability, is extensive and of long duration, and is addressed in Chapter 3. Evidence for context specificity in student learning (Chapter 4), under displaced feedback conditions (Chapter 5), in human error behavior (Chapter 6), in affective performance (Chapter 7), in social and team performance (Chapter 8), and in work performance as influenced by complex sociotechnical systems (organizations, institutions, and entire economies and nation-states) (Chapter 9), as a basis for performance variability observed in these domains, is less extensive. However, control systems concepts have been applied to an interpretation of the nature and basis of performance variability in all of these domains. Following an analysis of performance variability inherent to fracture-critical systems (Chapter 10), the book concludes by providing an evolutionary perspective on the origins and behavioral significance of human performance variability (Chapter 11).

Acknowledgments

We would like to acknowledge the support and encouragement of our wives—Jill, Sally, Martha, and Claudia in developing this book.

We acknowledge the contributions of the following students, Cory Jones and Jessica Atlas, in assisting Prof. Wade in developing Chapter 2.

A number of the ideas in this book stem from the insights of Karl U. Smith, whose inspiration—in memoriam—we gratefully acknowledge.

Authors

Thomas J. Smith has research and teaching experience and funding support encompassing many areas of human factors/ergonomics, including human performance variability, educational ergonomics, human error and hazard management, occupational ergonomics, occupational health and safety, surface transportation, occupancy quality, patient safety, work physiology, kinesiology, and ergonomics certification systems. He is a research associate with the School of Kinesiology, University of Minnesota. He is a Certified Human Factors Professional with over 100 publications. His honors include serving as senior editor for a recent special issue of *Ergonomics in Design* dealing with the topic of globalization of ergonomics, serving as chair of the Professional Standards and Education Standing Committee for the International Ergonomics Association (IEA), originator and committee member for the IEA K.U. Smith Student Award, editorial board member for the journal *Theoretical Issues in Ergonomics Science,* director and past president of the Human Factors and Ergonomics Society (HFES) Upper Midwest Chapter, general chair for the 2001 HFES Annual Meeting, member of the 2002 State of Minnesota Ergonomics Task Force, and member of the Dakota County (Minnesota) University of Minnesota Extension Committee.

Robert A. Henning is an associate professor of industrial/organizational psychology at the University of Connecticut. He holds degrees from the University of Wisconsin-Madison in psychology (BS), biomedical engineering (MS), and industrial engineering (PhD). He also received three years of postdoctoral training at the National Institute for Occupational Safety and Health (NIOSH) in a fellowship program administered by the National Research Council. He is a board-certified professional ergonomist. Since 2006, he has been an active researcher in the Center for the Promotion of Health in the New England Workplace where he conducts research on programs that integrate workplace health protection and promotion through the participatory design of workplace interventions by front-line employees. Other research areas include social psychophysiology of teamwork, work and rest patterns in computer-mediated work, augmented team cognition, and behavioral toxicology. Dr. Henning has served as a NORA panel reviewer for work-related musculoskeletal disorders, as a human factors reviewer on the Soldier Systems Panel of the National Research Council/NAS, as panel reviewer for the Information Technology Research Program, Collaborative Systems, Division of Information and Intelligent Systems, Directorate of Computer and Information Science and Engineering of the National Science Foundation, and on the NIOSH Peer Review Panel for the National Center for Construction Safety and Health Research and Translation. He also served as the secretary/treasurer and president of Psychophysiology in Ergonomics (PIE), a technical group of the International Ergonomics Association, and as a founding member on the Executive Committee of the Society for Occupational Health Psychology. He has codirected a NIOSH-supported graduate training program in occupational health psychology since 2005.

Michael Wade is professor of kinesiology and a faculty member in the University of Minnesota Center for Cognitive Science He is an internationally recognized scholar who has published extensively in two areas of movement science: developmental change across the lifespan, with an emphasis on individuals with motor difficulties, and a second focus on the effects of aging on motor skill performance. Dr. Wade holds Fellow status in the National Academy of Kinesiology, the American Academy of Mental Retardation, and the Research Consortium of the American Association of Health, Physical Education, Recreation and Dance. He is a past president of the North American Society for the Psychology of Sport and Physical Activity, and holds the Distinguished Scholar Award from NASPSPA.

Thomas Fisher is a professor in the School of Architecture and dean of the College of Design at the University of Minnesota, having previously served as the editorial director of *Progressive Architecture* magazine. With degrees in architecture from Cornell and intellectual history from Case Western Reserve, Dr. Fisher was recognized in 2005 as the fifth most published architecture writer in the United States, with 7 books, 47 book chapters or introductions, and over 325 articles.

1 Introduction

Thomas J. Smith

1.1 OVERVIEW

Variability in performance is ubiquitous to the human condition, and more broadly, to all life (Newell & Corcos, 1993b, p. 1). It represents both the basis and the consequence of evolution—our understanding of how and why evolution occurs dates back to Darwin's original account of the role of performance variability in the emergence of species, descent with modification. Performance variability is manifest in the functioning of all levels of human organization from the molecular to complex sociotechnical systems. All domains of human science—studies of anthropometry, cognitive or physical proficiency, chronobiology, learning, development, maturation and aging, health and well-being, disease susceptibility, language, social interaction, to name just a few—are grounded in the investigation of performance variability. In practical terms, analysis of performance variability leads to our understanding of the distribution of human populations and cultures across the planet and of the relative success or failure of learning and education, of social relationships, of health status and care, of athletic competition, of hiring and promotion, of organizational and institutional effectiveness and productivity, of community and societal functioning, of financial markets, and of entire state, nation, or economic systems. An entire branch of mathematics—statistics—has emerged to confront some of these analytical challenges.

It is relevant to cite earlier perspectives on human performance variability by Dodge (1931), author of one of the seminal modern texts on the topic:

> The psychophysical organism is in a perpetual state of flux. It changes not only from infancy to old age, from decade to decade and from year to year, but also from day to day, from hour to hour, and from moment to moment. (p. 5)
>
> It is possible that the variability of human reaction is not only its most conspicuous characteristic, but also a fundamental condition of intellective consciousness as we know it. (p. 7)
>
> There seems to be a widespread disinclination in healthy individuals to repeat the same behavior pattern in close succession. (p. 9)
>
> The more accurately observations are made the more conspicuous human variability becomes.... The scientific question is not the existence of variability, but how much, under what conditions, and with what consequences. (p. 10)

This chapter lays the groundwork for the remainder of the book by introducing major themes that will be developed in subsequent chapters. The following sections address definitions of the two terms in the book title (Section 1.1.1), the purpose and scope of the book (Section 1.1.2), the basic principles of human factors/ergonomic (HF/E) science (Section 1.1.3), key issues bearing on the properties and purpose of performance variability (Section 1.2), a control systems perspective on performance variability (Section 1.3), and a summary (Section 1.4).

1.1.1 DEFINITIONS

It is useful to review definitions of the two key terms in the title of this book to underscore the point that the meanings of these terms are not necessarily self-evident.

Variability. According to Webster (McKechnie, 1983) the biological definition of the term is, "the power possessed by living organisms...of adapting themselves to modifications or changes in their environment, thus possibly giving rise to ultimate variation of structure or function." This definition carries with it the idea that an intrinsic property of biological variability is context specificity; that is, susceptibility to influence by environmental design factors (see Section 1.1.2).

Modes or manifestations of variability considered in this book, as applied to individuals, groups, or complex sociotechnical systems, thus include differences at any given point in time, changes across time, differences across spatial locations, and/or differences within or across individuals, groups, or systems.

Performance. According to Webster (McKechnie, 1983) this term is defined as (1) the act of performing, (2) the carrying into execution or action, or (3) achievement, accomplishment, operation, or functioning. Perform, in turn, is defined as to act on so as to accomplish or bring to completion, to execute, to do (as a task or process), to carry out, to meet the requirements of, or to fulfill.

Two alternative operational views of the meaning of the term performance are provided by Meister (1995, p. 2) and by Murphy (1996a, pp. 260–261). The former author indicates that

> In these essays, I use the term 'human performance' rather than 'behavior' because the former suggests something more systematic, organized and measurable than the latter. The latter subsumes the former, of course; behavior is too comprehensive a term, because behavior—particularly when it refers to less rational, emotional, attitudinal states—is more appropriate in a non-technological context. This does not mean that behavior as defined above does not occur in a technological context, but in that context it should be viewed as a form of human performance.

The contrasting view of the latter author is as follows:

> *Performance* is synonymous with *behavior*. It is something that people actually do, and it can be observed... Performance is *not* the consequence or result of action; it is the action itself. Admittedly, this distinction is not always observable (for example, cognitive behavior, as in solving a math problem) and can sometimes be known only by its effects (for instance, producing a solution after much 'thought'). However, solutions, statements, or answers produced as a result of covert cognitive behavior and totally under the control of the individual are included as actions that can be defined as performance.

The present author adopts the view of Murphy that performance and behavior mean the same thing. However, the distinction that both of these authors draw between observable/measurable versus covert/less measurable behavior is dismissed in favor of the assumptions of both ecological psychology (Flach & Hancock, 1992; Flach et al., 1994; Gibson, 1966, 1979; Gibson & Pick, 2000; Jagacinski & Flach, 2003, pp. 1–7; Michaels & Beek, 1995; Neumann & Sanders, 1996) and behavioral cybernetics (K.U. Smith, 1972; T.J. Smith, 1993; K.U. Smith & Smith, 1966; T.J. Smith & Smith, 1987a, 1988a) that action, perception and cognition are inextricably, reciprocally linked (disclosure: K.U. and M.F. Smith are my parents). K.U. Smith and Smith (1966) introduce the term *homeokinesis* (p. 471) to underscore the view that it is through control of motor behavior that control of all modes and manifestations of behavioral expression is mediated. Recent empirical support for this view has been reported (Grens, 2013). A quote by Ahissar cited in this report encapsulates this idea: "If perception starts anywhere, it starts with the motor movement and not with the sensory reception." The assumption that control of motor behavior underlies all modes of behavioral expression is why Chapter 2 leads off the remainder of this book by addressing the kinesiology of human performance variability.

1.1.2 Purpose and Scope

This book is concerned primarily with *sources of* (i.e., factors contributing to) human performance variability and, less prominently, to its nature and characteristics. Earlier and contemporary books on the topic have tended to focus on the latter approach (Dodge, 1927, 1931; Harris, 2006; Miles, 1936; Newell & Corcos, 1993a). Moreover, in contrast to other works on the topic, the areas addressed here are as broad as the scope implied by the term "human performance" itself. In particular, different areas/dimensions of human performance variability are considered in the following chapters: Chapter 2, the variability in human motor and sports performance; Chapter 3, variability in cognitive and psychomotor performance; Chapter 4, educational ergonomics—variability in learning performance; Chapter 5, variability in human performance under displaced sensory feedback; Chapter 6, human error and performance variability; Chapter 7, variability in affective performance; Chapter 8, variability in social and team performance; Chapter 9, variability in human work performance—interaction with complex sociotechnical systems; Chapter 10, variability in fracture-critical systems; Chapter 11, human performance variability—an evolutionary perspective; and Chapter 12, summary and conclusions.

Perspectives on human performance variability adopted in this book rest on the idea of *context specificity* (alternatively termed design specificity) in performance. Context specificity refers to the general observation (documented in the remaining chapters) that variability in performance is closely referenced to design factors in the environment in which performance is occurring (T.J. Smith, 1998).

Goldhaber (2012, p. 2) calls attention to a terminology that distinguishes between the relative roles of context specificity versus innate biological factors in terms of their respective influences on performance variability. Specifically, the term *empiricist* is applied to those who believe that the environmental context (nurture) has a disproportionate influence on performance variability, whereas the term *nativist* is applied to those who believe that biological factors (nature) are more dominant. As Goldhaber points out, (1) these opposing viewpoints essentially are antithetical to one another, and (2) the intellectual and scientific tension between the two camps dates back at least to the nineteenth century if not earlier, and persists to this day. These terms are invoked throughout this book to refer to contrasting interpretations of how and why variability exists for the different performance domains under consideration.

It is fair to say that the idea of context specificity in terms of its implications for understanding performance variability merits greater attention on the part of the HF/E community than heretofore has been the case. This provides a rationale for the present book. A comment in Woods et al. (2010, p. 5) (with reference to human error) appears to support this observation:

> The results of the recent intense examination of the human contribution to safety and to system failure indicate that the story of 'human error' is markedly complex. For example...the context in which incidents evolve plays a major role in human performance.

There are few HF/E texts dealing in a dedicated manner with the topic of human performance variability. The motor behavioral basis of movement variability (i.e., the kinesiology of performance variability) is better understood than any other mode of performance variability (Chapter 2). Empirical evidence for the influence of context specificity on cognitive and psychomotor performance variability is extensive and long-lasting (Chapter 3). Evidence for context specificity in student learning, performance under displaced feedback, human error, affective performance, social and team behavior, macroergonomic and complex sociotechnical systems behavior, and fracture critical systems performance as a basis for performance variability in these domains is less extensive (Chapters 4–10). However, control systems concepts (Section 1.3) can be applied to an interpretation of the nature and basis of performance variability in all of these domains.

This book may appeal to readers across a broad range of disciplines with an interest in the general question of how and why human performance is influenced by design factors in the performance environment. However, the primary target audience for this book (published as part of the CRC Press Book Series on Human Factors and Ergonomics) is the HF/E community, with the following implications. First, the view of Meister (1995, p. 1) is favored here that there is, "no substantive distinction between Human Factors and Ergonomics, although the two are representative of slightly different traditions." Meister's conclusions (1995, p. 8) as to the essential nature of HF/E (i.e., a definition of the field, recapitulated by Dul et al., 2012) also are endorsed, as follows:

- HF/E is a distinct discipline. Its major antecedents have been psychology and engineering, yet it is not merely a variant of either of these antecedent disciplines.
- The primary factors that differentiate HF/E from psychology are that psychology is concerned with individuals and groups, but HF/E, in addition to

these, emphasizes a focus on the *design* of equipment and systems, whereas psychology has no such focus.
- HF/E is similar to engineering, but whereas engineering is concerned primarily with the functional aspects of technology, HF/E is primarily concerned with the *interaction* of the human with technology.
- As a distinct discipline, HF/E is not constrained by any limitations that may be inherent in the parent disciplines—it is free to consider any and all domains of performance/behavior, and how these interact with design.

The emphasis throughout this book on context specificity as a pervasive source of human performance variability is not meant to discount innate biological contributions such as genetic or physiological mechanisms. However, the rationale for such a context specificity emphasis may be summarized as follows:

- Both the HF/E field (the target audience) and the context specificity perspective feature a common emphasis on the importance of the nature and extent of design-performance interaction as key to understanding sources of variability in human performance.
- Unlike innate biological factors, design factors can be modified to evoke changes in performance (in accord with evidence from context specificity analysis), thus providing a practical and effective avenue for improving human safety, health, well-being and productivity through design interventions. Chapters 4, 6, 8, and 9 expand on these effects in more detail.
- Given this second point, design factors identified in this book contributing to adverse or debilitating modes of performance variability point to possible design intervention strategies that may mitigate or reverse such adverse effects, resulting in more positive performance outcomes.

1.1.3 Key Principles of HF/E Science

The purpose of this section is to elaborate on and extend the idea, introduced above, of context specificity in performance by articulating a set of key principles of HF/E science (T.J. Smith, 1993, 1994a, 1998). These principles offer a framework for the analysis of human performance variability provided in this book.

Principle 1: Performance and design are interdependent. In a general sense, one may argue that the status of HF/E as an integral scientific discipline rests upon the validity of this principle. If the nature and extent of the link between human behavior and performance and design characteristics of the environment within which performance occurs is weak (inconsistent or coincidental), then design and behavior may be dissected from one another, and HF/E becomes subordinate to the disciplines of engineering and psychology. Conversely, if there is functional interdependence between design and performance, such that neither can be realistically assessed in isolation from the other, then HF/E assumes a much stronger role as a distinct scientific discipline (as Meister [1995] and Dul et al. [2012] cited above and others have argued), inasmuch as neither engineering nor psychology traditionally have concerned themselves with design-performance interaction.

Principle 1, then, is the principle of context specificity, which assumes that the preponderance of observed variability in performance is attributable to design factors in the performance environment. Figure 1.1 represents a highly simplified depiction of this idea, with its suggestion that performance and design are reciprocally feedback integrated.

Support for this principle rests on at least six lines of empirical evidence, addressed in subsequent chapters, showing that variability in performance is specialized in relation to design of the performance environment: (1) variability in task performance is highly context-specific (Chapter 3); (2) three of the most robust "laws" in psychology—namely Fitts' law, Hick's law, and the law of training—each embody a highly reliable predictive relationship between a particular type of behavioral expression (movement time, reaction time, and the nature of incremental learning, respectively) and design conditions in which these behaviors are expressed (Chapter 3); (3) student learning performance is prominently influenced by design factors in the learning environment (Chapter 4); (4) displacements in sensory feedback originating with interface design factors cause immediate and dramatic decrements in human-machine performance (Chapter 5); (4) many occupational accidents, and more broadly the occurrence of human error or system failure in many different contexts, are linked to flaws in workplace or system design (Chapters 6 and 10); (5) variability in different modes of affective performance (emotion, motivation, attitude, etc.) has been shown to be context-specific (Chapter 7); and (6) social, team, organizational and complex sociotechnical system performance, and the design of environments in which such performance occurs, have been shown to be interrelated (Chapters 8 and 9).

Collectively, this evidence indicates that behavioral performance, by various analytic criteria, is context-specific. The contribution of design factors to total performance variance ranges from about 50 percent (some cognitive tasks, occupational accidents) to 90 percent or higher (on-the-job tasks; Fitts' law) (T.J. Smith et al., 1994a). Design-specific spatial or temporal displacements in sensory feedback cause immediate performance decrements ranging from two- to twelve-fold or higher. It appears that our performance is closely calibrated to specific design features of the environment in which performance occurs.

Principle 2: Behavior is organized as a control systems process. That performance is specialized in relation to design characteristics of the performance environment suggests that performance variability is feedback related or integrated with design in a systematic way, involving closed-loop, reciprocal links between behavioral expression and the environmental context of that expression. This is what is meant by "control systems process."

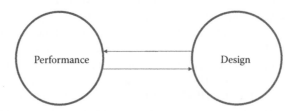

FIGURE 1.1   Basic scheme of performance-design interaction.

Figure 1.2 depicts this concept in more detail, extending the simple scheme in Figure 1.1 to a generic human-machine system. Arrows in the diagram symbolize closed-loop, feedback relationships linking patterns and levels of operator and organizational performance with organizational, psychosocial, and job, task, and interface system design factors. The cybernetic scheme shown in Figure 1.2 assumes that interdependence of performance and design involves three distinct modes of feedback integration. First, managerial and technological requirements and constraints mandate that organizational, psychosocial, job, task, and interface design features of the system be functionally interrelated (left side of the figure). Similarly, the performance of individual employees or operators of the system is feedback coupled with that of the overall organization (right side of the figure). Finally, individual and organizational performance becomes specialized through reciprocal feedback links with various system design factors.

There are three major implications of the cybernetic scheme shown in Figure 1.2: (1) there are not strict cause and effect relationships among the design and performance attributes shown in the figure, in that each serves simultaneously as both cause and effect of the others (Jagacinski & Flach, 2003, Figure 1.1), (2) consequently, design factors both originate from, and (3) in turn feedback influence the behavior and performance of the system.

Given that design affects performance (point 3), to what degree does performance, in a reciprocal manner, affect design (point 2)? In essence, the field of HF/E emerged as a consequence of scientific interest in this question. Because engineering design is a product of performance, it may be argued that specialization in the former is a product of specialization in the latter. More broadly, human history and prehistory are marked by a sustained and ongoing effort to customize the designs of technological, environmental, and organizational systems to meet human needs. The term "human factors compliance" gets at this idea, in the sense that successful realization of this effort since the dawn of the species has involved the development and application of designs, both functional and usable, that are aligned with (compliant

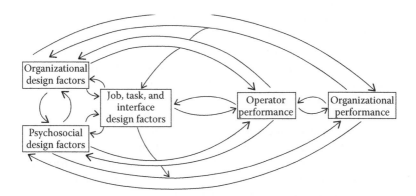

FIGURE 1.2 System performance-design interdependence. Individual and organizational behavior both are directed at closed-loop control of sensory feedback from system design factors, resulting in context specificity in performance.

with) the capabilities and limitations of human behavior and performance. From this perspective, viewed as the practice of integrating design and performance to support human well-being, progress, and survival, human factors may be considered as an essential aspect of the human condition (Flach, 1994) that played a fundamental role (long before its recognition as a formal discipline) as both a product and a determinant of human evolution (K.U. Smith & Smith, 1966, pp. 52–56).

Principle 3: The unit of analysis is the human-machine system. This principle is a logical extension of Principle 1. If performance and design are interdependent, neither can be understood without reference to the other. The appropriate analytical framework is human-machine or human-environment interaction considered as a context-specific, coupled system (Flach & Hancock, 1992). The principle is based on the concept that interdependence of performance and design arises as a biological consequence of the cybernetic properties of behavior (Principles 2 and 6); that is, human self-control of behavioral organization and expression (K.U. Smith, 1972; T.J. Smith 1993; T.J. Smith & Smith, 1987a). Self-control of behavior is contingent on self-control of environmental conditions that affect behavior, which serves to couple or yoke performance and design together as a feedback integrated, closed-loop system. Skepticism regarding this principle lies at the heart of debate over the scientific and practical utility of human factors. Efforts in psychology to study "pure" behavior, in isolation from the contaminating influence of design factors, define one end of the spectrum (Poulton, 1966). This may explain why engineers at the other end of the spectrum ignore human factors research findings in making design decisions because of their putative lack of relevance to solving real-world problems (Simon, 1987). The legitimacy of the latter approach is called into question by the record of complex systems failures linked to disregard for human factors (Hornick, 1987; Perrow, 1999). Chapanis (1991, 1992) has questioned the validity of both of these diametrically opposed approaches by noting that human factors research should distinguish itself with experimental designs that routinely feature a focus on design-performance interaction.

Principle 4: Better design cannot necessarily be deduced a priori *from performance analysis.* Better design is new design, which inevitably will evoke new modes and patterns of performance variability. As a result, consequences for performance evoked by the introduction of a new design are not necessarily predictable. Instead, as Gould (1990) advocates, better design should be developed iteratively by testing how successive design improvements affect performance. A further implication of this principle is addressed by Moray (1994), who reports on difficulties encountered in teaching design. His conclusion is that design of more and more complex systems is becoming increasingly dominated by context-specific features, and that effective teaching of design may require a contextual approach.

Principle 5: To change behavior, change design. Behavior is self-controlled. One individual cannot directly control the behavior of another. However, behavior does change as design conditions change (Principle 1). Therefore, the best strategy to effect behavioral change is to modify design characteristics of the physical or social environment in which that behavior occurs.

Principle 6: Tailor design to the control capabilities of behavior. Achieving "good" design is a widespread preoccupation of the human factors community.

Good design can make the difference between success or failure in usability, human-machine compliance, product acceptance, product quality, safety, health, and/or company survival. Literally thousands of guidelines for achieving good design have been developed. This principle provides a biobehavioral interpretation of what good design means.

The principle is grounded in behavioral cybernetic theory (K.U. Smith, 1972; T.J. Smith & Smith, 1987a). The theory assumes that performance is guided through execution of movements to control sensory feedback both from design factors and from movements themselves, and to thereby control perception, cognition, and patterning of subsequent movements. The underlying premise is that control and guidance of performance is both neurogeometric and neurotemporal (i.e., dependent on relative displacements in space and time between design- and movement-induced sensory feedback) (K.U. Smith & Smith, 1962).

From this perspective, task-specific variance in performance occurs as an inevitable consequence of the fidelity of dynamic feedback interaction between sensory feedback (design) and sensory feedback control (behavior) in addition to whatever contributions general ability and learning factors also may make. That is, in different contexts of human-environment or human-machine operation, performance and design become coupled as a specialized, interdependent system (Principle 1) through an ongoing process of behavioral control of sensory feedback originating with distinctive design factors of the system. Good design, therefore, may be defined as that giving rise to spatiotemporal feedback that can be effectively controlled through active behavior. Merken (1986) advances a similar viewpoint.

Compelling support for this principle is provided by findings from displaced sensory feedback research (Chapter 5) (K.U. Smith, 1972; K.U. Smith & Smith, 1962; T.J. Smith & Smith, 1987a). They demonstrate that the fidelity of behavioral performance is profoundly degraded by design-related displacements in the spatial or temporal properties of sensory feedback that must be controlled during task execution. This suggests that proficiency and skill in human-machine performance depend on the degree to which the operator can effectively control design-dependent sensory feedback generated across the interface. To ensure safe, effective performance, task and operational design factors must conform to the spatiotemporal imperatives of motor behavioral control.

What guidelines for good design can be deduced from this principle? A primary consideration is that sensory feedback from design factors be available for behavior to control. For example, the participatory approach often is effective because it enables workers to control sensory feedback from job-related decisions or working conditions that affect them (T.J. Smith & Larson, 1991). Conversely, performance difficulty during human-computer interaction often can be attributed to inadequate sensory feedback from the machine (T.J. Smith et al., 1994a).

As noted above, a second important consideration is that sensory feedback from the design environment be compliant with behavioral control capabilities and limitations. Whenever sensory feedback is displaced (i.e., spatially or temporally, modality conflict, magnification or miniaturization, inappropriate physical properties such as low or high extremes of light intensity in visual feedback), behavioral control inevitably is compromised and performance suffers.

A third consideration is that sensory feedback from the design environment supports the tracking control demands of the task or operation. During planning tasks or mobile equipment operation for example, the sensory feedback provided should enable projective or feedforward control of tracking behavior during task execution, in addition to whatever feedback (i.e., compensatory) control may be required. In behavioral cybernetic terms, "human-machine compliance" and "user friendliness" therefore refer to the degree to which properties of sensory feedback generated by the design environment during performance conform to the sensory feedback control capabilities of the operator. Meister (1994) also advocates an emphasis on behavioral considerations in design of complex automated systems.

Principle 7: From a behavioral perspective, the term "generalized performance" is an oxymoron. Performance varies in specialized ways in relation to the design context in which it occurs. Therefore, models of behavior and performance that do not accommodate phenomena of context specificity are irrelevant to HF/E science. Performance models that emphasize use of skilled movement behaviors to mediate interaction with design factors may be deemed most successful in dealing with the reality of context specificity (T.J. Smith, 1998).

Principle 8: Cause and effect work both ways. As performance modifies design, design influences performance. Linear causation models of behavior may be considered at best incomplete, and at worst invalid, in that they fail to deal with the cybernetic reality of reciprocal influence and mutual interdependence between behavioral performance and design.

Principle 9: Context specificity contributes to variability across all modes and patterns of human performance. Given that the role of context specificity in human performance variability has been empirically documented, with reasonable rigor, only for a limited number of performance examples (reviewed in Chapters 3 through 9), this principle must be considered almost entirely speculative. Yet it also is readily falsifiable. This book documents evidence for a role of context specificity in performance variability from the relatively basic level of motor behavior (Chapter 2) to the level of performance of complex sociotechnical systems (Chapter 9). Evidence for a role for context specificity in performance variability at the most basic levels of genetic expression and of evolution also may be cited (Chapter 11). Indeed, I am not aware of any type or mode of performance whose pattern of variability, based on available evidence, can be exclusively attributed to innate biological factors (the next section addresses this point in more detail in relation to the nature-nurture debate). In comments expressed during a panel on performance-design interaction (T.J. Smith, 1994b), Moray (1994) makes a similar point in stating that, "It is very doubtful whether there are any context free laws of behavior…" The broad implication of this principle is that context specificity may be universal across all levels and patterns of performance.

1.2 KEY ISSUES BEARING ON THE PROPERTIES AND PURPOSE OF PERFORMANCE VARIABILITY

This section addresses three issues—the role of variability in behavioral control, individual differences, and prediction of behavioral performance outcomes—all of which are relevant to understanding the properties and purpose of human performance variability. The latter two of these issues have prompted multiple publications in the form of both books and research reports by multiple authors over many decades. Consequently, only relatively brief perspectives on these issues are provided here. As for the role of variability in behavioral control, the past three decades have seen a substantial revision in scientific thinking about this question, prompted largely by motor behavioral research. Chapter 2 will address this topic in more detail.

1.2.1 INDIVIDUAL DIFFERENCES

It seems self-evident to note that individual differences between or within individuals or systems across time or space represent one of the basic phenomena giving rise to variability in human performance. Nevertheless, dealing with individual differences in terms of rigorous scientific inquiry has posed a challenge, and even a conundrum, for behavioral scientists for over a century. This section provides a brief overview of the nature and background of this challenge. Chapter 3 provides a more detailed analysis of how and why empirical investigation into individual differences offers insight into the sources of performance variability.

In his delineation of various examples of individual differences summarized at the beginning of this chapter, Dodge (1931, p. 6) points out that

> Individual variations delayed the development of a science of human nature for a long time. It was a fruitful methodological advance when psychology rescued individual differences from the scrap heap of scientific anomalies and began to study them. At the present time surfaces of frequency and the relative positions of individuals in continuous series express our knowledge of human nature vastly better than the hypothetical "average man" of a few years ago.

Meister (1995, p. 79) notes that in the first two decades of the twentieth century, American psychologists quickly began applying theory to practice in such areas as the study of individual differences, suggesting perhaps an even earlier appreciation by psychology in the scientific significance of individual differences.

However, it took over half a century before formal professional bodies dedicated to the study of individual differences appeared, with the establishment of the International Society for the Study of Individual Differences (ISSID) in 1983, and the Individual Differences in Performance Technical Group (IDPTG), a Human Factors and Ergonomics Society (HFES) technical group, in 1991. The research and applied topics addressed by the IDPTG include a broad range of individual difference variables such a performance ability, gender, intelligence and other abilities, education or training level, and anthropomorphic variables (http://tg.hfes.org/idtg/). The ISSID fosters research on individual differences in temperament, intelligence, attitudes, and abilities (http://www.issid.org/about/default.html).

A relatively early view on the somewhat conflicted approach to dealing with individual difference analysis on the part of psychological science is provided by Cronbach (1957). This author calls attention to the distinction between what he termed "experimental psychology" and "correlational psychology." He points out further that researchers in the former camp are interested in only the variability manifest in experimental designs, whereas correlationalists are interested in examining variability among individuals and groups. For experimentalists, he notes (p. 674) that individual differences are "an annoyance," whereas for the correlational psychologist, the question of interest is how the characteristics of individuals determine their mode and degree of adaptation.

It may be argued that the legacy of Cronbach's insight is that a clear consensus has yet to emerge in behavioral science generally, and HF/E particularly, regarding scientific treatment of individual differences. For example, in the first edition of *Handbook of Human Factors*, Christensen (1987, p. 7) calls for greater attention to individual differences based on increased respect for the individual (notably, a decade later, in the third edition of this handbook (Salvendy, 2006), no index entry for individual differences occurs). The sentiment of Christensen subsequently is echoed by Sanders and McCormick (1993, p. 5) who maintain that, "recognition of individual differences in human capabilities and limitations and an appreciation for their design implications…" represents the second "more or less established doctrine that characterizes the human factors profession"; and in panel comments by Hancock (Karwowski, 2003), who notes that

> In the last few decades, we have advocated for 'adaptive' systems…in which individual concerns have been raised to a higher level. I claim this is part of an evolution in which the single individual will be reified in design. The leitmotif might well be 'one size fits none.' Not merely simple adjustments or variations on limited themes, we shall see a true, life-span concern for the particular person.

Despite these noble aspirations for bringing study of individual differences into the mainstream of HF/E research and practice, it appears that personnel selection continues to represent a major focus of applied individual difference research. Thus, Eberts and Brock (1987, pp. 996–997) point out that traditionally, individual differences have been studied with the goal of selecting the right personnel for the appropriate job. Eight examples of individual differences that have been assessed in support of this goal are tabulated. Only four of these—time-sharing ability, information processing style, learning strategy, and foreign language learning approach—are claimed to have practical utility for personnel selection.

Much more extensively, the edited book by Murphy (1996a) documents a major emphasis (in each of the 14 chapters) by industrial and organizational (I/O) psychologists on the use of individual difference evaluation for purposes of personnel selection by organizations. In the first chapter, Murphy (1996b) points out that (1) I/O psychology has a long-standing concern for individual differences, (2) research on the roles of individual differences in understanding behavior in organizations has changed substantially in the last five to ten years, and (3) I/O psychologists are now considering a much wider range of individual difference variables and are linking them to a much wider range of potential outcomes. Yet this author (p. 5) goes on to observe that

> The field of I/O psychology has long exhibited a nearly schizophrenic attitude toward individual differences. On the one hand, the study of individual differences has always been a central component of I/O research and training… On the other hand, I/O psychologists have purposely avoided studying or even thinking about questions that fall outside of a very narrow range of concerns. In particular, I/O psychologists have often

ignored individual differences that have not seemed immediately relevant to predicting individual task performance.

A fitting wrap-up to this record of mixed, and occasionally conflicting, perspectives on how the fields of HF/E and psychology view the scientific significance of individual differences is provided in the panel summary by Karwowski (2003). In introducing the panel, this author notes that

> Too often, individual differences are treated as a nuisance variable, and are either controlled in the study or co-varied out in the statistical analyses of the results. Yet, to truly generate sound and useful human factors guidelines to facilitate the interaction between humans and systems, we need to fully understand how individual differences in aptitudes interact with the varying circumstances found in today's complex technological environments.

In her contribution to this panel, Weaver (Karwowski, 2003) goes on to observe that

> If one examines the human factors research published in the top journals of our profession, these [individual difference] variables are rarely considered as more than nuisance variables. Why [then] examine "individual differences?" It has been argued that individual differences contribute a significant amount of variance to many human factors related situations…and yet relatively few human factors studies make an attempt to systematically investigate these variables. This begs the question: why do most researchers either ignore or minimize the role that individual differences contribute? The answer might relate in part to the difficulty in determining what most people believe represents an "individual difference" variable? Obviously every possible variable cannot be studied in every situation, thus effort should be devoted to the determination of the most relevant variables within a particular context.
>
> Finally, human factors practitioners and researchers might be less willing or feel less capable to incorporate the study of individual differences variables into their work because human factors emphasizes performance. Perhaps we have over-emphasized performance as the dependent variable of interest at the expense of acquiring some general principles about the way that categories of individuals work most effectively relative to other groups under the same circumstances.

As the foregoing summary suggests, it may be argued that the role of individual differences as a central concern of HF/E science remains to be firmly established, However, for purposes of this book, (and in contrast to the suggestion of Weaver (above) that study of performance and of individual differences may not be compatible), an understanding of the nature and extent of individual differences is absolutely essential for delineating the sources of human performance variability. Chapter 3 will detail the empirical findings that address this question.

1.2.2 Performance Prediction

Performance prediction is at the heart of all modern economies in relation to such areas as financial, market, trading, gaming and betting, and investment systems. Collectively, across all such systems, valuations total in the many trillions of dollars.

In other words, performance prediction is integral to the effective functioning of today's economies, societies, states, and nations, indeed of global civilization itself.

In the behavioral domain, performance predication likewise plays a central role in such areas as education and training, personnel selection, athletics, and the design and management of organizations and complex sociotechnical systems.

A simplistic view of the perspectives of HF/E and behavioral science on performance prediction is categorization into the *empiricists* versus the *nativists*, terms introduced in Section 1.1.2 (Goldhaber, 2012). The former approach typically assumes that statistical inference and/or modeling methods can be applied to empirical observations of performance in order to arrive at predictive generalizations about performance variability in selected populations. As discussed above, this approach is at the heart of a disregard for, and sometimes conflicted relationship to, individual differences in much HF/E research. In contrast, the nativists assume that study of individual differences is key to understanding performance variability, yet analytical approaches to this avenue of study remain somewhat problematic (also noted above) even to this day.

Of course, the reality is that predicting the nature and extent of performance variability represents a more nuanced and formidable challenge than the foregoing characterization suggests. One straightforward way of putting it is the well-known saying, "It is difficult to make predictions, especially about the future" (attributed to 29 different sources, http://www.larry.denenberg.com/predictions.html). A more rigorous analysis is provided by Chapanis (1988), in one of the more well-known papers published in the HF/E literature, about generalizing findings from HF/E research. His major conclusions are as follows:

> Three fallacies about generalization are that so-called basic research is more generalizable than applied research, that general findings are immediately useful for design purposes, and that the use of taxonomies increases the generalizability of human factors studies. Some factors that limit generalizability are the use of unrepresentative subjects, insufficient training subjects receive before measurements are begun, inadequate sampling of tasks and situations, inappropriate selection of dependent variables, long-term changes in the world of work, and artifacts attributable to the measurement process itself. In designing a study to predict behavior in a specific application, the guiding principle is similarity. The study should be as similar as possible to the real situation.

Given the emphasis on context specificity in this book, Chapanis' last point bears rephrasing: if the goal of an HF/E study is to predict behavior with reasonable fidelity, the context specificity of the study design should be as close as possible to the design context of the real-world situation or system that the study purports to evaluate.

A key implication of the context specificity perspective that poses another challenge to performance variability prediction is that interaction with new designs of performance environments introduces new modes and patterns of behavioral variability that are not necessarily predictable in advance (this insight dates back well over three centuries to the philosopher John Locke, see Chapter 3). More detailed discussion of empirical evidence in support of this conclusion will be deferred to subsequent chapters. However, it is appropriate to cite the observation of Karwowski (2012, p. 983), namely, "behaviors of complex human–machine systems may be very sensitive to small changes in initial task performance conditions, leading to nonlinear and therefore often unpredictable—that is emergent—system responses."

A notable skeptic regarding the likelihood of making accurate predictions about the future was the economist and planner, Albert O. Hirschman (Adelman, 2013). As Gladwell (2013) puts it, "Hirschman was a planner who saw virtue in the fact that nothing went as planned." In one of his most famous essays, "The principle of the hiding hand," Hirschman (1967) articulated this viewpoint with the following contrarian observation:

> While we are rather willing and even eager and relieved to agree with a historian's finding that we stumbled into the more shameful events of history, such as war, we are correspondingly unwilling to concede—in fact we find it intolerable to imagine—that our more lofty achievements, such as economic, social or political progress, could have come about by stumbling rather than through careful planning… Language itself conspires toward this sort of asymmetry: we fall into error, but do not usually speak of falling into truth.

It is worthwhile to pause here momentarily to call attention to a notable exception to the hubris underscored in the observation of Hirschman that marks the unwarranted confidence of so many economists, politicians, and other private and public decision-makers in their ability to predict the future. In particular, Harry Truman, the 33rd president of the United States and notorious for his blunt and plain-spoken language, is quoted as saying (McCullough, 1992, p. 11), "We can never tell what is in store for us."

Further consideration of the implications of the premise that predicting the future is fraught with difficulty will be limited to a review of an engaging account of a number of examples of performance prediction failures and successes, and their interpretation, provided in a recent book by Silver (2012). Spectacular examples of performance prediction failures cited in the book include the recent "great recession" (the rating agencies indicted as a major culprit), the collapse of the Soviet Union in the late 1980s (no political scientist saw this coming), the lack of preparation on the part of the city of L'Aquila, Italy, for dealing with the effects of a severe earthquake that hit in April 2009 (attributed to complacency), and performance failures of a number of selected professional sports individuals and teams (various causes cited). An example of a prediction success cited by Silver is the collision of Hurricane Katrina with the city of New Orleans, predicted five days in advance by the National Hurricane Center (unfortunately, the warnings were ignored by many New Orleanians, leading to a substantial loss of life).

Based on his analysis of these and a number of other examples, Silver offers the following conclusions about performance prediction:

- If anything, human affairs are becoming less predictable.
- The volume of information is increasing exponentially, but the signal-to-noise ratio may be declining, making prediction more difficult. Better methods are needed to distinguish the two.

- Thinking probabilistically about possible outcomes may benefit prediction.
- How good we think we are at prediction, and how good we really are, may be inversely related. Our bias is to think we are better at prediction than we really are.
- The first twelve years of the new millennium have featured one unpredicted disaster after another. More modesty about our forecasting abilities may make it less likely that our mistakes will be repeated.

Silver's analysis provides a sobering counterpoint to an emerging body of thought that the growing availability of massive amounts of descriptive data pertaining to multiple factors—the so-called "big data" (Mayer-Schonberger & Cukier, 2013)—can be used through correlative analysis to predict the behavior of complex human systems. Brooks (2013) offers a skeptical perspective on this idea.

Interpretation. Although Silver and other observers of performance prediction never use the term, it can be argued that various examples of performance prediction failure, and the interpretation of their underlying causes, can be understood in terms of the key implication of context specificity, namely that new patterns of design (i.e., such as factors contributing to disaster or systems failure) evoke new modes of behavior that cannot necessarily be predicted in advance. Possible strategies for dealing with this conundrum are addressed under Principle 6 in Section 1.1.3.

1.3   A CONTROL SYSTEMS PERSPECTIVE ON PERFORMANCE VARIABILITY

The purpose of this section is to confront two fundamental questions regarding the *nature* of human performance variability. It is hoped that elucidation of these questions, in turn, will provide insight into sources of such variability.

The second of these questions—how and why so many human performance systems at different levels of complexity exhibit nonlinear, deterministic patterns of variability in line with the principles of dynamical systems theory (Harbourne & Stergiou, 2009; Karwowski, 2012; van Emmerik et al., 2004)—is addressed in Chapter 2.

The remainder of this section addresses the first of these questions; namely, why so many different human performance systems featuring dramatically different levels and modes of biological and sociotechnical complexity (elucidated by Karwowski, 2012) exhibit adaptive behavior. Harbourne and Stergiou (2009) suggest that motor behavioral variability is associated with flexibility on the part of the individual to adapt to variations in the environment. The views of West (2013) point to system complexity as a possible answer to this question, but they also underscore the fundamental challenge of predicting complex system behavior (as addressed in Section 1.2.2).

Complexity comes into play when there are many parts that can interact in many different ways so that the whole takes on a life of its own: it adapts and evolves in response to changing conditions. It can be prone to sudden and unpredictable changes—a market crash is the classic example. One or more trends can reinforce other trends in a "positive feedback loop" until things swiftly spiral out of control and cross a tipping point, beyond which behavior changes radically [Chapter 10 elaborates upon the propensity of complex systems for sudden, dramatic failure].

What makes a "complex system" so vexing is that its collective characteristics cannot easily be predicted from underlying components: the whole is greater than, and often significantly different from, the sum of its parts. A city is much more than its buildings and people. Our bodies are more than the totality of our cells. This quality, called emergent behavior, is characteristic of economies, financial markets, urban communities, companies, organisms, the Internet...and the health care system.

The view advocated here is that a control systems interpretation of the behavioral organization of human performance systems across all levels of performance complexity provides insight into this question (Section 1.1.3, HF/E Principle 2). The sections below elaborate this point of view by providing reviews of the historical development of a control systems perspective on human performance (Section 1.3.1), control systems concepts (Section 1.3.2), and the behavioral cybernetics of cognition and adaptive behavior (Sections 1.3.3 and 1.3.4).

1.3.1 Historical Perspective

The basic idea of a control system (the term implies an ability of the system to self-regulate its own behavior) is to link the output signal of the system to the input signal in what is termed a closed-loop relationship, such that variations in the output signal are "fed back" to adjust the input signal in order to bring about some sort of desired modification in output signal variations. In a classical negative feedback control system, the desired output signal modification is to minimize the difference between the output signal level and a target reference level. As Ramaprasad (1983) puts it in his paper, feedback is defined as, "information about the gap between the actual level and the reference level of a system parameter which is used to alter the gap in some way." This author goes on to emphasize that the information by itself is not feedback unless translated into action. The next section introduces a series of control system block diagrams to illustrate these points.

Appreciation of the fact that living systems somehow have the ability to self-regulate their own activities dates back well over a century. Pflüger (1875) provided an explanation for maintenance of steady-state or static conditions in living organisms by asserting, "The cause of every need of a living being is also the cause of the satisfaction of the need." Fredericq (1887) also called attention to the importance of biological self-regulation: "The living being is an agency of such sort that each disturbing influence induces by itself the calling forth of compensatory activity to neutralize or repair the disturbance." These early speculations regarding biological self-regulation were based on the work of Bernard (1865), who emphasized the importance of the internal environment in the establishment and maintenance of stable conditions in the body: "It is the fixity of the *milieu interieur* which is the condition of free and independent life, and all the vital mechanisms, however varied they may be, have only one object, that of preserving constant the conditions of life in the internal environment." This idea was elaborated upon by Cannon (1939), who introduced the familiar term homeostasis.

Contemporary with the advent of these seminal ideas about biological self-regulation were the first studies on the behavioral effects of spatially displaced visual feedback. This early work was one inspiration for the later conceptual and empirical development of the field of behavioral cybernetics (K.U. Smith, 1972), that will be addressed in Sections 1.3.3 and 1.3.4.

It was suggested in Section 1.1.2 that the idea of context specificity merits greater attention on the part of the HF/E community. It is interesting that the claim has been made that attention to control systems theory and analysis, based on the publication record, also has almost dropped off the radar of the contemporary HF/E community. In his survey of the "archeology" of ergonomics, Moray (2008) points out that, "interest in control theory appeared and flourished in the mesoergonomic period [late 1930s–1960], but appears to have become almost extinct in the neoergonomic period [1960–present]."

1.3.2 Control Systems Concepts

As suggested above, self-regulatory control in living systems is mediated by feedback control mechanisms. The term feedback control was used first by electrical engineers to describe automatic control of machine operations. The term referred to the return of a feedback signal of the effects of a machine's output to its input to correct its operation. Recognizing the similarity between such ongoing control in mechanical and living systems, Wiener (1948) coined the term *cybernetics*, from the Greek word "kybernetes" or "helmsman," to refer to the study of feedback-controlled guidance or steering in both living and nonliving systems. In the cybernetic analogy adopted by Wiener, behavioral and physiological mechanisms are seen as closed-loop control systems which, like self-controlled automatic machines utilizing feedback, generate action in following environmental changes and correct or modulate this action in terms of sensed feedback information. Jagacinski and Flach (2003) devote an entire book to the application of control systems theory for different modes of human performance.

The basic alternative to a cybernetic or closed-loop control system is a *top-down* or open-loop control system (Jagacinski & Flach, 2003, pp. 8–10). As depicted in Figure 1.3, with open-loop control there is no feedback from the response to modulate subsequent performance of the system. With behavioral systems, it is well recognized that open-loop control may result in an unstable pattern of performance variability. Jagacinski and Flach (2003, pp. 332–334) describe two examples of what happens when a negative feedback behavioral control system is converted, through experimental manipulation, to open-loop control. In the case of eye movement saccades under negative feedback control, successive saccades toward a stationary target angled off to the right from straight-ahead vision progressively reduce the difference

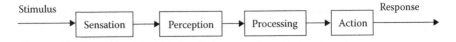

FIGURE 1.3   Block diagram of an open-loop control system.

between eye movement angle and target angle. If the control loop is opened, successive saccades progressively drive the target angle farther and farther to the right.

In the case of muscle contraction mediated (under normal circumstances) by negative feedback control of alpha motor neuron activation, removing the error signal causes muscular convulsion because of overstimulation of the alpha motor neuron.

In Chapters 9 and 10, which deal with variability in complex sociotechnical systems, examples will be cited for which top-down control also is associated with system inadequacy or dysfunction.

Wiener's cybernetic analogy provided the basis for the *servomechanism model* of behavior and physiological function. A servomechanism has three main components: an action (response) mechanism, a feedback (stimulus) detector, and a controller or computer that translates the feedback signal into corrective action. Thus any system functioning as a servomechanism has three main properties: (1) it exhibits goal-directed behavior oriented toward a target or defined path, (2) it detects error by comparing the effect of this action with the true path, and (3) it uses the error signal to redirect the action. This essentially is a negative feedback or compensatory action model, and it is the model emphasized in most human engineering texts (Kantowitz & Sorkin, 1983).

Figure 1.4 illustrates a block diagram model of a classical negative feedback, or servomechanism, control system (Jagacinski, 1977; Jagacinski & Flach, 2003, pp. 10–12). The external input of the system is called the reference signal, otherwise termed the set point. When one or more output variables of the system need to adjust to a certain set point over time, the controller manipulates the inputs to the system to obtain the desired effect on the output of the system. The sensor detects the gap or difference between the output signal and the reference signal and generates an error signal communicated to the controller. The usual objective of a control system is to calculate solutions for the proper corrective action from the controller that results in system stability; that is, the system will hold the set point and not oscillate around it. Thus, a negative feedback system is both self-regulatory and adaptive.

Negative feedback control, as in the servomechanism model, represents only one adaptive mechanism of behavioral and physiological self-regulation. Closed-loop control of behavioral and physiological functioning also includes positive feedback control and positive and negative feedforward control. In positive feedback control, an initial disturbance in a system actuates a series of events that tend to amplify the magnitude of the disturbance. For example, there are a number of metabolic and endocrine pathways in which production of a particular product enhances the subsequent rate of production of that product. It can be shown theoretically and experimentally that,

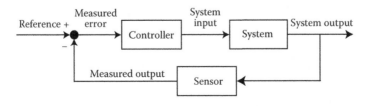

FIGURE 1.4   Block diagram of a closed-loop, negative feedback control system.

left unchecked, positive feedback control systems are inherently unstable and may exhibit wild oscillations. In behavioral and physiological systems, however, positive and negative control mechanisms are customarily coupled so that instabilities are ultimately brought under control.

In feedforward control, also termed pursuit control (Jagacinski & Flach, 2003, p. 71), the detection of an error or disturbance in a system is fed forward to control, either positively or negatively, the state of the system at some future point in time. A behavioral example is visual-manual tracking, such as a hunter shooting a moving target. If one treats the target as static (a negative feedback situation), one would always miss. Instead one uses the present trajectory of the target (the initial disturbance) to project its position in space some one or two seconds into the future and aims for that point instead. The most dramatic example of feedforward control is the ability of humans to control time itself by planning, projecting, scheduling, and predicting (Smith, 1966).

Figure 1.5 illustrates a control system block diagram that combines feedback, feedforward, and adaptive control elements. The feedback control subsystem, comprising the elements on the right side of the figure, is functionally identical to the scheme shown in Figure 1.4. That is, disturbances to the system are referenced against desired goals or targets, and the difference between these two signals is conveyed as an error signal command to a controller. The controller, in turn, actuates a controlled system (denoted as a forcing function in the figure), and the output from the controlled system (denoted as the controlled variable in the figure) is fed back to modulate controller output.

Superimposed on the feedback control subsystem is the feedforward control subsystem, denoted by the feedforward controller block to the left in Figure 1.5. The feedforward controller translates target, goal, and disturbance information into commands that are conveyed both to the feedback controller as well as directly to the controlled system via the feedforward path, bypassing the feedback control subsystem. In this manner the controlled system receives integrated input from both the feedback and the feedforward controller. The latter input enables the entire system to

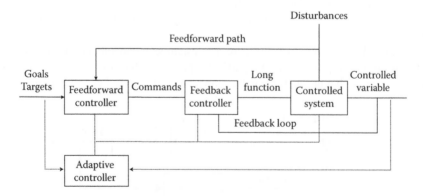

FIGURE 1.5 Block diagram of a closed-loop control system that combines feedback and feedforward control.

function in an anticipatory or projective fashion instead of in a purely compensatory fashion mediated by feedback control alone.

Finally, the adaptive controller is included as the third control subsystem in Figure 1.5 to illustrate the point made earlier that it is control system organization and properties of human performance systems that accounts for their adaptive behavior across a broad range of system complexity (Karwowski, 2012). The adaptive controller operates on a slower time scale than the other two control subsystems and functions by evaluating system performance relative to desired targets and goals (depicted by the two dotted input arrows to the adaptive controller) and, when warranted, altering the functioning of the other two control subsystems (depicted by the dotted output arrows from the adaptive controller).

1.3.3 BEHAVIORAL CYBERNETICS OF ADAPTIVE SYSTEMS BEHAVIOR

The purpose of this section is to introduce behavioral cybernetics as a comprehensive theoretical and empirical framework for the evaluation of human behavior as a closed-loop, adaptive control process. Other investigators, of course, also have provided closed-loop control interpretations of behavior (Annett, 1968; Barton, 1994; Carver & Scheier, 1998; Jagacinski & Flach, 2003; Sheridan, 2004). There are two distinguishing features of behavioral cybernetics however: (1) its emphasis on the role of movement—motor behavior—in mediation through feedback control of the organization and expression of all modes and manifestations of behavior and performance, and (2) its compilation of an extensive body of empirical evidence (some of which is summarized in Chapter 5) in support of point 1.

The implications of point 1 for this book may be summarized as follows: (1) variability is inherent to all modes of motor behavioral expression (Chapter 2), (2) all modes and patterns of performance are feedback controlled through motor behavior (assumption of point 1), and (3) variability in motor behavior therefore likely serves as the substrate for variability observed in different modes of performance.

The guiding force in the development and elaboration of the field of behavioral cybernetics was Karl U. Smith. His wartime observations on the devastating effects on operator tracking performance of temporal delays and spatial displacements in sensory feedback produced by new, semiautomated military hardware provided conceptual inspiration for subsequent study of behavioral feedback control. At the University of Wisconsin-Madison, he started the first postwar, nonmilitary human factors research program concerned with human factors design of work and performance. This program was centered on formulating and applying feedback concepts to research on all aspects of work performance (K.U. Smith, 1972; T.J. Smith, 1993; K.U. Smith & Smith, 1966; T.J. Smith & Smith, 1987a, 1988a). These concepts were believed to found a new theoretical approach to psychology based on human factors principles and were viewed as being opposed to stimulus-response, environmental determination, and open loop information processing doctrines of behavior as applied especially in the area of engineering psychology (Fitts, 1951). When plans for a new research laboratory were drawn up in the late 1950s, the term "behavioral cybernetics" was adopted to designate its research focus.

Figure 1.6 schematically illustrates a central idea of behavioral cybernetics that the organization and expression of behavior is controlled by motor activity (point 1 above).

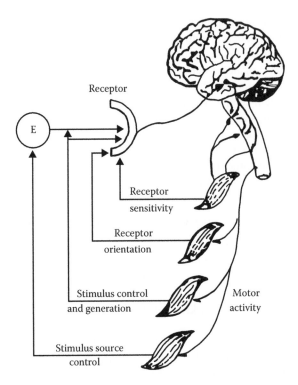

Receptor

E

Receptor
sensitivity

Receptor
orientation

Stimulus control   Motor
and generation   activity

Stimulus source
control

FIGURE 1.6   The motor system as a dynamic closed-loop control mechanism for feedback control of environmental, stimulus source, stimulus modulation, receptor orientation, receptor sensitivity, and neural parameters of perception.

In particular, at least five modes of feedback control by movement (the muscle icons to the right of the figure) are suggested in the figure, namely: (1) neural activation and integration, (2) receptor sensitivity, (3) receptor orientation, (4) stimulus control through modulation of stimulus input, and (5) stimulus source control via receptor selection. Two distinct targets of motor feedback control also are suggested in the figure.

Specific examples of motor activity can be cited that mediate the generalized feedback control effects specified in Figure 1.6. Thus, postural mechanisms serve to effect receptor-efferent neural integration across most if not all patterns of behavioral expression. Tremor helps to maintain receptor sensitivity. Transport movements govern receptor orientation, stimulus intensity at the receptor surface via stimulus modulation, and stimulus pattern formation via receptor selection. In addition, overt motor response can be transformed by tools, machines, and symbolic language—which is human-factored—in order to control interface and energy sources of environmental stimulation.

Figure 1.7 provides one example of how behavioral interaction with a tool increases both the scope and the demands of motor control of sensory feedback from the environment. Specifically, when a tool (the pen in the figure) is deployed by an operator to act upon the environment, three different modes of sensory feedback must be controlled through effector movement to achieve tool use objectives. These modes are: (1) reactive feedback from effector movement, (2) instrumental feedback

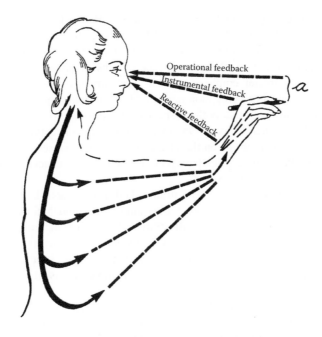

FIGURE 1.7   Sensory feedback control demands inherent to tool use. Use of any tool always entails control of reactive feedback from effector movement and control of instrumental feedback from movement of the tool. Tool use also may modify the external environment in a manner that persists after tool use has ceased, in which operational feedback from this action of the tool also must be controlled.

from tool movement, and (3) operational feedback from the action of the tool upon the environment (the written "a" in the figure) that may persist in time much longer after the duration of the first two feedback control modes. For example, that we can infer certain behavioral attributes of ancient hominids that lived well over 2.5 million years ago based on examination of their stone tools that have survived to this day underscores the unparalleled manner in which our environment has been human-factored over the millennia through use of tools, machines, and systems to generate operational feedback.

An important implication of the motor feedback control schemes in Figures 1.6 and 1.7 is that negative feedback control models are inadequate to deal with the many kinds of closed-loop control interactions that are involved in both primary and integrated levels of systems control in the human body on which significant phenomena of biological adaptation are based. Feedback control takes many systems forms and can be based on many modes of positive adjustment beyond detection of differences or errors. It may be involved in regulation of order, sequence, summing, separation, elaboration, and multiplication of biorhythms, integration, and/or serial interaction. It may interrelate movements, receptor functions, and physiological functioning, often to subserve human-tool or human-machine interaction, all on a positive active basis (homeokinesis), as contrasted with a purely negative feedback homeostatic basis with functional equilibrium as the end goal.

Ecological psychology (Flach et al., 1994; Flach & Hancock, 1992; Gibson, 1966, 1979; Gibson & Pick, 2000; Jagacinski & Flach, 2003, pp. 1–7; Michaels & Beek, 1995; Neumann & Sanders, 1996) shares this idea of an intimate link between *action* (movement) and *perception* (the consequence of sensory stimulation). Behavioral cybernetics places greater emphasis on self-regulatory (as opposed to ecological environmental) control of perception mediated by kinesiological mechanisms. The record indicates, however, that both perspectives have solid theoretical foundations and substantial empirical support. One of the important conceptual tenets of ecological psychology is emphasis on the idea that, given their closed-loop, interactive relationship, action and perception have reciprocal control influences on one another (Section 1.1.3, HF/E Principle 8).

This idea is illustrated in Figure 1.8 (transcribed from Jagacinski & Flach, 2003, p. 5), which depicts two different block diagram control models for understanding action-perception coupling. The top illustration depicts a standard feedback control model with system stimulation via environmental design factors (input) feeding through sensation, perception, decision, and action, resulting in a response signal that feeds back to modulate the input signal. This model is essentially identical to that shown in Figure 1.4.

The bottom scheme in Figure 1.8 turns the top model on its head. With this model, the "input" is action and the "output" is sensation. Yet functionally and operationally, the two models in Figure 1.8 are equivalent, and the control linkages in both of these models parallel those shown in more behavioral detail in Figure 1.6.

Observers in other system domains also have noted that the idea of reciprocal control coupling between input and output represents a viable conceptual alternative

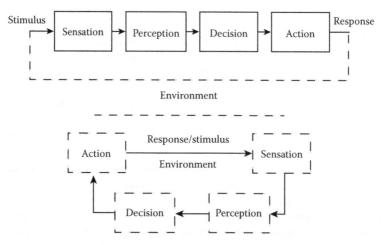

FIGURE 1.8 Two block diagrams of action-perception coupling illustrating the point that the action-perception closed-loop feedback link can be modeled in two ways. The conventional depiction (upper diagram) shows system output on the right side of the model feeding back to control system input on the left side of the model. The lower diagram turns this scheme on its head, with perception feeding back to control action. (Reproduced from Jagacinski, R.J., & Flach, J.M., *Control theory for humans: quantitative approaches to modeling performance.* Mahwah, NJ: Lawrence Erlbaum, Figure 1.1, 2003.)

to traditional cause-effect interpretations of performance variability. For example, as one economist is reported to have said (The Economist, 2013a) in an attempt to delineate the intimate feedback relationships between economic design factors believed to affect economic growth and the reciprocal influence of growth itself on those factors: "Result becomes cause and cause becomes result."

Implications for context specificity. Implicit in all control models of behavior is the role of context specificity in influencing variability in control system behavior. As the control system schemes in Figures 1.3 through 1.7 illustrate, the basic assumption of any control system model is that the behavior of the system is influenced by external environmental design factors; that is, by the context in which the system is operating. In the case of an open-loop system (Figure 1.5), this contextual influence is direct, with no feedback modulation by system output. In the case of closed-loop control systems of different designs (Figures 1.4 through 1.8), some type of feedback control from system output is deployed to modulate the nature and extent of such contextual influence. The control scheme in Figure 1.6 particularizes the generalized design-performance interactive schemes in Figures 1.1 through 1.5 and 1.8 by identifying specific motor feedback control mechanisms that can be deployed to actively and permissively control the influence of environmental design factors on system performance.

A further implication of the system control schemes depicted Figures 1.3 through 1.8 is that the *design* of the control system (open-loop, or negative or positive feedback or feedforward closed-loop) represents a context-specific influence on system performance variability. In summary, these considerations support the conclusion that systems control design characteristics, in and of themselves, represent a major source of variability in human performance.

1.3.4 Behavioral Cybernetics of Cognition

Chapter 3 reviews a substantial body of empirical evidence pointing to the conclusion that the preponderance of variability observed in task performance is context-specific; that is, attributable to design factors in the performance environment. The purpose of this section is to address the question of why this should be the case by providing a behavioral cybernetic perspective on how cognition is controlled.

Cognition in behavioral cybernetic terms emerges as a consequence of active motor control over sensory feedback, a concept that has been delineated from both a behavioral perspective (Figure 1.6; K.U. Smith, 1972; T.J. Smith & Smith, 1987b) and more recently from a neurobiological perspective (Jackson & Decety, 2004). The cybernetic view of cognitive behavior rejects the notions that environmental stimuli or external information sources can independently drive or control cognition or that cognition can occur separately from motor behavior. Attention and perception treated as independent processes under the information processing paradigm (Raley et al., 2004) also are understood in the cybernetic model as interdependent and integrated through motor-sensory control. Seemingly radical in the context of present-day cognitive science, the concept of motor-sensory integration is not new to psychology, having been introduced by functionalists William James and John Dewey over 100 years ago.

Cognition—perception and knowing—manifests itself through a variety of behavioral phenomena, denoted by terms such as thinking, problem-solving, understanding, insight, planning, situation awareness, mental workload, and so forth (T.J. Smith & Henning, 2005). In cybernetic terms, what these different manifestations of cognition have in common is predictive activity, which to us represents the essence of cognition. In other words, for effective guidance of behavior you have to be able to predict the sensory and perceptual consequences of your actions (Blakemore et al., 2001; Hawkins & Blakeslee, 2004). Based on empirical studies of brain function, neuroscientists now believe that such predictive guidance is based on what is termed a *forward model* (Miall et al., 1993), in which memory of sensory feedback from past action (the predictive model) is referenced against real-time sensory feedback from current action (perception) and the model updated (learning) based on any detected discrepancy. Recently, neurophysiological evidence has been reported documenting neuronal populations in the medial frontal cortex whose cognitive control functioning specifically subserves learned error prediction (Brown & Braver, 2005).

Figures 1.9 and 1.10 present two block diagram control schemes that illustrate the basic elements of a forward model. In both models, the idea is to enable the control system to predictively generate appropriate control output rapidly without waiting for possibly delayed feedback from this output. In Figure 1.9, sensory receptors (right side of diagram) integrate input from both external stimuli (i.e., design factors in the performance environment) and from effector movement (system output). Sensory feedback from these receptors, reflecting actual movement performance (state) of the system, is compared (C1) with predicted movement feedback (estimated state) from the forward model (assumed to be functionally generated by neuronal activity in the cerebellum and frontal cortex). The difference between these two sources of feedback serves as an input signal to a second comparator (C2) to be compared

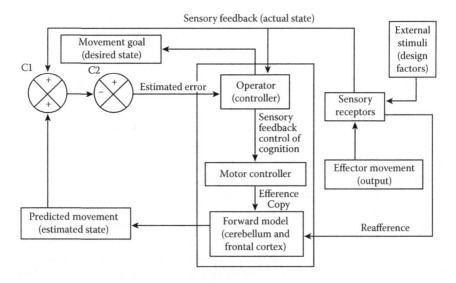

FIGURE 1.9   Block diagram of a forward model.

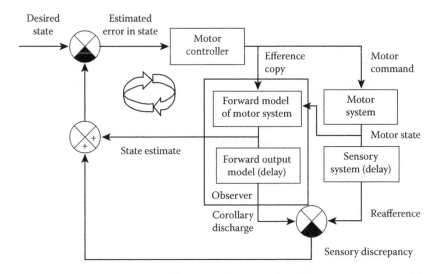

FIGURE 1.10 Block diagram of the cerebellum as a Smith predictor, an illustration of how the forward model might be incorporated into neuronal function to account for the planning, projective, and predictive capabilities of human behavior. (Reproduced from Miall, R.C. et al., *Journal of Motor Behavior, 25*(3), 203–216, Figure 2, 1993.)

with a movement goal signal (desired state) generated by the central nervous system (CNS) controller. The output signal from the forward model is referenced to input (efference copy) from both the motor controller (i.e., the system output or action) and reafference from the sensory receptors (actual state). Referencing sensory feedback from both current action and sensory receptors with predicted or estimated output enables the current state of the forward model (used to generate estimated output) to be updated with new sensory feedback from real-time action of the system. In this manner, the forward model is being continuously primed to reference sensory feedback information acquired from previous design contexts against novel sensory feedback acquired from newly encountered design contexts, thereby refining the predictive capabilities of the model (T.J. Smith & Henning, 2006).

Figure 1.10 is a more elaborate rendition of the forward model based on what is called the Smith predictor. Otto J.M. Smith invented the eponymous Smith predictor in 1957 as a strategy for dealing with delay in feedback control systems, with particular reference to industrial process control (VanDoren, 1996). That is, with its original process control application, the predictor model enables the process controller to estimate and execute appropriate process control moves without waiting for sensory feedback from the process system output to see how each control intervention turned out.

The scheme in Figure 1.10 represents an application of the Smith predictor to the cerebellum, one of whose neural functions is assumed (Miall et al., 1993) to embody a forward model of behavior. The behavioral objective in this case is to enable sensory consequences of motor activity to be predicted. Specifically, the cerebellar forward model is assumed to receive input from both the efferent signal sent to the motor effector (termed efference copy), as well as the signal from the actual output of the motor system (termed the reafference signal). These inputs enable the cerebellum both to predict what sensory feedback from system behavior should be (the state estimate), and to reference this predicted feedback to real-time sensory feedback from motor behavior that actually occurs. If there is a difference in these two sources of feedback (sensory discrepancy), the state estimate of the forward model is updated. Unless sensory feedback from motor activity is novel and relatively uncontrollable (as may be the case for displaced feedback conditions, described in Chapter 5), under customary circumstances the cerebellar forward model depicted in Figure 1.10 enables prediction with reasonable fidelity of the sensory consequences of ongoing motor activity.

This section concludes with a summary of a series of general behavioral cybernetic concepts of cognition based on the foregoing considerations. The first concept, from the perspective of the forward model, is that many cognitive demands arise from the need to control future behavior, a process termed feedforward control, anticipatory control, or projective tracking (Figure 1.7). Unlike compensatory control in which the human operator reacts to sensory feedback about system errors, feedforward control involves the cognitive projection of past memories and associations of feedback control dynamics to anticipate future events and the behavioral requirements for their control so that control actions can occur to prevent behavioral errors from occurring when these events transpire.

The second, related behavioral cybernetic concept of cognition is that effective cognitive performance depends on effective motor behavioral control of sensory feedback, as illustrated in Figure 1.6. Cognitive demands as they are commonly experienced can be understood as resulting from challenges to such feedback control caused by complex sensory environments (control limitations), lack of learning (poor understanding), lack of skill (poor training), poor human factors design, and/or displaced or perturbed feedback that is noncompliant with the control capabilities of behavior. In every case, the cognitive consequence is a compromised ability for projective guidance of behavior. Particularly dramatic manifestations of compromised sensory feedback control emerge under conditions of noncompliant displaced or perturbed sensory feedback from one's own behavior (Chapter 5).

The third behavioral cybernetic concept of cognition is that of context or design specificity in cognitive performance. A major emphasis throughout this book, as noted previously, the essence of this idea is that the preponderance of observed variability in such performance is attributable not to inherent biological factors or learning ability but to design of the task (Section 1.1.3, HF/E Principle 9). In other words, performance cannot be evaluated outside of its context—generalized models of performance have little scientific validity (Section 1.1.3, HF/E Principles 1 and 7).

The fourth behavioral cybernetic concept of cognition is that overt motor behavior and the behavior of neuronal populations subserving cognition collectively represent a closed-loop system mediated by respective functional mechanisms that act as both cause and effect of each other (Section 1.1.3, HF/E Principle 8). In this regard, self-regulation of cognition to support both goal realization and learning is recognized as integral to the development and education of children (Snow et al., 1996, p. 247).

The fifth behavioral cybernetic concept of cognition, derived from the fourth, is that control systems principles and methods (Section 1.1.3, HF/E Principle 9) can be applied to cognition generally (Hommel et al., 2002), and more particularly to the analysis of degradation in cognitive performance observed under high stress and/or complex interactive environments. The latter concept is aligned with the perspective of Sheridan (2004), who views driver distraction—resulting in a reduction in attention to the driving task—as a control systems problem, and applies control theory to its analysis. In the context of high stress and/or complex interactive environments in battlefield situations, we thus assume that compromised cognitive performance of the military operator under high/complex task demand conditions also represents a manifestation of distraction, where we define distraction as a battlefield process or condition that interferes with the attention of the military operator to the task at hand, thereby disrupting behavioral control of the operator (Sheridan, 2004). From this perspective, factors such as operator drowsiness, fatigue, and/or stress are not viewed as sources of distraction per se, but rather as factors that compromise brain activation and thereby alter parameters of cognitive task control such as gain, delay, and variability.

Implications for context specificity. Certainly since the ancient Greeks, and possibly going back much further in prehistory, humans have wondered about how knowledge about our world is acquired. This section introduces a control systems interpretation of cognition based on behavioral cybernetics as one solution to this question. Debate regarding the validity of this perspective relative to other, more popular, HF/E perspectives on cognition (such as information processing or situation awareness theories) will not be pursued here. However, the conclusion that control system design characteristics, in and of themselves, represent a major source of variability in human performance (Section 1.3) suggests that variability in cognitive performance in particular also is critically influenced by cognitive control system design features.

One particular example of this idea merits emphasis. As noted above, a distinctive feature of human cognition, perhaps not unique but certainly much more highly refined in the human species relative to any other subhuman species, is our ability to control time itself (T.J. Smith, 1966). This ability is based on cognitive capacities for planning, projecting, scheduling, and predicting behavior into the future, made possible by feedforward control mediated by the forward model. That cognition is controlled in this manner suggests that cognitive variability is a controlled process. The question is, does control of time through control of cognition imply that cognitive variability is context-specific?

The straightforward answer is that time, as a human construct for purposes of planning, scheduling, and predicting, is a design factor. To be sure, the year and the day reference the earth's rotation around the sun and around its axis, but the month, hour, and minute all are human design constructs. Furthermore, humans have been designing artifacts to measure and keep time for thousands of years. From these perspectives, variability in cognitive performance has distinct context-specific connotations.

1.4 SUMMARY

The purpose of this chapter has been to lay the groundwork for the remainder of the book. To this end, successive sections of the chapter discussed definitions of the terms "performance" and "variability" in the book title (Section 1.1.1), purpose and scope of the book, introducing two central points that the focus will be on sources of performance variability and that context specificity will be emphasized as a major influence on such variability (Section 1.1.2), key principles of HF/E science (Section 1.1.3), two issues bearing on the properties and purpose of performance variability (Section 1.2), and a control systems perspective on performance variability (Section 1.3).

2 Variability in Human Motor and Sport Performance

Michael G. Wade and Thomas J. Smith

2.1 INTRODUCTION

In typical human factors/ergonomics (HF/E) contexts, the human operator functions as the "redundant subsystem" in the person-machine interface. Most systems rely on an engineered operating system with the human operator in a monitoring role, ready to engage control of the system as appropriate. Engineered design seeks to reduce variability and maintain a high level of precision; such is the case for aircraft and ocean-going transportation. What is referred to as *unanticipated variability* (Vincente & Rasmussen, 1992) represents situations that are unfamiliar to the operator and not anticipated by design engineers who created the system. Turvey et al. (1978) refer to this as *context-conditioned variability,* whereby a different set of actions produce the same task goal.

In the sport sciences, and to some degree in the rehabilitation sciences, the HF/E influence has been primarily in equipment design of such things as golf clubs, tennis rackets, and equipment where product design has focused on maximizing the interface between performer and the actual equipment being used. For example, golf club shafts now have different flex characteristics made from both steel and graphite designed to maximize the different swing characteristics of the performer (e.g., swing speed). Club heads have weight distributions designed to optimize both launch angle of the ball and minimizing the effects of off-center shots that do not precisely impact the so-called sweet spot (the center of percussion). In the domain of rehabilitation equipment HF/E has influenced the design of wheelchairs, walkers, and other equipment assistive devices. The influence of HF/E has less of a presence when discussing variability in sport skill performance *per se*, or teaching or retraining of rehabilitation skills. Equipment design minimizes variability of the user by engineering into the design the different features of the performer to accommodate the known variability of the operator (i.e., strength and flexibility). This can improve the outcome of a golf shot or tennis stroke by minimizing errors produced by variation in response execution; one might say therefore that such designs assist in producing a more consistent (less variable) outcome, but have little or no influence on the inherent flexibility of the actor's coordination pattern in producing any specific skilled action, yet can support such performances. It is perhaps interesting to note that the almost incessant technology improvements in sports equipment (especially in golf and tennis) has produced only modest impact on the average performer but has enhanced the performance of the elite athlete. For example, the handicap index for the recreational golf enthusiast has remained essentially constant (18 hole scores of 88–90 for males), but that for elite performers has dropped due to longer drives, more stable ball flight, higher spin rates, and customized launch angles, to name a few. Advanced equipment design has generated higher profits ($4 billion in 2011), from advances in technology (endorsed by elite athletes!) but mixed results with respect to performance increases across the general population of users.

The focus of this chapter is on the motor skills related primarily to sport. When we speak about or describe an individual as "skillful," we imply a performance that appears both unhurried and smooth in its execution and with seemingly a minimum of observable variability as to outcome. It is this characteristic of variability that is the focus of this chapter. In everyday language skilled performance is assumed to be both reliable and consistent, and not given to wide deviations; consistency is the mark of a skillful performance. What we observe in an actual skilled response (e.g., an excellent golf shot or a superb tennis stroke) can differ markedly from the underlying coordination and control parameters that lead to the production of such consistent skillful responses. The skilled output, be it recorded as distance or accuracy, is an outward reflection of an array of coordination dynamics that have combined to produce this observed outcome. When we consider variability in skilled activity we are in fact talking about two separate aspects of this idea. First is the standard (traditional) view, represented by variability about a mean or average value, expressed either as variation between groups of individual performers or as a series of trials between or within individuals. Second is the variability that is present in the coordination and control parameters required to produce the actual skilled output. Variability represented by the first approach (standard deviation about a mean score) is seen as a mark of inconsistency with respect to skilled performance, while the second represents more the performer's capacity for flexibility with respect to a particular performance in an ever-changing array of environmental constraints.

Deviations in motor output around a mean score are viewed as error variance, which in standard statistical analysis is assumed to have a mean of zero and an unknown variance. Variability of this kind is often termed as error or noise in the motor system, represented by seemingly random fluctuations at some location in the system, be it anatomical, mechanical, or physiological. This linear modeling or deterministic perspective of motor variability has both a long research and theoretical history in the study of motor skills, medicine, and physiology.

By way of contrast regarding variability as error or noise is the notion that any skilled action is constrained by the dynamic interplay between the task, actor, and the environment (TAE). Formally proposed by Newell (1986) as a constraints model, skilled activity is viewed in the context of this triad of constraints. Given this assumption, intraindividual variability may be seen less as system noise but more as the individual's attempts to adapt performance to the changing array of constraints represented in the dynamic interplay of TAE. This casts variability in a much more positive light; no longer is it seen as a source of error or noise, but as playing an important functional role in skill performance representing the actor's (operator's) capacity to make ongoing adjustments to the ever-changing array of interactive constraints present in the dynamics of the TAE. This chapter reviews the two different perspectives on variability and the analytical approaches (linear and nonlinear), and the meaning attached to each; how each fits into an overall view of variability and its role in better understanding the expression of human skill, as observed in sporting and other functional contexts. As a starting point, we will consider the role of variability in behavioral control.

2.2   THE ROLE OF VARIABILITY IN BEHAVIORAL CONTROL

Given that skillful performance reflects a high level of stability, the question at issue is whether performance variability compromises or contributes to such stability. As noted above, traditionally the former view has prevailed. However, in the past three decades a growing body of research has emerged that suggests that variability is essential for consistent (stable) control of motor behavior. The late Jack Adams, a major contributor to human factors research and motor skills, once referred to the problem of individual differences as residing "in the cesspool of the error term" (Schmidt, 2011, p. 83). Adams and the majority of the motor skills research community, from the late 1950s and for the ensuing 30 years, regarded individual differences as almost unavoidable variation in performance rather than a mark of ongoing and necessary skill output. The idea that variability may be an important marker rather than an assumed cost of ongoing skilled performance emanates from the earlier insight of Bernstein (1967) that effective coordination of motor behavior requires the behaving system to level effective control of the multiple degrees of freedom manifest in the functioning of different joints, muscles, and motor units whose combined actuation is typically required for the execution of a coordinated behavior (skill). As Harbourne and Stergiou (2009, p. 268) note:

> ...multiple degrees of freedom of the body...combine with external forces during movement to produce countless patterns, forms, and strategies. The redundancy of the system allows for the use of multiple strategies to accomplish any given task. Logically, there are multiple performance variants for each movement...

The implications of this idea—that variability is essential for effective control of systems behavior—is drawn on analyses offered by the following authors: Newell and Corcos (1993a, b), van Emmerik and van Wegen (2000), Hamill et al. (2000), van Emmerik et al. (2004), Stergiou et al. (2006), Harbourne and Stergiou (2009), and Karwowski (2012). Points below offer relevant perspectives on these implications.

- Rather than considering variability as a reflection of the individual's ability (or inability) to accurately predict how degrees of freedom manifest in pattern expression should be controlled, biological systems are thought to self-organize in order to find the most stable solution for producing a given movement. This view is consistent with the observation that many behaviors that appear to be stable, including skilled behaviors, are performed in variable ways. This was first noted by Bartlett (1932) who among others referred to this kind of variation in skilled activity as evidence for what he termed a schema. Variability of this type increases during the establishment of stable, refined behavioral states (Stergiou et al., 2006).

More broadly, the emergence of new perspectives regarding the role of variability in biology has emphasized not only the importance of variability in biological systems processes (van Emmerik & van Wegen, 2000), but also its critical role subserving movement stability. Indeed, as Harbourne and Stergiou (2009) have suggested, variability is inherent in biological systems because it ensures survival. The crux of this idea is that variability permits choices among options, selection of strategies, and flexibility to adapt to variations in the environment, and in so doing optimizes success in both individuals and groups. A more behaviorally adaptive (more variable) individual may successfully challenge a more behaviorally rigid (less variable) individual.

For example, individuals who employ a high degree of variability in cognitive strategies at the beginning of task development have greater learning and eventual success in task performance. Harbourne and Stergiou (2009) also point out that variability exists at many system levels—thus a logical conclusion (in Darwinian terms) is that if variability is inherent to biological systems at multiple levels, and is pervasive across species, it cannot be considered error, but must be linked to survival.

- The introduction of nonlinear dynamics and chaos theory to the study of biological systems has opened the door to new ways to think about the role of variability (Hamill et al., 2000; Harbourne & Stergiou, 2009; Stergiou et al., 2006; van Emmerik et al., 2004).

- There is a strong rationale for applying nonlinear measures of variability (i.e., dynamical systems and/or chaos theory analysis methods) to the analysis of systems behavior, namely: (1) in averaging dependent measures of performance (i.e., linear analysis) collected over several trials, the temporal organization of the behavioral pattern is lost, (2) the application of linear analysis methods assumes that variations between repetitions of a task are random and independent; however, recent studies of many types of system performance have shown that such variations are neither random nor independent, but instead have a deterministic origin and are distinguishable from noise, and (3) linear methods yield different results from nonlinear methods when applied to the analysis of behavioral stability (Stergiou et al., 2006).

- The phenomenon of dynamical instability, first discovered by Poincaré (1900), demonstrated that two nearly indistinguishable sets of initial conditions for the same system can result in two final states or behaviors that differ vastly from each other. Newtonian physics assumes a predictable connection between cause and effect in nature, such that for any given system, the same initial conditions should always lead to identical outcomes.

However, such an assumption does not hold for all living systems, thus invalidating the underlying assumption that it is always possible to make accurate long-term predictions about any physical system (including human systems) as long as the starting conditions are well defined. Extreme sensitivity to initial conditions is called dynamical instability or deterministic chaos (Karwowski, 2012).

The notion of dynamic instability has important implications beyond the consideration of human motor skill. The field of medical science is fast recognizing that the behavior (function) of a variety of physiological systems exhibits variability as a critical feature of stability. Such behavior observed in both heart rate and respiration suggests that it may well be a characteristic of the many behavioral, physiological, and biochemical elements of all living systems. In a recent book entitled *Where Medicine Went Wrong—Rediscovering the Path to Complexity*, author B. J. West (2006, p. 89) writes:

Empirically there developed two parallel, but distinct, views of the physical world. One world was deterministic, in which phenomena are predictable and without error. The second world is stochastic, with uncertain predictions confined to a range of possibilities... What scientists found is that the principle of reductionism, which served so well in the building of machines, did not seem to work so well when biological systems were investigated.

West (2006) has written widely on both the history and fallacies of adhering to a strict reductionist approach to understanding the complex workings of the human body. A complex system requires a type of analysis that recognizes the ongoing interactions between the many subsystems in the human body, This is where a focus on nonlinear science takes both a different analytical approach and uses a different set of underlying assumptions. These contrasting methodologies are now reviewed.

2.3 METHODS FOR ANALYZING PERFORMANCE VARIABILITY

The challenges inherent to analyzing human performance variability discussed above considered the role of variability in behavioral control. This section compares and contrasts the various analytical methods that have been brought to bear for meeting these challenges. The review is somewhat brief and qualitative—the quantitative mathematics and statistics underlying these methods are described in the references cited. The report of Harbourne and Stergiou (2009) is particularly helpful in providing working definitions of relevant analytic terms (c.f., see p. 281), as well as key principles of nonlinear analysis (p. 282).

Although the methods addressed below have been applied to the analysis of variability in a number of modes of performance (Harbourne & Stergiou, 2009; Karwowski, 2012), it is in the study of variability of movement control that these methods have been applied most intensively, with a notable degree of refinement in recent decades.

2.3.1 LINEAR METHODS

As applied to variability analysis, this term refers to traditional, one-dimensional measures of dispersion of a cohort (N) of measured values around the central tendency (i.e., mean) of the sample. As applied to analysis of variability, the range, the standard deviation (s.d., the typical difference between each value (x) and the mean $\left(\text{s.d.} = \left(\left(\sum (x - \text{mean})^2 \right) \Big/ (N-1) \right)^{1/2} \right)$ and the coefficient of variation (cv = s.d./mean) are calculated. Linear measures of variability provide information about the quantity of a signal, but they are limited in their explanation of performance variability for several reasons (Harbourne & Stergiou, 2009). First, they contain no information about the time-evolving nature of the signal. That is, data from several trials generate a "mean" picture of variability, but the mean removes the temporal organization of the variability and thus masks the true structure present in the variability pattern. Second, the valid use of linear measures to study variability assumes that variations between repetitions of a task are random and independent (of past and future repetitions), an assumption that has been shown (through nonlinear analysis) to be false. Finally, linear measures provide different answers when compared with nonlinear measures, as pertains to the respective outcomes of variability evaluation.

2.3.2 NONLINEAR METHODS

2.3.2.1 Time Series Analysis

There are several different methods of time series analysis that have been applied to the evaluation of performance variability (Hamill et al., 2000; Harbourne & Stergiou, 2009). These methods fall under the aegis of what is termed dynamic systems theory (DST) because analysis with these methods reveals that variability in performance patterns exhibit what is termed complex dynamics (i.e., highly variable fluctuations that evolve over time). The complex dynamic variability patterns delineated by these methods also are termed nonlinear, because there is a between-data-point relationship, or dependency, for sequential data points throughout the time series. Nonlinear methods describe the patterns, or structure, of these relationships, not simply the quantity. Finally, complex dynamic nonlinear variability systems are termed deterministic and adaptive, because (1) for a given starting condition, the future state of the system is determined—randomness is not present, and (2) their adaptive nature features the combined action of different system components to carry out a coordinated behavior (in line with the early idea of Bernstein, 1967) directed at coping with (and thereby controlling) a change-initiating event or collection of events. Figures 2.1 and 2.2 (reproduced from Stergiou et al., 2006) illustrate the basic differences between less complex, less deterministic versus more complex, more deterministic variability patterns. Nonlinear methods demonstrate that temporal variability patterns across a wide variety of systems (numerous examples cited below), actually exhibit deterministic patterns even though they appear no different from random noise. These patterns have been defined, perhaps confusingly, as "chaotic" (Barton, 1994). Figure 2.1 illustrates this idea with a depiction (on the left side of the figure) of three variability patterns, a simple sine wave (top panel), a so-called chaotic pattern

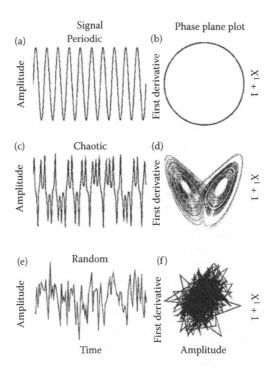

FIGURE 2.1 Contrasts between periodic, chaotic, and random time series (left side of figure) and the phase diagram plots for these different time series (right side of figure). Each phase diagram plots the amplitude of the time series versus its first derivative. The top panel plots a simple periodic function (i.e., a sine wave), the middle panel plots a chaotic system function, and the bottom panel plots a random number function (with a Gaussian distribution centered on zero and a standard deviation of 1.0). (Reproduced from Stergiou, N. et al., *Journal of Neurologic Physical Therapy, 30*(3), 120–129, Figure 1, 2006.)

(middle pattern), and a random signal (bottom panel). On the right side of the figure are plotted phase diagrams of each of these three variability patterns. The diagrams are generated by plotting, at each point in time (t), the amplitude (A) of the signal against its first derivative (i.e., dA/dt).

The phase diagrams in Figure 2.1 show a clear progression from a highly predictable (i.e., deterministic) sine wave signal to a random signal with very low predictability. The interesting result is the phase diagram for the chaotic signal in the middle panel of the figure. Even though the variability pattern for this signal appears random, the phase diagram demonstrates that the chaotic signal in fact has an entirely different temporal structure than the purely random signal. The distinctive butterfly shape of the phase diagram for the chaotic variability pattern in Figure 2.1 is the hallmark of a set of solutions for what is termed the Lorenz system of ordinary differential equations, otherwise termed the Lorenz attractor (Stergiou et al., 2006).

Stergiou et al. (2006) interpret these differences with the conceptual illustration in Figure 2.2, which plots increasing predictability on the *x*-axis against increasing complexity on the *y*-axis. Low amounts of predictability are associated with the

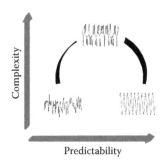

FIGURE 2.2 An illustration of the theoretical model based on the time series plotted in Figure 2.1. For each of the three time series, their predictability is plotted on the abscissa and their complexity on the ordinate. Both predictability and complexity are based on the regularity of the time series patterns. (Reproduced from Stergiou, N. et al., *Journal of Neurologic Physical Therapy, 30*(3), 120–129, Figure 2, 2006.)

random, noisy system, while high amounts are associated with the periodic, highly repeatable, rigid sine wave variability. The chaotic variability pattern features an intermediate level of predictability, wherein the system is neither too noisy nor too rigid. That is, the variability pattern of a chaotic signal can be considered to be moderately deterministic. On the *y*-axis of Figure 2.2, lesser amounts of complexity are associated with both the random and the highly periodic variability patterns, wherein the system is either too rigid or too unstable. Although the phase diagram for the random signal (Figure 2.1) appears on visual inspection to be complex, recall (above) that from a dynamic systems perspective a complex signal is defined as one that features highly variable fluctuations that evolve over time. Clearly, there is no indication in the random signal phase diagram of any evolution of the temporal structure of the signal over time.

In contrast, the phase diagram for the chaotic signal in Figure 2.1 strongly suggests an evolution of the temporal structure of the signal over time, which equates to a high complexity of the signal (Figure 2.2). Stergiou et al. (2006) believe that greater complexity in system variability is associated with a rich behavioral state for which system output is characterized by a chaotic structure. These authors conclude that chaotic temporal variations in the steady state output of a healthy system represents an underlying capability to make flexible adaptations to everyday stresses placed on the system. Accordingly, a reduction or deterioration in the chaotic nature of the temporal variability of a system represents a decline in healthy flexibility of the underlying control system associated with emergence of behavioral rigidity and inability to adapt to stress.

As noted above, a number of analytical tools have been developed to ascertain the temporal structure of a given variability pattern of interest. For example, Hamill et al. (2000) describe four different dynamical systems methods for quantifying a given variability pattern, methods that have been applied to movement variability analysis by van Emmerik and colleagues (van Emmerik & van Wegen, 2000; van Emmerik et al., 2004). Harbourne and Stergiou (2009) cite a total of eight specific methods of nonlinear analysis and allude to the other additional methods that can be applied.

A favored approach in this and in publications by Stergiou et al. (2006) and Karwowski (2012) cited above is on applying chaos theory to the analysis of the temporal structure of performance variability observed in various types of systems.

In addition to movement control discussed above, what other modes of performance exhibit variability patterns that can be classified as dynamic, complex, nonlinear, and deterministic? The answer appears to be a large number that encompasses impressive variety. For example, May (1976) cites variability patterns observed in population dynamics, prey-predator systems, ecological systems, economic systems, business cycles, learning, and social interaction as possible candidates that meet these criteria. In a similar vein, Karwowski (2012) cites 11 different complex sociotechnical system domains (with some overlap with May's candidates), for which the theory of complex adaptive systems can be applied, namely (1) biology, including ecosystems, (2) business management, (3) social interactions, (4) economics, (5) disease prevention, (6) health care management, (7) human service systems, (8) information and software engineering, (9) manufacturing operations and design, (10) medicine, nursing, and medical practice, and (11) occupational biomechanics. For each of these domains, multiple references are cited in support of the complex adaptive system claim.

2.3.2.2 Fourier Transform Analysis

Fourier transform analysis (Boashash, 2003; Jagacinski & Flach, 2003, pp. 120–136) is a method of assessing the distribution of different frequencies in a given variability pattern. Therefore, like linear analysis methods, but unlike dynamical systems methods of analysis described above, Fourier transform analysis does not reveal the temporal structure of a variability pattern. However, unlike these other two methods, it does reveal the relative distribution of different frequencies in a time-varying signal, and thus represents another, and potentially useful, approach to time series analysis.

With regard to human performance variability, one interesting and useful application of Fourier transform analysis is to assess the limits of a behavioral control system in the frequency domain. The putative objective of the system is to control variability in system output in the face of input disturbances imposed on the system. Fourier transform analysis can be used to assess the frequency distribution of both the input and output signals, and thereby to show how the power of the dominant frequency of the output signal varies in relation to the power of the dominant frequency of the input signal (the term "power" refers to the relative proportion of different frequencies across the frequency spectrum) as the input signal is varied across a selected frequency range. In other words, the outcome of this analysis is an indication of the control system response across the frequency bandwidth of the input signal. This outcome typically is depicted as a Bode plot (Jagacinski & Flach, 2003, pp. 137–157; see Flach, 1990, and T.J. Smith et al., 1994b, for two examples of the application of Bode plots for analysis of behavioral control systems). This method plots the gain of the control system on the ordinate against the range of different frequencies of the input signal imposed on the system on the abscissa.

A readily demonstrable movement example illustrates the utility of Fourier transform analysis and the Bode plot. In mammals, the proprioceptive system is highly refined to detect joint displacement of a muscle-joint system with input from muscle spindle proprioceptors conveyed to the central nervous via relatively high-velocity IA afferent fibers. In contrast, processing of visual input by the visual system is much slower. If a subject is asked to track the back-and-forth hand movements of a partner, with the hands of the two partners placed in contact palm-to-palm, the subject can essentially track the partner's hand movements with complete fidelity (i.e., gain = 1) at the fastest back-and-forth hand movements the partner can generate. The Bode plot for this condition (derived from Fourier transform analysis of the hand movement signals of both partners in the frequency domain) thus would show a flat horizontal line at gain = 1 for subject hand movements across the entire range of frequencies of partner back-and-forth hand movements.

In contrast, if the palms of the subject and the partner are separated by one inch, the ability of the subject to move her or his hands at the same frequency as the partner's back-and-forth hand movements rapidly diminishes as the frequency of the partner's hand movements increases. For this condition, therefore, the Bode plot would show a horizontal line at gain = 1 for subject hand movements across a relatively low range of frequencies for partner hand movements, but as partner hand movement frequencies continue to increase, the gain for subject hand movements will fairly rapidly decrease to a much lower value.

Adherence to linear methods of analysis when investigating skilled movement have limited applicability to both the development and expression of human skill and more generally biological systems. Both the traditional motor program perspective and its notion of variability as system noise, and the more recent DST ideas of coordination as self-organizing, have both regarded decreased variability as a mark of effective movement expression, but DST posits the systems to transition to a "new" stable state when variability increases to a chaotic state. Stergiou et al. (2013, p. 94) point out that both the traditional motor program view (general motor program [GMP]) and DST both adhere to linear measures. They note

> As motor learning occurs, the magnitude of variability continuously decreases and will reach a plateau. At that time we have a very stable behavior according to DST, or an appropriate selection of parameters to correctly execute the motor program according to GMP.

Earlier reviews of the role and potential importance of motor variability (Newell & Corcos, 1993b; Newell & Slifkin, 1998) suggested that *variability* and *not* invariability is likely the key to better understanding of how the coordination and control of movement is expressed in the observed response output. More recently Riley and Turvey (2002, p. 120) have stated this more succinctly when they note

> In particular an explicit focus on variability promises insights and lessons of potentially broad scope, for example, that more variable does not mean more deterministic.

Figure 2.3 illustrates the point they make by comparing both linear and nonlinear forms of analysis. Figure 2.3 shows both the linear and nonlinear methods of analysis; reading from top to bottom periodic, chaotic, and random signals are presented. The two columns to the right of the figure show the range and value of the Lyapunov exponent (LyE). The conclusion illustrated by this figure is that signals can have the

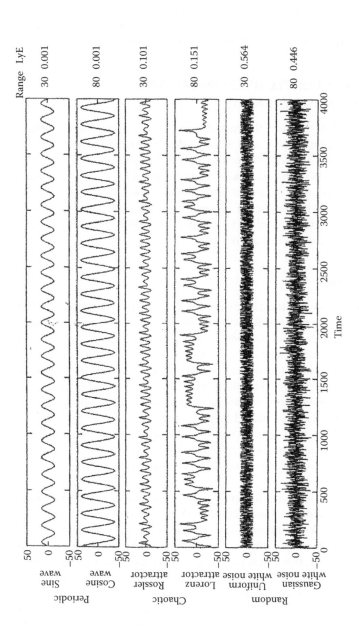

FIGURE 2.3 Complementary linear and nonlinear measures from different signals; six signals are displayed, with the respective values for range and largest Lyapunov exponent (LyE). The first two time series are periodic and have been generated using the sine function 15sin(*t*/24) and the cosine function 40cos(*t*/24). The following two time series are chaotic and have been generated using the Rössler and Lorenz systems, respectively. The final two time series are random and correspond to uniformly and Gaussian distributed white noise, respectively. All time series contain 4000 data points. The figure demonstrates that signals can have the same range but differ in terms of temporal structure (LyE) or they can have different ranges but the same LyE. (Reproduced from Stergiou, N. & Decker, L.M. *Human Movement Science, 30*(5), 869–888, Figure 1, 2011.)

same range but differ in terms of LyE, or vice versa. Nonlinear analytical methods suggest that the development and learning of both phylogenetic and ontogenetic movement skills rely on an optimal state of variability. In similar fashion Stergiou et al. (2013) and others have proposed that atypical motor behaviors may be represented by either inflexible behavior states that occupy a narrower range of functional stability, or by more random, unfocused, and unpredictable behavior.

2.4 THEORETICAL PERSPECTIVES ON VARIABILITY AND SPORT SKILL LEARNING

In the field of motor skill research, the past 36 years have seen a growing contrast between representational models of skill learning that invoke the traditional metaphors of motor programs and the assumed computational resolution of such programs, compared to the DST approach that views movement skills as self-organizing, essentially emerging from the dynamic interplay between the constraints present in the environment, task, and the abilities of the performer. The contrasting theoretical approaches are quite different regarding the assumptions, yet both have relied primarily on analytical approaches that are linear.

2.4.1 SCHEMA THEORY

Perhaps the most well known in the motor (sport) skills literature is Schmidt's (1975) schema theory of discrete motor skill learning. In a recent retrospective Schmidt (2003) reflected on the impact and resilience of this contribution, especially in light of the more recent dynamical systems account of skill expression. While there is little doubt that this contribution triggered a storm of empirical work among movement scientists, the central tenet of the theory that practice variability would enhance skill learning especially in producing novel (unpracticed instances of a skill) variations of a skill produced mixed results. Based largely on the earlier ideas proposed by Head (1926) and Bartlett (1932), schema theory addressed two persistent problems in motor skill learning, namely how one accounted for learning a "novel" skill (not present in memory), and how the storage problem is similarly addressed in representational memory.

Variability of practice was seen to better prepare the learner when confronted with a novel skill event, and the assumption of a GMP capable of being modified to address a changing set of environmental and other circumstances went a long way to addressing the difficulty of assuming each and every skill activity was stored in memory, much like a video store! Schema theory was in the tradition of cognitive psychological theory in that it relied on some form of representational metaphor that argued for some kind of computational activity to solve the particular movement problem requiring a response. This was in direct contrast to the dynamical systems theorists who argued that much of the skilled activity was "self-organized" and "emerged" as a consequence of the constraints placed on the dynamic interplay between TAE. In a commentary on Schmidt's (2003) "Reflections on Schmidt's theory after 30 or so years," Newell (2003) delivered almost the Shakespearean-like eulogy spoken by Marc Antony…"Friends Romans and Countrymen I come to…" (Julius Caesar, Act III scene II), when he remarks:

> These changing times lead naturally to the suggestion that Schmidt's (1975) schema theory will have the additional distinction of representing the last occasion in which an exclusively behavioral psychological theory for studying motor learning and control will receive as much scholarly recognition (p. 387).

Schema theory placed the spotlight on a variability of practice hypothesis that has not produced the kind of empirical support initially proposed and perhaps anticipated by Schmidt's (1975) theory. In a review of the empirical support for schema theory, Van Rossum (1990) concluded that practice variability was relatively weak both for adults and children. What is of interest, however, is the notion that variability in the *coordination patterns of movement production* has in fact, retrospectively, garnered more support than the learning and performance of discrete motor skills, which was Schmidt's initial theoretical proposal in his 1975 paper. The variability reported in the analysis of the actual movement properties of skill learning and performance is somewhat more supportive of the schema idea.

2.4.2　Dynamical Systems Theory

The central idea of DST is that performers exhibit three essential characteristics. First is the idea that preferred rates or periodicities (attractors) abound in performers. Second, these preferred rates are resistant to a wide range or perturbations, and third, if the perturbations become too disruptive the system can reorganize (phase shift) to a new level of stability. Examples of this would be the four biomechanical states of horse locomotion (walk-trot-canter-gallop), or humans transiting from walking to running. With both examples, the locomotor state is reorganized as a consequence of velocity demands, each producing stability as a function of the speed vector. In the developmental literature Adolph et al. (2008), in an analysis of the process of developmental changes observed in infant locomotion, argued that inappropriate sampling intervals failed to detect the process of change. Her data demonstrated clearly an increase in variability prior to a transition—a phase shift to a more stable state with respect to the development of locomotion. Standard sampling rates of three months failed to detect this important aspect of transition from early to stable locomotion in the development of infant walking. This is a good example of how the development of coordination and control reflects the laws of thermodynamics in that prior to a system transitioning to a new stable state, a rapid increase in variability (chaos) is observed.

With the case of infant development, we would do well to remember that variability has more to do with flexibility and responding to environmental change rather than reducing variability as a mark of skill and consistency. The paradox would appear to be that the observed consistency (minimal standard deviation) of performance is a product of the actor being able to continually adjust to changing contexts, which is a mark of inherent variability in the coordination and control required of any particular skilled response. Bernstein (1967) referred to this as "repetition without repetition," and Gibson (1966) as "regular not regulated." The British psychologist Frederick Bartlett's concept of motor schemata was influenced by his lifelong passion for cricket. Specifically, he marveled at the ingenuity with which batsmen shaped their strokes within a certain "range of anticipation" of the bowler's intentions. One might also argue that no single swing of the bat is precisely the same in baseball.

Variability is an inherent characteristic of skill expression with no two instances of a skilled action being identical. Stergiou and Decker (2011, p. 2) make a similar observation that we all share the high-level skills of Michael Jordan (basketball) and Yo-Yo Ma (cello) when we successfully navigate through crowds or cross diverse and challenging terrains. Variability is equated with rich behavioral states in which we are all capable of flexibly managing changing environmental contexts while maintaining optimal stability and performance.

2.5   VARIABILITY IN LEARNING AND PERFORMING SKILLS

Returning once again to the holy trinity (TAE) of movement skills (Newell, 1986), variability is present in both the outcome of any performance action and in the underlying coordination dynamics contributing to that performance. Clearly, producing a skilled action requires and reflects several aspects of the human condition, both biological and behavioral. Hubbard (1976, p. 6) noted:

> Perception and understanding of the world around us also depend, in part, on our ability to move. Even memory, thinking, and concept formation may well depend on sub-overt movements of speaking and writing…it is little more than a convenient evasion to assume that thinking is completed entirely within the nervous system. This is no more logical than to assume that action results from muscular activity alone.

Hubbard's (1976) comments reflect essentially the theory of direct perception, proposed a decade earlier by James Gibson (1966, 1979) and his claim of the critical link between perception and action. As a function of the ongoing interplay between TAE, it not hard to understand that "variability" is lurking in each of the three components and in their interactions. With respect to the actual TASK to be performed one might ask, is it slow or fast with respect to execution? What is the level of accuracy or precision required for success? Is the coordination pattern required to complete the skill complex or relatively simple? What about the actor? What is his or her level of conditioning; how much practice has been devoted to the activity? What is the actor's mental state (stress level, anxiety level, arousal level)? With respect to the environment we might ask, is it conducive to permitting a high level of performance, or are environmental circumstances seriously impeding optimal performance (windy, rainy and cold outdoors, noisy and uneven surfaces, etc.)? All of these factors can produce changes in the variability in both performance and optimal levels of coordination and control. Figure 2.4 (from Davids et al., 2003) illustrates how Newell's (1986) constraints perspective of TAE resonates with the perception–action synergy to produce the desired skilled performance.

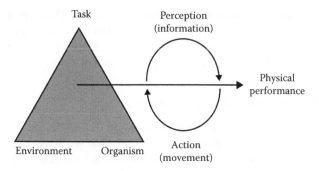

FIGURE 2.4 Newell's model of interacting constraints adapted to illustrate the resulting effects on variability of physical performance. (From Newell, K.M., Constraints on the development of coordination. In M.G. Wade & H.T.A. Whiting (eds.), *Motor Development in Children: Aspects of Coordination and Control*, Dordrecht, Netherlands: Martinus Nijhoff, pp. 341–360, 1986.)

2.5.1 Variability between Expert and Novice Performers

From a deterministic perspective, discrete sport skilled responses will likely be more variable for novices and less so for elite performers, other things being equal (a closed versus an open environment). Nonlinear analysis will show the expert performer exhibiting higher levels of variability with respect to making adjustments in the actual coordination dynamics demanded by changes in the environmental context in which the performance is required. Such an example would be adjustments made for wind gusts during a tennis serve or a golf shot. Again the contrast is perhaps best described as follows: given changes in the TAE dynamics, the expert will typically demonstrate a lower rate of "within" variability in terms of accuracy (error) of the response outcome, but increased "between" variability expressed as flexibility in managing the array of changing environmental demands. This likely will be more apparent in an open as opposed to a closed skill setting.

A good example of this is cited by Bauer and Schöellborn (1997; cited in Davids et al., 2003) with respect to the coordination patterns demonstrated by discus throwers. Filmed during both different training and actual competitive sessions, data were collected on both a decathlete and a discus-throwing specialist. Biomechanical analysis of body positions and joint velocities revealed differences between training sessions and competition for both athletes. These results confirm that the invariance often ascribed to expert performance is in fact a *between*, not a *within* feature of their skill set. Experts do not produce identical performance characteristics each and every time they perform, but make the necessary ongoing adjustments as a function of the particular constraints present in any training or competitive situation.

There are several other examples of this so-called functional variability, such as pistol shooting (Arutyunyan et al., 1968; Davids et al., 2003; Koenig et al., 1994). As expertise develops over time, the motor patterns of elite performers do not reveal an increase in invariance for the motor pattern required of a particular skill. Davids et al. (2003) refer to this as "coordination profiling," whereby an individual performer meets the specific demand characteristics of producing an optimal performance given the constraints present at any particular point in time. This approach to understanding the role of variability in performance applies not only to sport but also to medicine (see West, 2006, discussed above) and to a range of therapeutic interventions by both physical therapy (PT) and and occupational therapy (OT) clinicians.

2.5.2 VARIABILITY IN REHABILITATION SETTINGS

Based on the research findings, variability is a key element in the development of the phylogenetic skills of postural stability, locomotion, reaching, and grasping, all of which are fundamental to the later development of the ontogenetic skills, whether they are activities of daily living or recreation and sport-related activities that reflect our cultural norms. Variability with respect to gait has a substantial extant research literature, both for typical development and for those with particular physical and neurological disabilities. This research literature has demonstrated that nonlinear analysis of variability, measured as a time series (stochastic), shows that the effects of aging per se, and trauma such as a neurological event, shows a reduction in the typical step-by-step gait patterns. In such cases, a reduction in this kind of variability can influence the incidence of falls in the elderly and instability in patients with physical deficits as a result of trauma or disease.

In this context, a reduction in variability is tantamount to a loss of flexibility to respond to changes in the demand characteristics of the environment in which the gait (navigation) is occurring. van Emmerik and colleagues (van Emmerik, 2007; Heiderscheit et al., 2002) report gait changes in individuals, both typically aging and elderly with orthopedic or neurological constraints. In a similar vein, Stergiou and colleagues (Harbourne & Stergiou, 2009) report how stochastic aspects of variability using nonlinear analytical techniques can be an important guideline for the retraining of limb use in knee and hip repair and in those learning to use artificial limbs.

2.6 SUMMARY

By way of summary, the discussion of variability in the preceding sections provides little insight with respect the underlying *sources* of variability, a major focus of this book. However, the analysis above does raise two basic questions:

1. First, what is the basis of the adaptive nature of variable human performance systems across a broad range of movement based system complexity?
2. Second, what are the origins of the nonlinear and chaotic properties of the temporal structure of movement system variability?

Both questions address what essentially represents the origins of the nonlinear, chaotic structure of performance variability. Chapter 1 addressed the first of these questions and attributes adaptive behavior to control system designs. Here we address the second of these questions, namely *how* and *why* so many human performance systems, at different levels of complexity, exhibit nonlinear, deterministic patterns of variability in line with the principles of dynamical systems theory (Harbourne & Stergiou, 2009; Karwowski, 2012; van Emmerik et al., 2004). Stergiou et al. (2006) broach this question with the following observations regarding movement behavior:

- Analysis of selected types of movements (such as tremor and gait) shows that chaotic temporal patterns of movement variability are associated with healthy movement behaviors.
- A reduction or deterioration of such patterns is associated with a decline in healthy flexibility of movement behaviors, producing behavioral rigidity and the inability to adapt.
- These observations support the conclusion that for healthy movement behavior there is an optimal amount of temporal variability characterized by a complex chaotic structure. A reduction in such variability equates with increased behavioral rigidity, leading to an increase in noise and instability.
- Tononi et al. (1998) propose that subserving the chaotic temporal structure associated with overt behavioral variability is a comparable chaotic temporal pattern in brain signaling. The substance of this proposal is that (1) signals from large numbers of functionally specialized groups of neurons distributed across brain regions are integrated with signals from environmental stimuli to generate a coherent, meaningful percept, and (2) chaotic temporal complexity of both neuronal signals and environmental signals underlies this integrative process.

As noted earlier, the evolutionary rationale offered for the ubiquitous prevalence of performance variability in biological systems is that it ensures survival. The above observations, and those cited in preceding sections, support the proposition that both healthy movement behaviors and integrative brain functioning are mediated by signaling variability that embodies nonlinear, deterministic, chaotic temporal structures. Similar temporal structures are assumed to characterize variability in more complex performance systems (Karwowski, 2012; May, 1976). Collectively, this body of information and evidence about the temporal patterning of variability across diverse modes of performance appears to offer insight into the question of why such patterning, under normal circumstances, typically is characterized by nonlinear, deterministic, chaotic temporal structures.

This leaves the question of how these types of temporal structures are generated. Arguably, none of the foregoing information and analysis offered in this chapter really answers this question. As Harbourne and Stergiou (2009) point out for the movement domain: (1) over time complexity may be hidden within the time series of a movement sequence, (2) a nonlinear, deterministic temporal pattern of movement variability is generated when a movement that occurs at one moment affects, and in turn is affected by, movements that occur either before or after that given movement, and (3) nonlinear dynamics methods based on chaos theory, can be deployed to analyze this complexity. Stergiou et al. (2006, p. 121) offer a hint as to how this type of temporal movement patterning might be generated:

The plurality of human movement patterns and the multitude of motor control feedback loops make movement similar in many respects to other physiological life rhythms (e.g., heart beat), for which variability has been described as exhibiting deterministic dynamics.

The hypothesis on the "how" question offered here extends the observations of Stergiou et al. (2006) with reference to a behavioral cybernetic assessment about behavioral control systems made earlier such (Chapter 1, Section 1.3): "Feedback control takes many systems forms and can be based on many modes of positive adjustment beyond detection of differences or errors. It may be involved in regulation of order, sequence, summing, separation, elaboration and multiplication, integration, and/or serial interaction." Phrased in control systems terms, temporal variability observed in the performance of any complex system is assumed to originate as a consequence of some combination of both negative and positive feedback and feedforward control. Varying the input of a closed-loop performance system with modes of feedback control that can vary in both type and level on a continuous basis has the potential to generate a system output pattern with a complicated temporal structure. The question remains as to why this pattern should be nonlinear, deterministic, and chaotic.

As a possible answer, consider the chaotic temporal structure of the variability pattern in the left-hand column, middle panel (Illustration c), of Figure 2.1. Assume that input to the system that generated this output pattern was influenced by a combination of negative and positive feedback control. The objective of a negative feedback control system is to generate solutions for the proper corrective action that maintains system stability. That is, under perfect negative feedback control, the system will hold the set point and not oscillate around it (Jagacinski & Flach, 2003, pp. 10–12). In contrast, unconstrained positive feedback control will cause the system to swiftly spiral out of control and cross a tipping point into instability (West, 2013).

In functional complex systems, however, positive and negative feedback control mechanisms are customarily coupled so that ultimately instabilities are brought under control. Thus, generalizing the conclusion of Stergiou et al. (2006) above, there likely is an optimal amount of temporal variability associated with functional system performance at many levels of system complexity. A reduction in such variability (arising out of an imbalance in system control favoring negative feedback) equates with increased behavioral rigidity of the system. An inappropriate increase in such variability (arising out of an imbalance in system control favoring positive feedback) equates with an increase in noise and instability in system behavior.

Now consider the chaotic variability pattern in Figure 2.1 in light of the foregoing summary. Under the hypothesis advanced here, the periodic temporal deviations featuring higher amplitudes are assumed to be associated with a more intense influence of positive relative to negative feedback control on system performance. In contrast, the periodic temporal deviations featuring lower amplitudes are assumed to be associated with a more intense influence of negative relative to positive feedback on system performance. If the influence of these two modes of feedback control on system performance exists, then the overall pattern of variability in system output resembles the chaotic pattern in Figure 2.1, with alternating higher and lower amplitudes in temporal deviations. Moreover, this analysis is aligned with the Harbourne and Stergiou (2009) description noted above, in a more generalized sense, that a nonlinear, deterministic temporal pattern of system variability is generated when system output that occurs at one particular moment both affects, and in turn is affected by, system output that occurs either before or after that particular output signal.

Admittedly, this analysis is both qualitative and highly speculative—it is unknown whether or not other patterns of chaotic variability in system performance, arising perhaps out of other combinations of feedback control influence, can be subjected to a comparable interpretation. Also, we are not aware if other authors have provided more definitive accounts of how nonlinear, deterministic, chaotic patterns of performance variability are generated. What seems reasonable to suggest is that the above analysis provides a credible interpretation of the origins of the chaotic variability pattern illustrated in Figure 2.1.

Nonlinear techniques such as fractal geometry provide a new perspective to view the world. For centuries, "the line" has been the basis whereby we have comprehended objects around us. Chaos science uses a different geometry called fractal geometry. Fractal geometry is a relatively new language used to describe, model, and analyze complex forms found in nature. Human variability that is present in dynamic coordination and control is just such a venue to apply this analytical process.

It would seem that the assumption that invariance is a characteristic of better motor control is not especially valid when analysis of the temporal structure of performance variability is undertaken. In fact, it would appear that nonlinear analysis has clearly established that variability is akin to flexibility of responses in dynamic contexts where environmental and other constraints operate in ways that cannot always be anticipated. As we noted at the start of this chapter, there exists a continuum of unanticipated events in complex human-machine systems from the familiar and anticipated to the unfamiliar and anticipated, and finally to the *unfamiliar and unanticipated* (Vincente & Rasmussen, 1992). It is this latter condition—not anticipated by either the operator or the system architecture—that requires improvisation by the operator. In human factor consideration of sport-related contexts, the onus is almost always on the operator (athlete) to extract a solution to a motor problem without the aid of an *a priori* system design.

The unique characteristic of sport-related activity, be it individual or team play, is that advances in equipment design has enhanced the ability of the performer to produce skill responses that seek to minimize outcome errors. In a similar fashion, medications have been shown to aid both strength and endurance of athletes, often in violation of the rules of a particular sport. That being said, it falls to the performer to utilize the technological advances afforded by better equipment design, impacted by human factors considerations, which have produced a continuous improvement in performance, especially at the elite level, where athletes appear to be more able to exploit the effects of improved equipment design. In sport and similar motor skill contexts there is clear evidence that performance variability is task-dependent (Vaillancourt & Newell, 2002). As such, the approach to better understand both the *how* and the *why* of the observed performance requires analytical measures that are sensitive to levels of scale. The complexity (fractal measure) of skilled activity can be useful to assess and monitor the output from the motor system across a range of skill sets, of which sport is a rich context to explore.

There remain a number of unresolved issues with respect to performance variability with regard to sport and other modes of performance. In a recent review, Preatoni et al. (2013) highlight the need for more complete data sets with respect to the types of movement skills under investigation, the relationship between "causes and effects" (Preatoni et al., 2013, p. 84), styles of inquiry, and the need to generate feedback that can be translational in order to foster application of these analytical techniques to real-world settings.

Clearly, analyzing and interpreting variability can be regarded as both good and bad. "Good" in the sense that the inherent flexibility of skilled responses is seen as the performers' ability to make ongoing adjustments to the output as a function of changing environmental demands. This response flexibility is present in the underlying dynamics of the coordination pattern and is not seen when standard linear analytical techniques are employed. In fact, increases in variability at this level of analysis reflect output errors, but do not reveal how the ongoing dynamics are functioning, a key element of the observed output. The latter approach to variability (linear) casts variability as system noise, which is regarded as "bad." Both approaches provide important insights into performance and in their own way allow the learning process to take advantage of two key aspects of the skilled output: performance results and the coordination dynamics that support performance.

3 Variability in Cognitive and Psychomotor Performance

Thomas J. Smith

3.1 OVERVIEW

This chapter deals with sources of variability in inter- or intraindividual performance in the execution of particular cognitive and psychomotor tasks (portions of this chapter expand on material in T.J. Smith [1993, 1998]). The chapter thus extends the analysis of movement variability in Chapter 2 to a consideration of variability in the application of different movements to carry out these types of tasks. A recurrent focus will be on context specificity as a prominent influence on variability in cognitive and psychomotor performance. Sections below expand on this theme with discussions of a historical perspective (Section 3.2), sources of variability in transfer of training and differential learning performance (Section 3.3) and in psychomotor performance (Section 3.4), and qualitative observations about task performance variability (Section 3.5).

3.2 HISTORICAL PERSPECTIVE

Issues addressed in this chapter have their origins in the basic question, "How do we know what we know?" Debate regarding this question dates back many centuries and has centered on alternative views of whether human knowledge represents an innate attribute or arises through experience. In ancient Greece, contrasting views of Plato (437–347 B.C.E.) and Aristotle (384–322 B.C.E.) possibly represent the earliest historical record of this debate (Goldhaber, 2012, p. 14). Plato argued that experience was insufficient to account for all the knowledge and abilities that humans possess, and that since these cannot be taught, they must be innate. In Goldhaber's terms (Chapter 1, Section 1.1.2), Plato thus can be classified as one of the earliest nativists. In contrast, Aristotle (Plato's student) held that maturity occurs not simply as a summation of earlier characteristics but rather proceeds through a novel synthesis of them. Again using Goldhaber's terminology, Aristotle thus can be classified as one of the earliest empiricists.

Fast forward some 2000 years, and the next great development in the nativist-empiricist debate emerged with the turmoil of the intellectual revolution of the 17th and 18th centuries (Kors, 1998). The nativist champion was René Descartes (1596–1650) (1996), one of the most esteemed philosophers of the age, who argued that all human ideas are innate (Kors, 1998, Lecture 5). A less well-known advocate of the nativist point of view, from a religious perspective, was Pierre Bayle (1647–1706) (1985), a devout Christian who held that neither evidence nor reason, but faith alone, supports Christianity (Kors, 1998, Lecture 12).

In the opposing empiricist corner were Thomas Hobbes (1588–1679) (2012) (Malcolm, 2002) and John Locke (1632–1704) (1974), both of whom championed the idea of experience as the source of human knowledge (Kors, 1998, Lectures 6, 10, and 11). Hobbes held that all ideas enter the mind through the senses. Locke, perhaps the strongest empiricist of his or any other age, echoed and extended the epistemology of Hobbes in insisting that all of our ideas and knowledge are acquired only through, and are limited to, our experience of the world. For Locke, there are no innate ideas—we have no knowledge of matters that are not acquired through experience. In Locke's view, the mind begins as a blank slate upon which experience imprints ideas via the senses and reflection—we cannot know what is not within our experience, and because experience is not logically determined, our knowledge of the world is merely probable. In a very real sense then, Locke may be considered the father of context specificity.

Another empiricist perspective on the role of experience in human understanding, as Chapanis (1988) notes in his seminal article on generalization, is that of David Hume (1711–1776) (Lindsay, 1911, pp. 91 and 94). Hume maintains that, "It is impossible for us to satisfy ourselves by our reason, why we should extend that experience beyond those particular instances which have fallen under our observation... there can be no demonstrative arguments to prove, that those instances of which we have no experience resemble those of which we have had experience." In other words, extending Locke's point of view, and as an early exemplar of the challenge confronting attempts to generalize from observed results (Chapanis, 1988), Hume's position is that there may be no logical basis for generalizing or extrapolating from particular observations.

The first definitive scientific support for the empiricist perspective on the influence of experience on the acquisition of knowledge was provided by Helmholtz (1856–1866) in the mid-nineteenth century. Helmholtz recognized that there is great behavioral variability in the manner in which humans (and animals) respond to particular conditions of stimulation, and that experience affects the way in which situations are perceived and how behavior is organized through learning. Based on this insight, accompanied by supporting research, he developed what came to be known as the empirical doctrine of space perception, based on the assumption that space perception is learned, not innate. Today, we remember Helmholtz as the first to experimentally document the phenomenon of brain plasticity, the modification of brain function through experience.

One of the first major programs of psychological research on cognitive variability was carried out by Thorndike and Woodworth (1901a, b, c), over a century ago. The purpose of this program was to ascertain if skill training of subjects on one type of task (the control task) predicted performance on other tasks of similar or dissimilar design (the test tasks). This approach is termed a transfer of training paradigm. The prevailing view at the time (that persists to some degree down to today) favored the idea of generalized transfer of learning such that improvement in levels of performance on the control task during training was predictive of performance improvement on the test tasks, as measured at control task pretraining and posttraining stages. The actual result was that performance improvement in the test tasks between testing sessions at the control task pretraining and posttraining stages was irregular and undependable and seldom as great as the improvement in that observed for the control task.

The conclusion drawn by the above authors from this research, that aroused considerable dissent at the time, is summarized in Woodworth (1938, pp. 194–195) as follows:

> Improvement in any single mental function need not improve the ability in functions commonly called by the same name. It may injure it. Improvement in any single mental function rarely brings about equal improvement in any other function, no matter how similar, for the working of every mental function-group is conditioned by the nature of the data in each particular case... There is no inner necessity for improvement of one function to improve others closely similar to it, due to a subtle transfer of practice effect. Improvement in them seems due to definite factors, the operation of which the training may or may not secure.

Thorndike and Woodworth (1901a) clarify use of the word "function" in this conclusion as follows: "The word 'function' is used for all sorts of qualities in all sorts of performances from the narrowest to the widest. It may refer to the mental basis of such things as spelling, multiplication, discrimination of size, force of movement, marking *a*'s on a printed page, observing the word *boy* in a printed page, quickness, verbal memory, chess playing, reasoning, etc."

This research set the stage for an extensive series of studies of variability in cognitive performance over the ensuing century based on experimental transfer of training and differential learning protocols accompanied by a number of analytical reports in the psychological and HF/E literature that examine the scientific meaning and implications of the results. This work has been reviewed from different perspectives by Ackerman (1987), Adams (1987), K.U. Smith and Smith (1991), and Jones (1966, 1969). The next section summarizes the key experimental findings and the perspectives offered by these authors.

3.3 VARIABILITY IN COGNITIVE PERFORMANCE

This section reviews the extensive series of investigations into variability in cognitive performance carried out over the last century, inspired by the seminal studies of Thorndike and Woodworth cited above. Collectively, this work is aimed at exploring the nature and implications of variability in individual differences (Chapter 1, Section 1.2.1) in transfer of training and differential learning across different tasks. In contrast to the use of movement variability and control systems analysis to investigate the temporal structure of performance variability that was addressed in Chapter 1, the analytical approach to exploring cognitive variability reviewed here is based on either transfer of training or correlational analysis methods. As we shall see, these approaches feature both methodological strengths and shortcomings.

Arguably, the implications of findings from this research are central to the scientific characterization of HF/E. That is, HF/E science continues to struggle to define itself. Since emergence of the field some seven decades ago, the HF/E community has engaged in a more or less continuous debate regarding the scientific essence of the discipline (Dul et al., 2012). As Durso (2013) notes, even today HF/E science is not recognized by some as a legitimate candidate for inclusion in science, technology, engineering, and mathematics (STEM) education. The widely accepted proposition that HF/E is intimately linked to design (Meister, 1989; Chapanis, 1991, 1992) raises the obvious challenge of precisely defining the nature of this link.

The answer to this question, introduced in Chapter 1, is that of context specificity—the idea is that the preponderance of observed variability in performance is attributable not to innate biological factors or learning ability but to design of the task (T.J. Smith, 1993, 1998). Thus, if design factors critically influence human behavior and performance such that the latter cannot be realistically assessed in isolation from the former, then HF/E assumes a much stronger role as a distinct scientific discipline (Chapter 1, Section 1.1.3, Principle 1), and performance cannot be evaluated outside of its context—generalized models of performance have little scientific validity.

In the modern era, the debate in the field of psychology regarding the importance of context in understanding behavior has been ongoing for some time. One perspective, dismissing the significance of context, is exemplified in the observation of Poulton (1966):

> ... A man's reaction to a complex equipment is now not always considered part of the appropriate subject matter of experimental psychology, since the performance measures are a function of both the man and the equipment...

The opposing perspective, based on the assumption that context always contributes to variability observed in performance, is exemplified in the assertion of Taylor and Birmingham (1959): "All measures of performance are partly a function of equipment."

Additionally, despite broad agreement that better design represents a key objective of human factors, there are differing viewpoints on reaching this goal. The statement of Poulton (1966, p. 178) defines one perspective: "The aim of engineering psychology is...to specify the capacities and limitations of the human, from which the choice of the better design should be deducible directly." Gould (1990, p. 762) summarizes the opposing viewpoint: "You can't rely upon descriptive data. Even if a description of the intended users were as complete as it possibly could be it would not be an adequate basis for design. It would not substitute for direct interaction and testing."

On balance, findings from the cognitive variability research reviewed below—categorized in relation to evidence from transfer of training and differential learning studies, coupled with critical analysis of this evidence—favor the positions of Taylor and Birmingham and of Gould. First, across a broad range of cognitive tasks, this evidence consistently points to a role of context specificity in cognitive performance variability. The logical implication of this conclusion, noted in Section 3.2, is that interaction with a new design introduces new modes and patterns of performance variability that cannot necessarily be predicted in advance. This conclusion underlies the account by Chapanis (1988) about limitations in generalizing findings from HF/E research. Gould's insight (1990) also helped launch the entire field of usability analysis, predicated on the idea that how users interact with new designs of products or software interfaces can only be characterized through iterative usability testing rather than predicted *a priori*.

3.3.1 EARLY TRANSFER OF TRAINING EXAMPLE

One of the first studies to follow up that of Thorndike and Woodworth cited above was that of Poffenburger (1915). This author studied a variety of judgmental and arithmetical tasks, using control groups and training groups, to assess the direction and amount of transfer of training for seven pairs of tasks. Results are summarized in Table 3.1. Training group subjects first were administered a series of learning trials on a training task (leftmost column). Then, their performance on a transfer task was assessed and compared with that of control group subjects, whose performance was tested only with the transfer task. Improved performance of the training group subjects on the transfer task, relative to that of the control group subjects, was judged to represent positive transfer of training.

As the table shows, positive transfer occurred for only one of the pairs of tasks, in which the numbers 3 and 5 were canceled in the training task and groups of numbers containing 3 and 5 were canceled in the transfer task. In all other cases, the transfer effect was either negligible or negative. The results confirm those of Thorndike and Woodworth (1901a) in indicating that the specific nature of the tasks defines training transfer. The remarkable feature of these findings is that context specificity in transfer of cognitive performance persists as the prevailing result.

Subsequent research on transfer of learning in cognitive as well as other types of tasks has not added much to early findings. Some investigators have drawn conclusions regarding the role of structural or situational similarity in determining positive transfer. Overall, findings suggest that the direction and degree of transfer are

---

TABLE 3.1
Transfer of Training in Cognitive Tasks (Results of Poffenburger, 1915)

Transfer Experiment	Type of Transfer
1. Color naming → Form naming	None
2. Opposite test → Adjective noun test	Negative
3. Two digit cancellation → Group cancellation	Positive
4. Two digit cancellation → Different two digits	None
5. Addition → Subtraction	None
6. Addition → Multiplication	Negative
7. Addition → Division	None

Source: Poffenburger, A.T. (1915). *Journal of Educational Psychology, 6,* 459–474.

determined by task makeup or structure, terms usually used by investigators to refer to the physical design features of particular tasks.

3.3.2 DIFFERENTIAL LEARNING RESEARCH: EARLY STUDIES

Differential learning research is concerned with determining the degree to which initial performance in learning is predictive of later performance. With the initial studies summarized here, the procedure involved obtaining initial measures of performance in a learning task for a sample of subjects and thereafter correlating the initial measures with final measures of performance or improvement on the same task. Such research addresses two questions: (1) To what extent can initial measures of performance predict overall improvement? and (2) To what extent do individual differences in performance increase or decrease during the course of learning? Findings related to both questions are relevant to the issue of the role of learning in the specialization of cognitive performance.

Kincaid (1925) summarized the results of 24 of the early studies of learning predictability based on correlations of initial and final performance (tabulated by Hunter, 1934). Twenty-two of these studies dealt with cognitive performances, predominantly arithmetical tasks. Among these 22, correlations between initial performance measures and percentage gain after practice were negative for all but two studies, indicating that initial performance is markedly inefficient in predicting learning. In 13 cases, the standard deviations of performance measures decreased after learning, but in 9 cases they increased (i.e., individual variation was affected by learning differently for different tasks).

Across the cognitive studies reviewed by Kincaid (1925), initial performance levels show highly variable correlations with overall performance gains after practice and predominately negative correlations with percentage gains. The latter finding may reflect statistical bias in that the higher the level of initial performance, the more limited the possibility for increased performance (as Ackerman, 1987, points out).

Overall, the experimental results of Kincaid, coupled with those summarized in Table 3.1, suggest that some factor other than individual differences or general learning ability is the major determinant of variability in cognitive performance. If such variability depends primarily on initial individual differences in skill or knowledge, then such skill or knowledge should transfer from one task to the next, and practice should affect the degree but not the pattern of variability among subjects after learning. Conversely, if cognitive performance variability is determined primarily by learning, practice should improve performance without changing the relative distribution of scores among subjects. The findings confirm neither of these predictions. The clear implication is that the extent of change in variability in a given task between the beginning and end of training is dependent largely on the nature of the task itself.

3.3.3 DIFFERENTIAL LEARNING RESEARCH: ANALYTICAL INNOVATIONS

Just before and after World War II, the questions posed above about cognitive performance during learning were addressed in two new ways. First, Perl (1934) originated a novel experimental attack on the problem of predictability of cognitive learning by intercorrelating performance scores of all possible pairs of trials, giving a complete array of predictive relationships during learning. Using four different intelligence test tasks as learning situations, she had subjects practice such tasks for 20 trials, and correlated scores for each trial with each later trial. Her findings were surprising for learning theorists, who generally believed that learning increases the predictability of performance. Perl found that instead of increasing with practice, intertrial correlations decreased with practice. That is, the correlations for trials 1–2, 1–3, 1–4, or 2–4 were larger than those for trials 1–20, 1–19, 1–18, or 18–19, 18–20, or 19–20. Jones (1966, 1969) has discussed Perl's work as the forerunner of extensive later research establishing task design as the prime determinant of variability in cognitive learning.

Perl's research laid the experimental foundation for that of Woodrow (1938, 1939), who originated an entirely new approach to the question of cognitive performance in learning. Although his work has not been widely recognized for its striking originality, Woodrow established a new theoretical basis for research on learning and cognition by subjecting performance scores in cognitive learning tasks to analysis of variance. This technique enabled him to identify the principal sources of variability in performance during learning and to determine their relative contribution to improvement. He defined experimentally the concept and phenomena of *specific variance* as related to the factors of practice, individual differences, and task structure or makeup.

Woodrow's (1938) first study involved extensive research on 56 subjects, using nine learning tasks: digit cancellation (three different tasks), adding, anagrams, digit-letter substitution, judgments of length, spot pattern identification, and speed in tracing gates in a visual maze. The subjects practiced each task during nine sessions. Woodrow computed the variances in performances for each task at the beginning and at the end of practice and expressed the ratio between the final and initial variance (final variance/initial variance) for each task. The results were as follows: anagrams, 1/21; digit-letter substitution, 2/69; spot pattern, 0/32; adding, 6/25; digit cancellation tests, 3/50, 2/28 and 2/89; speed in tracing gates, 0/9; and length judgments, 0/96. These results, along with those of Perl, argue against individual differences or generalized learning effects as major contributors to cognitive performance variability. That is, if individual differences contributed in a major way to performance variability with these tasks, then the consistent low/high pretest versus posttest variance ratios should not have been observed with every task. If generalized learning contributed in a major way to performance variability, then variance after training should have been *lower* than that before training. Instead, for every task, the opposite is observed.

Collectively, these findings suggest instead that it is task structure predominantly that determines the specific variance in performance during learning.

In a second major study, Woodrow (1939) practiced 82 subjects for six trials in four cognitive tasks: horizontal adding, substitution, and 2- and 4-digit cancellation. The same subjects performed 21 other tasks without practice. For each of the four practiced tasks, correlations between improvement scores (final minus initial scores) and initial scores were low and barely significant. Correlations between improvement scores on practiced tasks and scores on unpracticed tasks also were low and more often than not were negative. Woodrow was attempting to determine whether there is a general factor common to improvement in cognitive learning task that would be indicated by significant correlations between improvement scores and initial scores for either the practiced or the unpracticed tasks. Since he found only the most limited correlations for both relationships, he concluded that there is no general factor that accounts for improvement in cognitive learning. This experiment poses a major barrier to the concept of a general mental decision-making process in cognitive learning.

As another aspect of his 1939 study, Woodrow determined the relative degree to which initial and final scores on the practiced tasks could forecast performance on the unpracticed tasks. Correlations of the initial and the final scores for the four practiced tasks with those for the 21 unpracticed tests are as follows: horizontal adding −0.30, 0.19; substitution −0.22, 0.13; 2-digit cancellation −0.15, 0.10; and 4-digit cancellation −0.18, 0.16. The results indicate that the specific variance of the practiced tasks increases with practice, and that this increase is a function of the nature of the task situation.

Arguably, in empirically delineating context specificity as a prominent and pervasive feature of performance-design interaction, Woodrow's research established the experimental foundation of HF/E science and also defined a primary tenet of cognitive theory. He demonstrated for the first time that the critical determinant or source of variability in cognitive performance and learning is the design of the task rather than practice or individual differences among subjects. In a later report, Woodrow (1946, p. 157) concluded that practice gains are task-specific and are determined by task designs that subjects interact with during performance to achieve gains. To my knowledge, this is one of the earliest usage of the term task specificity in the psychological literature. (As an aside, my father, K.U. Smith, was personally acquainted with Woodrow. He told me once that Woodrow died an embittered individual because of the dedication of the psychological community at the time to generalized learning theory and their consequent refusal to recognize the validity of the idea of task specificity in learning.)

After military research in World War II had more firmly established HF/E science (K.U. Smith, 1987), Adams (1953) conducted an extensive experiment to examine possible differences in the variability of learning of different types of tasks. He administered 32 printed (cognitive) tests, 13 simple psychomotor tests, and 7 complex psychomotor coordination tests (CPTs) to 197 airmen, who practiced only the CPTs, with extensive practice on one in particular, termed the Complex Coordination Test (CCT). Initial and final scores for this CCT were correlated with overall mean performance scores for each of the three types of tests and also with overall mean final scores for the other six CPTs. This procedure yielded eight coefficients indicating the degree of correlation between both initial (I) and final (F) learning scores on the CCT and cognitive (printed test) scores, simple psychomotor test scores, and initial and final CPT scores. These correlation values are: for the printed test scores, [+]0.38 (I) and [+]0.25 (F); for the simple psychomotor task scores, [+]0.24 (I) and [+]0.27 (F); for the initial CPT scores, [+]0.45 (I) and [+]0.38 (F); and for the final CPT scores, [+]0.41 (I) and [+]0.49 (F).

These results on psychomotor learning agree generally with earlier findings on cognitive learning, that principal source of variance is in the design of the task. Adams concluded that the design of specific performance situations determines their relationships with initial and final scores for practiced performances. Learning reduced the level of correlation with dissimilar tasks, whereas with similar tasks the correlation levels increased.

Fleishman (1954, 1966) and Fleishman and Hemple (1954) measured the specific variance during learning of various complex psychomotor tasks, most of which could be classed as having significant judgmental (cognitive) components. In summarizing this work, Fleishman (1966, p. 159) states:

> In general, these studies, with a great variety of practice tasks, show that: (a) the particular combination of abilities contributing to performance changes as practice continues; (b) these changes are progressive and systematic and eventually become stabilized; (c) the contribution of "nonmotor" abilities (e.g., verbal, spatial) which may play a role early in learning, decreases systematically with practice, relative to "motor abilities"; and (d) there is also an increase in a factor specific to the task itself.

Bilodeau (1953) and Bilodeau and Ryan (1960) studied two relatively simple cognitive tasks: micrometer reading and judgment of the length of lines. They found that the specific variance due to task makeup increases with practice, even in the case of such limited tasks. It has been speculated that the more complex the task, the greater the variance related to its specific makeup during and at the end of learning (Jones, 1966, 1969).

After reviewing all studies of the determinants of sources of variability in cognitive and other learning tasks, Jones (1966, 1969) concluded that numerous additional studies have confirmed the work of Woodrow (1938, 1939), Adams (1953), Fleishman (1954), and Bilodeau (1953), and have shown that the amount of specific variance associated with most cognitive learning tasks is even greater than that found in these original studies. He concluded also that in the case of actual work performance, the variance attributable to the specific design of the task situation is even greater than that found for experimental test situations, amounting usually to at least 75% of the total variance, frequently to as much as 90%.

3.3.4 CRITICAL ANALYSES OF ACKERMAN

The most definitive critical analysis of the transfer of training and differential learning studies discussed above is that of Ackerman (1986, 1987, 1988). In the 1987 paper, he reanalyzed data from the 24 studies of Kincaid (1925) (noted above), converting all scores to reaction times to introduce consistency in comparing practice effects on performance variability (σ) across studies. Ackerman's results show that for 21 of the 24 studies, the level of σ after training was significantly lower than that before training, concomitant with an improvement in mean performance across all studies.

This suggests that individual differences in cognitive performance decrease or converge with practice. However, this conclusion may not apply to cognitive tasks lacking an attribute termed *consistency*. Specifically, Ackerman (1986, 1987) altered the consistency of a series of verbal and spatial memory tasks by changing the mapping between category items to be memorized and exemplars of those categories. Consistent tasks display practice effects (mean performance improvement with reduced σ), which mirror those found for the studies reviewed by Kincaid (all of which also employed consistent cognitive tasks). Inconsistent memory tasks also display mean performance improvement with practice, but σ levels show little or no change. These observations on differences in the effects of task consistency versus task inconsistency on the contribution of task specificity to performance variability (also applied to the results of Fleishman; see below) represent a highly distinctive contribution of Ackerman to our understanding of sources of variability in cognitive performance.

Ackerman (1987, 1988) also reanalyzed the results of Fleishman (Fleishman, 1954; Fleishman & Hemple, 1954). In particular, he subjected findings from Fleishman's original individual difference studies of three tasks—complex coordination, discrimination reaction time, and rotary pursuit—to more rigorous statistical analysis. Based on the results, Ackerman challenged the validity of conclusion (d) in the above quote by Fleishman (1966) (who had concluded from his 1954 results that there was an increase in a factor specific to the task itself reflected in performance changes observed with practice). Specifically, Ackerman's reanalysis shows that task-specific variance increased during training only for the complex coordination task. However, for the three tasks, task-specific factors (as opposed to factors related to prior abilities or intelligence) still accounted for 40, 65, and 69 percent (respectively) of total performance variance before training, and 54, 62, and 67 percent after training.

In an approach parallel to that described above for his evaluation of the Kincaid results, Ackerman (1987) followed up his reanalysis of the Fleishman data by conducting studies of cognitive tasks that also differed in their consistency. For three consistent cognitive tasks, task-specific variance increased from the range 41 to 58 percent before practice, to the range 62 to 83 percent after practice. For four inconsistent cognitive tasks, task-specific variance decreased from the range 44 to 67 percent before practice to the range 18 to 34 percent after practice.

Figure 3.1 illustrates in graphical form the contribution of task specificity to performance variability with cognitive tasks before and after practice, with consistent versus inconsistent task designs. The percentage of task-specific variance before practice is plotted on the abscissa; the percentage of task-specific variance after practice is plotted on the ordinate. The open circles are results for three consistent cognitive tasks, and the + signs are results for four inconsistent cognitive tasks evaluated by Ackerman (1987) and described in the preceding paragraph. The closed circles are results for three consistent cognitive tasks evaluated some three decades earlier by Fleishman (1954, 1960) and Fleishman and Hempel (1955).

Results in Figure 3.1 show that the contribution of task-specific variance to total performance variance before practice is less than 50 percent for three consistent tasks, but greater than 50 percent for three other consistent tasks; after practice is greater than 50 percent for all six consistent tasks; before practice is less than 50 percent for three of the four inconsistent tasks; and after practice is markedly less than 50 percent for all four inconsistent tasks. Ackerman's (1987) interpretation

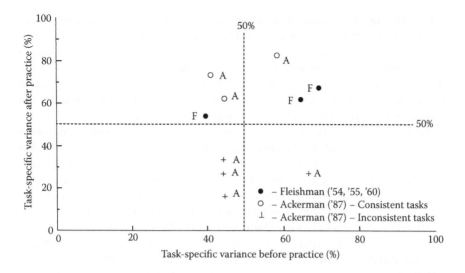

FIGURE 3.1   Variance in performance before and after practice, attributable to task-specific factors, for 10 cognitive tasks. (From Ackerman, P.L., *Psychological Bulletin*, *102*(1), 3–27, 1987; Fleishman, E.A., *A factorial study of psychomotor abilities* [USAF Personnel & Training Research Center Research Bulletin No. 54–15]. Dayton, OH: Wright-Patterson USAF Base, 1954; Fleishman, E.A., *Journal of Experimental Psychology*, *60*, 162–172, 1960; Fleishman, E.A., & Hempel, W.E., Jr., *Journal of Experimental Psychology*, *49*, 301–312, 1955.)

of the large body of evidence from transfer of training and differential learning research over most of the twentieth century, with an emphasis on the results of Kincaid (1925) and those of Fleishman (1954, 1960) and Fleishman and Hempel (1954, 1955), favors an integration of individual difference and information process-ing theories. From this perspective, his key conclusion is that whatever the contribu-tion that task-specific variance may make to total variance observed in cognitive task performance, individual differences in postpractice performance are associated with innate cognitive and intellectual abilities. Nevertheless, as results in Figure 3.1 show, task-specific variance persists in contributing to total performance variance even in the face of cognitive performance challenges posed by tasks with inconsis-tent designs.

3.3.5 SUMMARY

This section has traversed a century of transfer of training and differential learn-ing research, whose implications for documenting the role of context specificity in cognitive performance are well recognized, epitomized in the reviews of Ackerman (1987, 1988) and Adams (1987). However, in the field of HF/E today, this material is largely ignored. For example, in the first, third, and fourth editions of *Handbook of Human Factors and Ergonomics* (Salvendy, 1987, 2006, 2012), there are no chapters devoted to performance variability, and no index entries for the terms "performance variability," "task specificity," or "variability."

The experimental findings summarized in this section are highly consistent with regard to theme of context specificity. The results of early research clearly indicate that cognitive performances are task specialized in relation to transfer effects between different tasks, as well as in relation to the predictive significance of initial performance in forecasting final performance. Results of subsequent research confirm and extend these conclusions by demonstrating that no general factor can account for the variability observed in cognitive performances.

The range and variety of cognitive tasks and types of cognitive performance that have been explored in this work merit emphasis. For example, the 24 studies reviewed by Kincaid (1925) dealt with cognitive demands requiring knowledge/concept skills (arithmetic, naming opposites), visual perception (canceling numbers or letters, color naming), and perceptual-motor dexterity (typing, sorting, maze tracing). The research of Fleishman and colleagues involved over 200 tasks using some 20 different testing devices, with 52 different ability factors (all but seven of which have cognitive attributes) specified in the effort to link cognitive task performance to generalized ability (Fleishman, 1966; Fleishman & Quaintance, 1984).

With the possible exception of observations pertaining to practice effects on inconsistent tasks made by Ackerman (1987), research spanning nine decades leads to the conclusions that cognitive performance and learning are highly task-specific, vary primarily in relation to task design characteristics, differ in no significant way from other types of motor-sensory performance in regard to task specificity, and are specialized during and at the end of learning in terms of the specific variance attributable to task makeup. Thus, the proficiency of cognitive performance and learning cannot be attributed to stimulus determination of learning, inherent ability, or training effect, but rather to the human factors design of the task itself (Jones, 1966, 1969).

3.4   VARIABILITY IN PSYCHOMOTOR PERFORMANCE

Although the emphasis in the preceding section was on evidence documenting the role of task specificity in cognitive performance variability, studies of Adams (1953) and of Fleishman (1954, 1966) and Fleishman and Hemple (1954) were cited indicating that this evidence also is applicable to selected psychomotor tasks evaluated by these authors. This section first introduces an interesting example of transfer of training evidence in the psychomotor domain. The example recapitulates the pattern of findings from the much earlier cognitive study of Poffenburger (1915) (Table 3.1) in implicating task specificity as major contributor to performance variability. The section then goes on to call attention to three of the most robust and reproducible "laws" of psychology—Fitts' law, the Hick-Hyman law, and the law of practice— each of which predicts variability in performance that is prominently influenced by context specificity.

3.4.1   Transfer of Training among Different Balance Tests

Table 3.2 lists cross-correlations of performance for six different tests of static and dynamic balance (Schmidt & Wrisberg, 2014, Table 7.2). *A priori*, a reasonable

TABLE 3.2

Cross-Correlations among Six Tests of Static and Dynamic Balance

	Stork Stand	Diver's Stand	Stick Stand	Sideward Stand	Bass Stand	Balance Stand
Stork stand	–	.196	.144	.0676	.0400	.0009
Diver's stand		–	.144	.0009	.0049	.0196
Stick stand			–	.0016	.0484	.0361
Sideward stand				–	.0961	.0361
Bass stand					–	.0324
Balance stand						–

Source: Reproduced from Table 7.2 of Schmidt, R.A., & Wrisberg, C.A. (2014). *Motor Learning and Performance: From Principles to Application* (5th edition). Champaign, IL: Human Kinetics.

postulate might be that once an individual has mastered the motor skill of balancing, this skill should be transferable to different tests aimed at assessing proficiency in balancing. In fact however, results in Table 3.2 show a remarkable similarity to the cognitive task results of Poffenburger (1915) (Table 3.1) in indicating that performance proficiency with a given balance test does not predict proficiency with any of the others. The highest correlation coefficient (r) in Table 3.2 is 0.31 (between the bass stand and the sideward stand balance tests), which means that variance in performance for one of these tests accounts for only 9.6 percent ($r^2 = 0.31 * 0.31$) of performance variance for the other. The lowest correlation coefficient in Table 3.2 is .03 (for two different pairs of balance tests), which means that performance variance for one test of either pair accounts for only 0.1 percent of the variance for the other member of the pair.

In summary, results in Table 3.2 indicate that the vast preponderance of variance in performance proficiency for any of the six balance tests listed in the table is attributable to some factor or factors other than balance performance proficiency with any of the other tests. This research did not assess what these factor or factors might be. Nevertheless, as with the Poffenburger results (Table 3.1), results in Table 3.2 implicate task specificity as a major contributor to observed variability in balance test performance.

3.4.2 Fitts' Law: Context Specificity in Movement Time Performance

Fitts (1954) originally observed that the time required for reciprocal hand-arm movement between two targets can be predicted from two task design factors: distance between the targets and their diameter. Equation 3.1 expresses Fitts' law in its original formulation.

$$MT = a + b*(ID) \tag{3.1}$$

where

$$ID = \text{Index of Difficulty} = \log_2(2A/W) \tag{3.2}$$

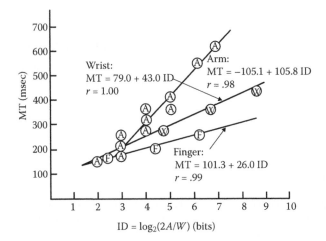

FIGURE 3.2 Compliance with Fitts' law by three upper limb effectors—finger, wrist, and arm. (Reproduced from Langolf, G.D. et al., *Journal of Motor Behavior*, 8(2), 113–128, Figure 6, p. 120, 1976.)

and where A = distance between targets, and W = target diameter (the assumption is that both targets have identical diameters).

Fitts' law has proved to be extremely robust, accounting for well over 90 percent of the variance in movement time performance under a variety of experimental paradigms for a number of different muscle-joint systems (Langolf et al., 1976). Figure 3.2 reproduces Figure 6 in the report by these authors, showing compliance with Fitts' law for three different effector movements, namely finger, wrist, and arm movements. In the figure, ID is plotted on the abscissa, and movement time (MT), in msec, on the ordinate. For each of the three effectors, the regression of MT on ID features markedly high r values, ranging from $r = 0.98$ (for the arm) to $r = 1.00$ for the wrist. Given that ID is defined by the task design factors of target width and target separation, results in Figure 3.2 underscore the point that most of the variability in a behavioral parameter—movement time—is attributable to task design features. In general terms, the predictions of Fitts' law are highly context-specific.

3.4.3 Hick-Hyman Law: Context Specificity in Choice Reaction Time Performance

In the early 1950s, Hick (1952) and Hyman (1953) documented a robust relationship between the number of choices that an individual was asked to react to and the time that elapsed for a reaction time decision to be made. Mathematically, the Hick-Hyman law is expressed as the following linear equation:

$$\text{Choice RT} = a + b[\log_2(N)] \qquad (3.3)$$

where RT = reaction time, N = number of reaction time choices, b is the slope of the line relating RT with $\log_2(N)$, and a is the y-axis intercept of this line.

Figure 3.3 illustrates results for the Hick-Hyman law for four different subjects originally evaluated by Hyman in his 1953 paper (p. 192) (subject initials are in the upper left corner of each panel). With each subject, three different experiments were conducted to generate the results in each panel of the figure. For each panel, the $\log_2$ of the number of reaction time choices (labeled as stimulus information in bits) is plotted on the abscissa (actual choices thus range from 2 to 8), and the reaction time in msec is plotted on the ordinate. Results in Figure 3.3 point to the following key conclusions about the Hick-Hyman law: (1) choice RT is linearly related to $\log_2$ of the number of reaction time choices (the linear equation is listed in each panel of the figure), (2) the tight regression between these two variables for each subject, with relatively little scatter, indicates that the number of reaction time choices is highly predictive of the time that a subject takes to react to the choices, and (3) as shown by the linear equations for the results for each subject plotted in the figure, there are marked individual differences in the RT-N relationship (the same conclusion applies to Fitts' law and the law of practice). There is a substantial body of evidence

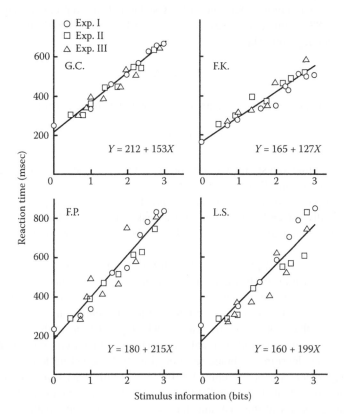

FIGURE 3.3 Compliance with the Hick-Hyman law based on three repeated stimulus-response studies of four different subjects. (Reproduced from Hyman, R., *Journal of Experimental Psychology*, 45, 188–196, with approval by the American Psychological Association, Figure 1, p. 192, 1953.)

bearing upon the Hick-Hyman law that supports these conclusions (Schmidt, 1988, pp. 80–88).

The number of reaction time choices that a subject may be asked to react to in a choice reaction time condition is a design factor. Studies of the Hick-Hyman law document a strong predictive relationship between choice reaction time and this design factor. Thus, as with Fitts' law, results for the Hick-Hyman law demonstrate a high degree of context specificity.

3.4.4 THE LAW OF PRACTICE: CONTEXT SPECIFICITY IN LEARNING

In the previous section, evidence from differential learning research was summarized to support the conclusion that task specificity makes a substantial contribution to variability observed in cognitive learning performance. This review of the law of practice extends these results in indicating that the actual temporal course of learning (i.e., the learning curve) also is highly context-specific.

The law of practice specifies that incremental improvements in learning continue to occur as the number of repetitions of the task being learned progressively increase (Schmidt, 1988; Schneider, 1985). Mathematically, this relationship is expressed with the following power law:

$$T = B * N^{-\alpha} \tag{3.4}$$

Equation 3.4 is expressed as a linear equation by logarithmic conversion:

$$\log(T) = \log(B) - \alpha\log(N) \tag{3.5}$$

where T is the time to respond to or complete the task being learned, N is the number of task repetitions, and B and α are constants.

As Equation 3.5 indicates, the law of practice predicts that performance will improve rapidly for early repetitions of the task being learned, but will also continue to improve at a decreasing rate as the number of task repetitions progressively increase.

Studies of the law of practice show high linear correlations, based on Equation 3.5, between learning performance ($\log(T)$) and task repetitions ($\log(N)$) across a broad range of cognitive, psychomotor, and motor tasks (Schmidt, 1988, pp. 459–460), a consistency that has prompted Newell and Rosenbloom (1981) to label the relationship as ubiquitous. Specific tasks for which predictions of the law of practice have been documented include manual cigar rolling, adding digits, editing text, playing card games, learning a choice reaction time task, detecting letter targets, and performing geometry proofs (Schneider, 1985). In a university-level motor learning course that I have taught, the predictive power of the law of practice also has been demonstrated for a knitting task and for the task of cursively writing one's name upside down and backwards.

Figures 3.4 and 3.5 illustrate two examples of the predictive power of the law of practice. Figure 3.4 is a plot, on a log-log scale, of results from a mirror-tracing learning task evaluated by Snoddy (1926, p. 11) and replotted by Schmidt (1988,

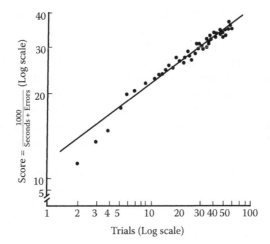

FIGURE 3.4 Results from a 100-trial study of improvement in performance on a mirror-tracing task illustrating compliance with the law of practice. (Reproduced from Snoddy, G.S., *Journal of Applied Psychology*, *10*(1), p. 11, 1926.)

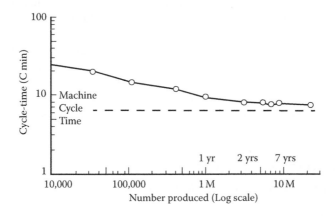

FIGURE 3.5 Seven-year study of improvement in manual cigar assembly task performance illustrating compliance with the law of practice. (Reproduced from Crossman, E.R.F.W., *Ergonomics*, 2, 153–166. Taylor & Francis Ltd. http://www.tandfonline.com, 1959.)

p. 458). The logarithm of the number of trials (100 total) is plotted on the abscissa and the logarithm of the task performance score (= 1000/(seconds + errors)) on the ordinate. The task performance score increases progressively as the number of task repetitions increases. The tight log-log linear regression demonstrates the predictive power of the law of practice.

Figure 3.5 illustrates the result of one of the most heroic studies in the history of psychology (Crossman, 1959). The study involved measurement of the elapsed time required for a single subject to manually assemble cigars, starting with novice performance; tracking task performance over the ensuing seven years, and comparing time required for manual cigar-making with that for a cigar-making machine. In Figure 3.5, the log of the number of cigars produced manually is plotted on the abscissa (10 million total produced at the end of seven years), and the log of the number of cigars produced manually per minute (the cycle time) on the ordinate. The manual cycle time dropped from about 25 sec/cigar at the manual production stage of 10,000 cigars, to about 7 sec/cigar at the end of seven years, close to the machine cycle time. In accord with predictions of the law of practice, the log-log relationship displays two linear phases, one featuring more rapid improvement in manual task cycle time out to two years, and a second phase featuring less rapid improvement between two and seven years.

The number of task repetitions presented to a subject during a task-learning condition represents a design factor. Studies of the law of practice document a strong predictive relationship between the number of task repetitions and incremental improvement in task performance. Thus, as with Fitts' law and the Hick-Hyman law, results for the law of practice demonstrate a high degree of context specificity.

3.4.5 SUMMARY

The analysis in this section has barely scraped the surface regarding the broad variety and modes of psychomotor performance that have been evaluated by the motor performance and psychology communities (Schmidt, 1988). The objective rather has been to call attention to the prominent role of context specificity in explaining the predictive power of three of the most robust and reproducible laws in psychology, namely Fitts' law, the Hick-Hyman law, and the law of practice. The possible role of context specificity in contributing to performance variability observed with other modes of psychomotor performance merits closer study. Transfer of training results for balance test performance reviewed at the outset of this section suggests that such study might be rewarding.

3.5 QUALITATIVE OBSERVATIONS ABOUT TASK PERFORMANCE VARIABILITY

There are numerous examples of qualitative, nonempirical observations of the influence of design factors in the performance environment on variability in performance. This section focuses on three such illustrative examples relevant to cognitive and motor performance: language, situation awareness, and a series of context specific observations by Gladwell (2008).

3.5.1 LANGUAGE

Few more dramatic examples of variability in cognitive performance can be cited than the proliferation of human languages. An estimated 6000–7000 languages are thought to exist today, many on the verge of disappearing, and only about 1000 of which have been analyzed in any detail (Wuethrich, 2000). K.U. Smith and Smith (1966, pp. 84–107) briefly review the origins of language and its behavioral manifestations of speech, writing, and reading. The emergence of language in the course of hominid evolution is now thought to date back hundreds of thousands of years—current archeological and genetic evidence suggests that Neanderthals may have been capable of both symbolic behavior and language.

There is a vast literature on the sources and nature of language, from behavioral, biological, evolutionary, and linguistic perspectives, whose analysis is well beyond the scope of this section—at least five major books on language evolution alone have been published in the past fifteen years (Calvin & Bickerton, 2000; Fitch, 2010; Hurford, 2007; Larson et al., 2010; Lieberman, 2006). Instead, the focus here will be on the narrow question of whether context specificity plays a role in language expression.

The basic answer to this question is that despite decades of research, the question remains unresolved. The two sides of the debate can be characterized using the nativist versus empiricist framework of Goldhaber (2012). The nativist champion is Chomsky (1965), who has maintained for decades that humans have a species-specific, innate knowledge of universal grammar, a mechanism for generating all possible human linguistic structures across all languages (Cook & Newson, 2007). Holden (2005) cites findings from a study of three Nicaraguan deaf children that appear to support the Chomsky perspective. These children had no contact with anyone who knew formal sign language, yet they developed, apparently based on an innate mechanism, a "home" sign language with a conventional subject-verb linguistic structure.

As summarized below, however, a number of observers have taken issue with the Chomsky point of view by emphasizing the likely role of learning in the developmental mastery of language.

- Stetson (1951) favored a motor phonetic theory that assumed that language learning occurred as a consequence of deployment of motor behaviors required for language expression through speech.
- Enfield (2010) points out that in favoring a nativist perspective, Chomsky has ignored decades of linguistic research that calls this perspective into question, and that instead favors an empiricist interpretation of language development.
- Enfield (2010) also points out that social interaction likely plays a major role in language learning.
- Fitch (2010) acknowledges the possibility that a "narrow language facility" (such as a universal grammar) may not exist. Hurford (2007) discusses the idea—attracting growing support among cognitive scientists—that language expression relies on a constellation of cognitive capacities that individually may subserve other functions as well. Evidence introduced in Section 3.3 that task specificity contributes to variability observed in cognitive performance supports the hypothesis that context specificity may contribute, at some level, to the behavioral modality of language expression.
- Calvin and Bickerton (2000) point to the likelihood of a link between motor skills and use of grammar.
- Sign language relies on motor behavior for purposes of communicating language meaning, and the mastery of sign language reflects a motor skill. Hundreds of sign languages have been developed around the world, often

to serve the needs of local deaf cultures. Professional linguists have studied many sign languages and have concluded that they exhibit the fundamental properties that exist in all languages and are as rich and complex as any spoken language (Fox, 2007; Sandler & Lillo-Martin, 2006). Sign languages, like spoken languages, organize elementary, meaningless units (analogous to phonemes in spoken language) into meaningful semantic units. Identical areas of the brain are activated during expression of both spoken and sign language. More generally, both sign and spoken languages share the following common features that linguists have found in all natural human languages: communicative purpose, semantics, and cultural transmission.

Therefore, with sign language we have a motor behavioral skill that meets all the criteria of human language and whose mastery requires motor learning. In Section 1.3 of Chapter 1, motor behavior was characterized in control systems terms, and it was noted that the role of context specificity in influencing variability in control system behavior is implicit in all control models of behavior. Sign language thus represents a bona fide language whose mastery relies upon experience (learning) and whose expression is prominently influenced by context specificity.

- In a study not related to human language but highly suggestive nevertheless, most songbirds have been shown to learn their species-specific song by listening to others of their species who have mastered such singing; they develop abnormal song patterns if deprived of such experience. More recent investigation of birdsong reveals detailed and intriguing parallels with human speech (Marler, 1970, 1997).

Collectively, the above points suggest, but do not definitively prove, that context specificity plays a role in the acquisition and expression of spoken language. Nevertheless, the idea that there are innate neural mechanisms that mediate language ability, as championed by Chomsky (1965), remains a viable thesis in the linguistic community. Arguably, an objective interpretation of both available evidence and prevailing linguistic and cognitive science opinion is that language ability requires both context-specific and innate mechanisms. Indeed, in a recent coauthored article, Chomsky himself implicitly appears to support this point of view (Hauser et al., 2002). In this analysis, arguments for both broad and narrow components of language are advanced. The narrow component is assumed to rely entirely on internal (innate) computation. The broad component is assumed to rely on the narrow component as well as sensory-motor and learning behaviors. Given evidence already adduced of context specificity associated with variability observed in these latter two behaviors, the analysis of three of the most distinguished linguists in the field arguably favors a composite theory of language acquisition and expression.

3.5.2 Situation Awareness

It is fair to say that few concepts have captured the imagination of the HF/E community over the past two decades more than the notion of situation awareness. A search of this term on Google Books (http://www.google.com/advanced_book_search)

retrieved just over 1000 instances of use of the term in different books for the period of January 1990 through May 2013. In contrast, for the period of January 1970 through December 1989, only 250 such instances were retrieved.

The undisputed champion in the HF/E community for the concept of situation awareness is Endsley (Endsley, 1988, 1995a, b; Endsley & Garland, 2000; Endsley & Jones, 2011), whose consulting company (http://www.satechnologies.com) eponymously references the term. According to Webster (McKechnie, 1983), "situation" is defined as, "being located or positioned with reference to the environment," and the definition of "awareness" has cognitive connotations, "having knowledge or cognizance." Endsley (1988) expands on these specifications with a broader conceptual definition of situation awareness, "the perception of the elements in the environment within a volume of time and space, the comprehension of their meaning, and the projection of their status in the near future." Her theoretical model of situation awareness (Endsley, 1988, 1995a) essentially represents a modification of the traditional feedback control model (Figures 1.6 and 1.8), with perception (awareness) of environmental conditions (the situation) interposed as an additional control systems stage between environmental stimulation of sensory receptors and controller decision-making.

The idea of situation awareness has attracted its detractors, who point to such concerns as the vagueness of the whole concept (the term "situation" can refer to practically any condition, and "awareness" as a behavioral construct is equally hard to pin down), and the consequent implication that the existence of awareness of a situation essentially is nonfalsifiable. As it was gaining in popularity, Wiener (1993, p. 4) referred to the concept as a "buzzword of the '90s." Sarter and Woods (1991) maintain that developing a definition of situation awareness is futile and not constructive. Finally, the putative role of situation awareness for purposes of projective behavior ("projection of the status of the environment into the future") is muddled in the Endsley model, in that the model does not actually specify the behavioral stages thought to be required for behavioral feedforward control (Figure 1.7). Flach (1995) offers a thoughtful perspective on the meaning of situation awareness:

> The argument that situation awareness (SA) is valuable as a phenomenon description draws attention to the intimate interactions between human and environment in determining meaning (or what matters) and reflects an increased appreciation for the intimate coupling between processing stages (e.g., perception, decision, and action) within closed-loop systems. However, I caution against considering SA as a causal agent. When SA is considered to be an object within the cognitive agent, there is a danger of circular reasoning in which SA is presented as the cause of itself. As a causal explanation, SA is a simple, easy-to-understand wrong answer that, in the end, will be an obstacle to research.

As noted above, the dictionary and the Endsley definitions, as well as Flach's perspective, all maintain that situation awareness arises as a consequence of the influence of environmental conditions on a cognitive process. In other words, the phenomenon of situation awareness embodies critical elements of context specificity. In maintaining that it is futile to try to determine the most important contents of situation awareness because the significance and meaning of any data are dependent on the context in which they appear, Sarter and Woods (1995) appear to share this conclusion.

Furthermore, in her advocacy of the concept, Endsley (Endsley, 1995a, b; Endsley & Jones, 2011) has emphasized the importance of situation awareness for understanding how and why operator behavior varies during interaction with dynamic systems of different designs. These include process control, military command and control, systems mobilization to deal with pandemics, aircraft operation, air traffic control, remote vehicle control, and other complex sociotechnical systems, most requiring team coordination. The putative role of context specificity in situation awareness, argued above, implies that context specificity makes a prominent contribution to variability observed in the performance of complex sociotechnical systems (presumably based in part on situation awareness). This thesis will be addressed in more detail in Chapter 9.

Endsley and Jones (2011) devote an entire book to a discussion of how application of situation awareness theory and methods can benefit user-centered design. Their assertion is (p. 12) that by developing design methods that keep users in control and at high levels of situation awareness, user-centered designs will be created. This assertion ignores the extensively documented conclusion of Gould (1990), discussed in Section 3.3, that it is not typically possible to develop effective user-centered designs by making *a priori* assumptions about how these designs will influence user performance (Gould advocates iterative usability testing as the preferred approach to user-centered design development).

Despite shortcomings outlined above, the concept of situation awareness is firmly ensconced in the HF/E firmament. The scientific credibility of the concept undoubtedly would be enhanced if these shortcomings were addressed. Furthermore, from the perspective of this chapter, the current assumption that situation awareness relies exclusively on information processing mechanisms (Endsley, 1995a) requires modification to acknowledge the likelihood that context specificity represents a major source of variability in any performance based on situation awareness.

3.5.3 CONTEXT SPECIFIC OBSERVATIONS OF GLADWELL

In his popular book, *Outliers*, Gladwell (2008) relates a series of interesting stories of success based on differential performance. I describe them here because the common denominator is the prominent role that a design factor plays in contributing to the performance success described by Gladwell with each story. A synopsis of the Gladwell stories is provided in Table 3.3, with the different types of performance success listed in the first column and the design factor contributing to each success listed in the second column. For the different types of performance detailed by Gladwell, practice (training), socioeconomic background, duration of classroom exposure to learning, and birth year effect are the design factors implicated as major contributors.

Some may question the assumption in Table 3.3 that birth year effect represents a legitimate design factor. To be sure, accidental or unanticipated pregnancy is a common occurrence. On the other hand, planned pregnancies are not uncommon either. Thus, it is likely that a planned birth year effect explains the elite hockey or

TABLE 3.3

Stories by Gladwell (2008) about Variability in Different Modes of Performance that Appears to Be Attributable Primarily to Design Factors

Type of Performance Success	Design Factor
Elite (i.e., NHL) Hockey Performance/Skill	Birth Year Effect
World Class Musical Performance (e.g., The Beatles)	Practice, Practice, Practice
Achieving Extraordinary Wealth (e.g., 20% of richest individuals in history)	Birth Year Effect
Career Success	Family Socioeconomic Background
Becoming a Highly Successful Lawyer	Practice, Practice, Practice
Academic Achievement by Children	Length of Time in Classroom
Lack of Predictability of Future Performance (i.e., in job interview and hiring decisions, talent assessment)	Variability in Performance is Most Prominently Influenced by Design Factors in the Performance Environment, Not by Innate Ability or Skill

extraordinary wealth performance success of a subset of the individuals described by Gladwell.

3.5.4 SUMMARY

Many qualitative examples of context specificity featuring variability in different types of performance influenced by design factors could be cited. The three examples discussed above are selected because language represents one of the most dramatic examples of human performance variability, the concept of situation awareness enjoys wide popularity among the HF/E community, and the stories of Gladwell (2008) are consistent in attributing performance success to design factors, and frankly also are fascinating. Unlike quantitative evidence for task specificity derived from research on transfer of training, differential learning, or selected examples of psychomotor performance (Sections 3.3 and 3.4, respectively), evidence for context specificity with the three examples discussed above is not definitive. Nevertheless, persuasive arguments are presented to suggest that design factors represent a key source of performance variability observed in each case.

3.6 CONCLUSIONS

Underlying the discussion throughout this chapter is the scientific and philosophical debate between those who believe that innate biological factors are most important in influencing human performance variability (the nativists), versus those who believe that design—context—is most important (the empiricists). This debate dates back to the ancient Greeks, but it was John Locke in the latter half of the seventeenth century who staked out a pronounced empiricist position that galvanized philosophical and scientific support and opposition during the intellectual ferment in Europe during the seventeenth and eighteenth centuries. As noted earlier, Locke's epistemology maintains that (1) all knowledge is derived through experience, (2) we cannot know what is not within our experience, and (3) because our knowledge is derived only through our senses, based on experience, knowledge is not logically determined and consequently is only probabilistic in nature. Based on these assumptions, Locke went on to conclude that (1) we are products of our environment, which, if changed, would change our very nature, and (2) our character and our knowledge therefore are relative to time, place, circumstance, and experience (Kors, 1998, Lecture 11).

I have rehashed this Lockean agenda to emphasize the point that all the evidence regarding the role of context specificity in human performance variability, accrued in ensuing centuries, serves to validate Locke's original assumptions. For example, Locke's premise that all knowledge is probabilistic helps us understand why prediction of systems performance can be so problematic, particularly as system complexity increases (Chapter 1, Section 1.2.2). Indeed, in his book on this topic, Silver (2012) emphasizes the point that statistical analysis (based on Bayesian statistics) can improve the reliability of systems performance prediction.

In this and other chapters of this book, my emphasis on context specificity is not intended to minimize or ignore the undeniable contribution that innate biological factors make to observed variability in human performance. However, it is a relatively straightforward proposition to modify the design of the performance environment in order to modify performance, preferably in a more favorable direction. Modifying innate biological factors is far more problematic. This is why I have chosen to emphasize context specificity as a key source of performance variability in this chapter and throughout this book. The following summary of key implications of context specificity thus represents an appropriate way of concluding the chapter.

1. In many cases, the preponderance of variability in performance is attributable to design factors in the performance environment.
2. Therefore, to change performance, change design.
3. New designs evoke new modes and patterns of behavioral variability that cannot necessarily be predicted in advance. This implication provides the rationale for usability testing.
4. For an individual or group to effectively control their behavior, they must be able to predict the sensory consequences of their actions.
5. If a user can predict the sensory consequences of interaction with a given design, then behavior during such interaction can be effectively controlled—we conclude that such designs have good usability.
6. If a user cannot predict the sensory consequences of interaction with a given design, we conclude that such designs have poor usability.
7. A law is a design factor aimed at evoking/promoting comparable patterns of variability in behavior across multiple individuals.
8. However, any given law always has unintended consequences (implication 3).
9. Implications 5 and 6 provide insight into the basis of the observation that good usability of a given design has positive economic consequences.

4 Educational Ergonomics
Context Specificity in Student Learning

Thomas J. Smith

4.1 OVERVIEW

Chapter 3 reviewed evidence from a century of transfer of training and differential learning research supporting the conclusion that context specificity represents an important source of variability in cognitive, learning, and psychomotor performance. This chapter extends this analysis to learning by K-12 students (i.e., kindergarten [typically age 5] through high school senior [typically ages 17–18] students) (portions of this chapter adapted from T.J. Smith [2007, 2013]). The goal is to discuss and delineate design factors in K-12 learning environments that have been shown to influence academic achievement on the part of K-12 students, a field of study termed educational ergonomics (Kao, 1976; Legg & Jacobs, 2008; Lueder & Berg Rice, 2008; K.U. Smith & Smith, 1966; T.J. Smith, 2007, 2009, 2011, 2012, 2013; T.J. Smith et al., 2009; Stone, 2008). To recapitulate the debate highlighted in Chapters 1 and 3 (Goldhaber, 2012), empiricist evidence for the influence of context—design—on student learning will be contrasted in this chapter with the nativist assumption that innate cognitive mechanisms account for much of the variability observed in student learning, an idea that continues to persist in educational psychology.

It should be emphasized that attribution of the role of context specificity in student learning addressed in this chapter is based on observational evidence rather than the more stringent experimental analysis of differential learning described in Chapter 3. However, it can be argued that the nature and extent of this observational evidence is substantial enough to provide strong support for the conclusion that context specificity plays a significant role in variability observed in the learning of children and young adults, as also demonstrated for more mature individuals.

4.2 INTRODUCTION

Remedies and nostrums proposed to mitigate the maladies undermining school performance in the United States and elsewhere are almost as numerous as the number of children targeted by these proposals. This chapter introduces a conceptual framework and empirical evidence for the conclusion that a common theme underlies almost all of these proposed solutions, namely their reliance on the design—that is, the HF/E—of educational system features and operations (Hourcade, 2006; Kao,

1976; Legg, 2007; Lueder & Berg Rice, 2008; K.U. Smith & Smith, 1966; T.J. Smith, 2007, 2009, 2011, 2012, 2013; T.J. Smith et al., 2009; Stone, 2008). As noted above, the focus of this analysis is on students in K-12 classrooms.

It is worth noting at the outset that good ergonomic design of learning environments has potential economic as well as learning benefits. Total employment of elementary, middle, and secondary school teachers in the United States exceeded 3.5 million in 2012, representing 2.7 percent of total U.S. employment, the thirtieth largest occupational group in the country (http://www.bls.gov/oes/#data). The job performance of this important group of workers, as well as the learning performance of the some 48 million K-12 students (Picciano & Seaman, 2007) whose development is so critically influenced by these workers, each should be positively influenced by application/adoption of good learning environment ergonomics.

As noted in an earlier paper on the ergonomics of learning (T.J. Smith, 2007), the educational community has yet to generally recognize the contributions that H/E science might make to improving student learning and the performance of educational systems. That is, of all the sociotechnical systems areas whose design features received attention from HF/E, education has been least affected (the term sociotechnical system refers to a conceptual approach [Trist, 1981] that emphasizes the need to integrate social behavioral, technical engineering, and organizational requirements in the design of complex systems). The application of HF/E principles and practices has achieved proven success in improving performance, productivity, competitiveness, and safety and health in most occupational sectors. In contrast, in the educational arena, the benefits that the application of HF/E science to the design of educational systems features and operations might bring to promoting student learning have yet to be widely recognized. This is particularly true for K-12 education.

Thus, the relevance of HF/E to education receives little or no attention in analyses over the past four decades devoted to problems with U.S. K-12 school systems (Berliner & Biddle, 1995; Carnegie Foundation for the Advancement of Teaching, 1995; Davison, 2013; Denby, 2010; Grissmer et al., 2000; Rhee, 2010; Ripley, 2010; Silberman, 1970; U.S. Department of Education, 1983; Wilson & Daviss, 1995). This disregard likewise is true of the content of the 2010 issues of three U.S. news magazines—*Newsweek* (December 13), *Time* (September 20), and *U.S. News and World Report* (January)— each of which has a series of articles devoted to the need for reform in U.S. K-12 education. In the 1996 and 2006 editions of *Handbook of Educational Psychology* (Alexander & Winne, 2006; Berliner & Calfee, 1996), with 33 and 41 chapters, respectively, and over 1000 pages each largely devoted to K-12 education, HF/E design of the learning environment is nowhere mentioned as a key aspect of the psychology of learning.

Despite this state of affairs, an objective look at available evidence for factors or strategies that have been shown to improve performance and learning of K-12 students leads to the conclusion that these interventions, in almost every case, are based on changes in the design of learning environments. In other words, school districts, administrators, teachers, educators, and even students themselves are practicing and applying HF/E in all but name. The remainder of this chapter first provides a brief review of the traditional view of the nature of student learning from the perspective of educational psychology. The alternative HF/E view of learning as a context-specific process based on an array of evidence alluded to above then is explored in more detail.

4.3   THE NATURE OF STUDENT LEARNING FROM THE PERSPECTIVE OF EDUCATIONAL PSYCHOLOGY

Why have K-12 educators largely ignored the relevance of HF/E to understanding the nature of student learning? A possible reason was suggested by K.U. Smith and Smith (1966, p. 1): "Factors of human design long have been ignored in experimental psychology. It has been believed that learning could be studied as a general process." Although a large body of evidence regarding context specificity in learning can be cited (T.J. Smith, 1994a; T.J. Smith et al., 1994a; reviewed in Chapter 3) to contradict generalized learning theory, there is no question that the latter viewpoint still plays an influential role in shaping the understanding of K-12 educators regarding the nature of learning.

The primary influence today on how K-12 school teachers and administrators view the nature of student learning originates with theoretical and empirical perspectives of the field of educational psychology. A brief synopsis of these perspectives is offered here to illustrate the point that the idea of context specificity in learning receives tacit, if not explicit, recognition by the field. For this purpose, the first (1996) and second (2006) editions of *Handbook of Educational Psychology* (Alexander & Winne, 2006; Berliner & Calfee, 1996) are used here as the primary reference sources.

As noted above, the terms "ergonomics" and "human factors" nowhere appear in either editions of the *Handbook*. In the 1996 edition (Berliner & Calfee, 1996), two of the chapters largely are concerned with the nature of learning, that of Greeno et al. (1996) dealing with cognition and learning, and that of Gustafsson and Undheim (1996), dealing with sources of individual differences in cognition and learning.

After an extensive review of various lines of evidence bearing on the idea of generalized intelligence (the "g" factor), the latter authors go on to point out that both genetic and environmental factors have been shown to make essentially identical contributions to observed variance in cognitive abilities. They then address research dealing with the relationship between cognitive ability and learning ability, with the observations that empirical evidence linking cognitive ability and learning ability is weak, and substantial evidence has been compiled pointing to the role of context specificity in learning.

Greeno et al. (1996) observe that with regard to theoretical perspectives on learning, three viewpoints dominate educational psychology, termed the behaviorist/empiricist view, the cognitive/rationalist view, and the situative/pragmatist-sociohistoric view. The first of these views draws on Pavlov's reflex conditioning and Skinner's operant conditioning theories with the assumption that learning emerges as a consequence of the student forming, strengthening, and adjusting associations between stimuli and responses. The second of these views represents a cognitive information processing theory of learning and assumes that learning is equated with conceptual understanding on the part of the substantive mind of the student. The third view emphasizes that learning requires the student to actively participate in the learning process through interaction with learning environments, with both social intercourse and practice representing key prerequisites for learning to occur.

The influence of design on student learning, a topic that receives abbreviated treatment in the 1996 *Handbook* edition (Berliner & Calfee, 1996), is accorded dramatically heightened attention in the 2006 *Handbook* edition (Alexander & Winne, 2006). Thus, comparing the numbers of index topic entries in the two editions with direct relevance to the role of design in student learning, there are 27 index topic entries, plus six entire chapters, in the 2006 edition devoted to the significance of the *context* of the learning environment for learning, relative to three index topic entries only in the 1996 edition; seven index topic entries, plus one entire chapter, devoted to learning environment *design* in the 2006 edition, relative to two index topic entries only in the 1996 edition; and 46 index topic entries devoted to student *self-regulation* of their interaction with learning environments in the 2006 edition, relative to four index topic entries only in the 1996 edition. These differences in index content are reflected in the substantially greater number of pages devoted to the HF/E-relevant topics cited above in the 2006 relative to the 1996 edition of the *Handbook*. The key themes relevant to the influence of design on student learning addressed in the 2006 edition of the *Handbook* may be summarized as follows:

- Bredo (2006) offers a biting critique of three cherished psychological theories of student learning and performance—namely, stimulus-response (input-output), trait/aptitude (including IQ), and substantive mind (i.e., mind as computer) theories—all of which he argues represent manifestations of conceptual confusion. This author observes that over a century ago, Dewey pointed out that stimulus (generated by design factors in the learning environment) and response (performance and learning) are mutually interdependent, a view that essentially represents a cybernetic interpretation of behavior as a feedback process, and one that more recently has been reincarnated in ecological psychology. As for aptitude and IQ, Bredo points to emerging perspectives that thinking skills reside in the student-situation interaction, not solely in the mind of the student, and that the idea of generalized student aptitudes should be replaced by the idea that different students have different context-specific aptitudes critically influenced by the designs of different environments in which they perform and learn. Finally, Bredo dismisses computational models of the mind as irrelevant to understanding the neurobiological basis of cognition and learning, phenomena that should be more realistically viewed as outcomes of student-environment dynamic interaction.
- In their chapter dealing with learning theory and education, Bransford et al. (2006) pay lip service to the generalized concepts of human thought and learning (outlined above) that Greeno et al. (1996) had defended in the 1996 *Handbook* edition, but then go on to point out that concepts such as early childhood learning mediated via neuroplasticity (modification of brain structure and function by experience), social learning, informal learning mediated by student interaction with learning environmental contexts

outside of the formal classroom, and adaptive learning (changes in student aptitudes mediated by changes in contexts of learning environments), provide a more realistic perspective on how student learning actually occurs.

- Schunk and Zimmerman (2006) address the importance of the level of student perceptions of their own capabilities (competence beliefs) and of their ability to self-control their behavior and environment (control beliefs), mediated primarily by social experience, relationships, and interactions, as a key influence on student learning.
- Martin (2006) and O'Donnell (2006) emphasize the central significance of student interaction with the social and cultural substrates of learning environments, and of group learning, as key influences on student learning.
- de Jong and Pieters (2006) discuss design features of powerful online learning environments, with an emphasis on student self-direction, social collaboration, and student engagement in the design process itself as key strategies for supporting student learning.
- In reviewing the influence of technology-rich environments on student learning, Lajoie and Azevedo (2006) emphasize the importance of both the design and social contexts of such environments, as well as that of student self-regulation of interaction with technology, in supporting learning.

In summary, to a limited degree in the 1996 *Handbook* (Berliner & Calfee, 1996), and much more prominently in the 2006 *Handbook* (Alexander & Winne, 2006), the influence of a series of design (i.e., context) factors—social and group learning, informal learning, learning environments that support both student self-regulation and participation in learning, technological interfaces, and the amount of practice—are delineated. In addition, older and more traditional theories of learning as a generic process are dismissed as irrelevant to understanding how students actually learn (Bredo, 2006). In the series of chapters in the 2006 *Handbook* outlined above that deal with the influence of design on student learning, the role of context as an important factor in learning recurrently is emphasized as an emerging theme in educational psychology.

Given this emphasis, what is the point of proceeding with the present review of a series of HF/E design factors that have been shown to influence student learning. A series of answers to this question may be cited. First, this chapter offers an appreciably broader perspective on the array of such design factors. Second, the 2006 *Handbook* treatment of these factors is piecemeal—that is, there is no evidence of a coherent synthesis of the concept of the central role of design in student learning. That the field of HF/E, whose central emphasis is on the interaction of design and performance, is nowhere referenced in either edition of the *Handbook* provides telling support for this conclusion. Finally, there is ample reference by other authors and chapters (not referenced here) in the 2006 *Handbook* and earlier editions, to generic learning theories that assume innate cognitive mechanisms—termed nativist perspectives on learning using the terminology of Goldhaber (2012)—that the authors and chapters cited above tend to refute. In other words, one is left with the impression that the field of educational psychology, as of 2006, can't quite come to grips with what is going on with student learning, a condition of cognitive dissonance if you will.

Finally, and perhaps most importantly, the various claims in the 2006 edition of the *Handbook* that recognition of the impact of design factors on student learning represents an emergent theme in learning psychology portrays only ignorance, not insight. Fully forty years earlier, K.U. Smith and Smith (1966) reviewed the role of many of the design factors cited above in terms of their influence on student learning, presented and defended the argument that generic learning theories have little relevance to education, and went on to point out that, "The main challenge in the science of human learning is to understand the requirements of educational design at all levels" (1966, p. 478). Tellingly, this work is nowhere cited in either edition of the *Handbook*—one is reminded of Hegel's admonition that people (in this case psychologists) have neither learned anything nor acted on principles derived from history. In their seminal works on the critical contribution of design to observed variability in learning performance, both Kao (1976) and K.U. Smith and Smith (1966) delineate an extensive series of design factors that should be considered for supporting student learning (Kao was a graduate student of K.U. Smith).

Let me close this section by noting that the field of educational psychology, despite an evident growing acceptance of the idea that learning and design are related, largely continues to display an ignorance (or a disregard) of the contributions that HF/E science could make to supporting further elaboration of this idea. In addition, some devotees of the field continue to espouse traditional ideas about the nature of learning, ideas that others in the field dismiss. Collectively, these considerations motivate the HF/E perspective on educational ergonomics offered here.

4.4 THE NATURE OF STUDENT LEARNING FROM THE PERSPECTIVE OF HF/E

Conceptual understanding of how and why learning performance and educational design are interdependent—termed an empiricist perspective on learning using the terminology of Goldhaber (2012)—is informed by four theoretical models of human-system interaction: (1) behavioral cybernetics (K.U. Smith & Smith, 1966; T.J. Smith & Smith, 1987a), (2) sociotechnical systems (Emery, 1969), (3) macroergonomics (Hendrick, 1986), and (4) balance theory (M.J. Smith & Carayon-Sainfort, 1989). All of these models are dedicated to the application of HF/E principles and practices to promote human-factored design of complex human-technological-social-institutional systems. Given the evident complexity of educational systems (Caldwell, 1992), it is to be hoped that educational performance can benefit from the application of HF/E systems principles and practices.

All four theories cited above assume that participants in a system should have some degree of control over their interaction with system design features. Sociotechnical and macroergonomic theories both stress the participatory approach—individual participation in decision-making governing system design and direction—as key to effective system performance. Balance theory views people as the center of the system and advocates that other system elements be designed (human-factored) to enhance human performance.

Behavioral cybernetic theory (Chapter 1, Sections 1.3.3 and 1.3.4) views self-control of behavior as a biological imperative and assumes that context specificity in performance arises as a biological consequence of closed-loop behavioral control of sensory feedback from design factors in the performance environment. System designs that assume that participant interaction with the system can be controlled externally (by managers or administrators through limiting information and/or involvement) are judged as fundamentally flawed from a biological perspective because they ignore the essential self-regulatory nature of behavior. From this perspective, the ability of system participants to self-govern the nature and extent of their interaction with the system is the linchpin of successful system performance.

The relevance of these concepts to educational systems is direct. The behavioral cybernetic theory of learning espoused by K.U. Smith and Smith (1966) assumes that as a behavioral mechanism, learning relies on self-regulation by the learner of feedback from design factors in the learning environment. The role of self-regulation in learning has achieved evident recognition in educational psychology (Snow et al., 1996), and Paris and Cunningham (1996) observe that self-regulatory behavior is one of the essential learning skills that emerges during childhood development. McCaslin and Good (1996) point out that this idea can be extended to the social aspects of learning through what they term coregulation by team members of their joint interaction with the learning environment during the learning process.

In spite of these observations, the prevailing state of affairs in most educational systems today arguably features a learning process that is open- rather than closed-loop. That is, as discussed in the next section, most design-related sources of sensory feedback expected to influence learning cannot be effectively controlled either by those targeted by, or by those guiding, the learning process. In behavioral cybernetic terms, it is unrealistic to expect optimal learning performance under such open-loop conditions.

4.5 ORIGINS AND SCOPE OF LEARNING ERGONOMICS

The scope of educational ergonomics encompasses all modes and levels of learning–design interaction that may occur in educational environments and systems. To illustrate this point, Table 4.1 specifies 26 different classes of educational system design factors that potentially may influence student learning (listed in the first column of the table), identified by seven different observers who have investigated this question at some systematic level (specified in the rightmost seven columns of the table). The check marks in selected cells of Table 4.1 for each given observer denote that the observer has associated student academic achievement with the specified educational ergonomic design factors. There is considerable overlap among observers cited in Table 4.1 regarding these associations. However, no two observers specify exactly the same set of design factors.

The design factors listed in Table 4.1 frame the two key questions pertaining to learning ergonomics: is the learning process affected by ergonomic design factors, and if so, which factors have the most critical influence?

Scientific attention to these questions has its origins in an extensive body of differential learning research dating back over a century, demonstrating that much of the variability in cognitive performance and learning (whose development and refinement is a primary focus of education) is attributable not to innate biological

TABLE 4.1

Classes of Educational System Design Factors that May Influence Student Learning, Identified by Different Observers

	Observer						
Ergonomic Design Factor	K.U. Smith, 1966	Kao, 1972	T.J. Smith, 2007	Stone, 2008	Legg and Jacobs, 2008	T.J. Smith, 2011	T.J. Smith, 2013
Class design and scheduling		√	√		√		√
Classroom design		√	√		√	√	√
Classroom furniture and equipment		√		√	√	√	√
Classroom technology—machine instruction	√	√	√	√	√		√
Community design factors			√			√	√
Cooperative learning		√				√	√
Curriculum design			√		√		
Design of educational resources (libraries, laboratories, etc.)		√	√				
Duration of exposure to classroom learning		√				√	√
Early childhood education						√	√
Family design factors			√				
Homework							√
Informal learning							√
Multimedia—study aids	√	√	√	√			
Online learning environments							√
Organizational design and management of educational system			√		√		
Physical environmental design features (air quality, lighting, noise, thermal comfort, etc.)		√		√	√		√
School building design			√		√	√	√
School choice							√
School funding							√
School size							√
Student personal factors (fitness, health, etc.)			√			√	√
Teacher training and pay	√	√					√
Teaching style and quality		√	√			√	√
Testing design	√	√					
Textbook design	√	√	√				

factors but rather to specific design features of the learning environment (reviewed in Chapter 3). The seminal text in the field is that of K.U. Smith and Smith (1966), a work that represents the first comprehensive effort to apply a well-defined HF/E perspective to learning and education (the first chapter is entitled, "Human Factors in Learning Science"). As specified in the second column of the table, these authors evaluate a range of design factors (such as audiovisual techniques, textbook design, training program design, and programmed instruction methods) that can be expected to influence learning and educational performance. The authors also advance a behavioral cybernetic theory of learning that maintains, based on evidence for context specificity in learning performance, that learning must be understood not as a generalized, innate phenomenon, but rather as a closed-loop process mediated by interaction between behavior of the student and learning environment design factors. Given that publication of this work was almost five decades ago, the time is long overdue to intensify efforts to demonstrate that student learning and educational systems today can benefit from the application of HF/E principles and practices, as has been the case with many other human systems and areas of human performance.

Nevertheless, when it comes to the actual practice of education, there is a deep chasm between the ideal of self-regulation of learning proclaimed by educational psychology and educational systems design of learning environments that typically prevail in most classrooms and educational institutions. Figure 4.1 illustrates this point with a depiction of a behavioral cybernetic model of the educational system from the perspective of the student. The outer two rings of the figure specify a series of different educational system design factors using factors identified by T.J. Smith (2007) listed in Table 4.1 that have the potential to influence the nature and extent of student learning. The inward-pointing arrows in the figure symbolize that this influence is mediated by sensory feedback to the student from different design factors. However, the figure also dramatizes a fundamental difficulty that students in most educational systems face.

As noted above, from a behavioral cybernetic perspective, learning is most effective if the learner not only is provided with, but is able to control, sensory feedback from design factors in the learning environment (K.U. Smith & Smith, 1966). It may be argued that most students are unable to achieve any meaningful level of control over sensory feedback from most educational system design factors. Indeed, Figure 4.1 suggests that the only learning-relevant design factors that the student can control to a substantial degree are of a personal nature: native language, health status, substance abuse, and nutritional status. For some students, because of peer pressure (such as gang influence), different household and school languages, and/or impoverished family status, it is possible that even these personal factors cannot be effectively controlled.

A behavioral cybernetic model of educational system performance comparable to that shown in Figure 4.1 also could be developed from the perspective of the teacher. A further fundamental difficulty confronting most educational systems is that the latitude of most teachers for control of sensory feedback from system design factors does not extend much beyond that of their students. Of the design factors specified in Figure 4.1, it may be argued that teachers are only able to levy direct control over

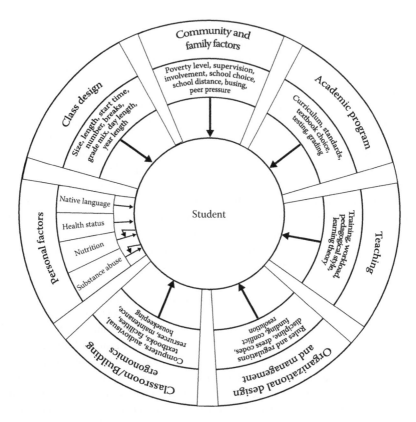

FIGURE 4.1 Behavioral cybernetics of learning ergonomics. Sensory feedback from a diverse range of educational system design factors has the potential to influence the nature and extent of student learning (inward arrows). However, the student can effectively control sensory feedback (outward arrows) from only a limited subset of these design factors.

teaching factors (pedagogic style, learning theory employed, etc.) and perhaps some elements of classroom ergonomics.

4.6 THE INFLUENCE OF HF/E DESIGN FACTORS ON STUDENT PERFORMANCE AND LEARNING

This section surveys a range of different HF/E design factors whose possible influence on K-12 student performance and learning in the classroom has been investigated through empirical analysis (T.J. Smith, 2011, 2013). Table 4.2 summarizes the design factors that will be addressed in this section, a subset of those listed in Table 4.1. In Table 4.2, the design factor category is listed in the first column, the type of design factor is listed in the second column, and a brief synopsis of the demonstrated impact of the specified design factor on student performance and learning is listed in the last column. It should be noted that some of the design factors specified in the second column of Table 4.1 conceivably could be classified in more than one of the design categories specified in the first column. The categorization scheme is

TABLE 4.2

Design Factors Whose Possible Effects on Student Performance in K-12 Classrooms Have Been Investigated

Category	Design Factor	Impact on Student Performance
Classroom and school building design factors	Environmental design of classroom and building facilities	Student academic performance influenced by level of classroom and school building design quality
	Classroom technology	Equivocal effects of computer use and mobile devices
	Online learning environments	No systematic analysis yet available
Educational system design factors	Smaller class size	Positive for lower grades but inconclusive for higher grades
	Longer exposure to learning	Strongly positive
	School choice	Varied effects in different countries and U.S. states
	School funding	No relationship
	School size	Varied results
	School start times	Nonacademic benefits, but academic impact unproved
	Levels of teacher training and teacher pay	No relationship
Learning strategy design factors	Cooperative learning	Strongly positive
	Early childhood education	Strongly positive, with possible limitations related to cost and long-term impact
	Amount of homework	Varied results
	Teaching quality	Necessary, but not sufficient
Student character and emotional status	Student perseverance, curiosity, conscientiousness, optimism, self-confidence, and level of control over the learning context	Largely positive
	Emotional well-being	Strongly positive
Design factors influencing student health	Nutritional adequacy	Mostly positive, but not definitive
	Good physical fitness levels and participation in physical activity	Largely positive
Community system design factors	Informal learning	Promising but limited and not definitive positive results
	Community socioeconomic status and school-community integration	Strongly positive

introduced in an attempt to impose some degree of order among the different design factors addressed in this section.

Also worth noting is that for many of the design factors listed in the second column of Table 4.2, thorough documentation of the evidence supporting the impacts cited in the third column for these factors would require a full report, or in some cases an entire monograph. Such a level of analysis is beyond the scope of this chapter. The purpose rather is to provide a more cursory series of analyses in order to drive home the point that, from a broad perspective, understanding how to human factor the design of learning environments is absolutely essential to understanding how and why students learn.

Finally, following presentation of findings for the design factors listed in Table 4.2, an overview perspective is offered on the implications of these findings.

4.6.1 IMPACT OF CLASSROOM AND SCHOOL BUILDING DESIGN FACTORS ON STUDENT PERFORMANCE AND LEARNING

Design factors addressed in this section encompass design features (Table 4.2) of classroom environments and school building facilities, classroom technologies, and online learning environments.

4.6.1.1 Design Features of Classroom Environments and Building Facilities

Both interior design features of the classroom itself, as well as the facilities design of the entire school building, are addressed here. A relatively early judgment regarding the possible impact of classroom design factors on academic performance is that of Caldwell (1992). His research on college classrooms identified chair design, air quality, and noise as primary classroom design factors needing improvement, and provided an estimate that poor classroom design and maintenance can lead to decrements of 10–25 percent in student performance. It is tempting to conclude that Caldwell's findings also apply to classrooms in K-12 schools as well.

This conclusion is supported by the more recent analyses of Schneider (2002) and Horrell (2009), whose reports summarize evidence that the academic performance of K-12 students is adversely affected by poor control of classroom environmental indoor air, room ventilation, temperature, humidity, thermal comfort, lighting, and acoustic quality. Regarding the latter, exposure to excess noise levels has been shown to degrade acquisition of reading skills in first and second graders as well as language acquisition in preschool children (Evans & Maxwell, 1997; Maxwell & Evans, 2000). On the other hand, there is little evidence that the age of the school building has much effect on student academic performance (Schneider, 2002).

Another classroom ergonomic design feature that has received attention in relation to student behavior and performance is that of classroom furniture. Various studies have shown that classroom furniture properly designed for children improves on-task behavior, promotes better sitting and standing postures, reduces back pain and other musculoskeletal complaints, increases trunk muscle strength, and improves overall academic marks (Horrell, 2009; Knight & Noyes, 1999; Koskelo et al., 2007; Legg & Jacobs, 2008; Mandal, 1982; Stone, 2008). Other emerging classroom furniture trends include a movement away from straight-row ranks of student desks to clustered or U-shaped desk arrangements that favor group discussion and cooperative learning (Horrell, 2009), and replacing standard classroom chairs with innovative seating designs that promote both better student posture and more active movement, such as inflatable stability ball chairs, which are reported to improve student attentiveness and productivity (Matheson, 2013).

Yet a third classroom design strategy receiving recent attention is the introduction of appropriate colors, full-spectrum lighting, plants, and background music in school classrooms to promote a more relaxed, calming, and attractive learning environment (Draper, 2005). It is claimed that these design innovations reduce discipline problems, but their possible effects on academic achievement are as yet unproven.

Finally, it should be noted that some architects are moving away from standard cookie-cutter approaches to school design to the application of an array of innovative classroom and school building design features aimed at achieving more effective and beneficial learning environments. For example, Nair and Fielding (2005) enumerate 25 design patterns claimed to represent a fairly complete range of various design principles that define school design best practice.

Perspective. Attention paid by the findings cited above to the influence of classroom environment and school building design factors on student performance has been accompanied by the emergence of various occupancy quality guidelines and standards pertaining to ergonomics for children (Legg & Jacobs, 2008; Lueder, 2010; Lueder & Berg-Rice, 2008). The necessity for this guidance has been prompted by recognition that, when it comes to performance effects of interaction with environmental design features, children are different from adults.

However, our understanding of the influence of environmental ergonomics on human performance is much more extensively and systematically informed by research on adult workplaces, particularly office environments (Aardex Corporation, 2004; American Society of Interior Designers, 2005; Brand, 2008; Leaman & Bordass, 1999, 2006; Orfield et al., 2006; Vischer, 1989). The study of Dainoff (1990) represents a specific example of such research. This author demonstrated that upgrading suboptimal office workstations with design improvements (ergonomic chairs, adjustable keyboard supports, document holders, and better monitor positions) resulted in a 17.6 percent improvement in the performance of computer-based data entry and editing tasks.

One of the more important findings growing out of office ergonomic research is reported by Leaman and Bordass (1999, 2006). Based on responses collected from workers in over 150 office buildings located in a number of different countries, these authors demonstrate that office worker perceptions of the *comfort of their buildings* account for 70 percent of the variability observed in worker self-perceptions of their *productivity on-the-job*. Comparable findings, based on a much smaller sample of buildings, later are reported by Smith and Orfield (2007). To my knowledge, no other design factor has been shown to have such a strong predictive relationship with the productivity of office workers.

It is tempting to extrapolate these findings to student academic performance. In reports noted above, Caldwell (1992) found that poor environmental design of college classrooms is associated with decrements in college student performance and Matheson (2013) observes that introducing inflatable stability balls in K-12 classrooms improves student productivity. Generally however, the putative benefits of recent architectural exploration of new school design innovations, aimed at bolstering academic achievement (Draper, 2005; Nair & Fielding, 2005), remain to be systematically investigated and definitively demonstrated.

4.6.1.2　Classroom Technology

One concomitant of the computer revolution has been a dramatic rise in the abundance and types of technology, particularly computers, in K-12 classrooms (Wilson, 2010). K.U. Smith and Smith (1966, Chapters 10–12) survey the earlier history of the development and application of teaching machines and programing for self-instruction. Somewhat later, Eberts and Brock (1984, 1987) provided major reviews of experience with computer-aided and computer-managed instruction. Their analysis also anticipated HF/E issues that have more recently emerged with the design of online learning environments (Section 4.6.1.3).

Over the past three decades, school districts in the United States (and elsewhere) have spent millions of dollars in technology enrichment of their classrooms. Nevertheless, the putative benefits of this expenditure for student academic performance have been (and continue to be) hotly debated, and remain unresolved, with ardent proponents on both sides of the issue (Carr, 2008; Draper, 1998; Healy, 1998; Hofferth, 2010; Kelly, 2000; Lewis, 1988; Meltz, 1998; Mervis, 2009; Solomon, 2010).

To arrive at an informed judgment on this question, I take as my guide the work of Straker and colleagues (Straker et al., 2009, 2010), who have provided some of the world's most definitive ergonomic research regarding the influence of classroom computers on student performance. The Straker et al. 2009 paper reviews current exposure data and the evidence for positive and negative effects of computer use by children. The background for this analysis is that computer use by children at home and school is now common in many countries, and that the majority of children in affluent communities now have substantial exposure to computers. Key conclusions from this review are: (1) child computer exposure varies with the type of computer technology available and the child's age, gender, and social group, (2) potential positive effects of computer use by children include enhanced cognitive development and school achievement, reduced barriers to social interaction, enhanced fine motor skills and visual processing, and effective rehabilitation, (3) potential negative effects include threats to child safety, inappropriate content, exposure to violence, bullying, Internet addiction, displacement of moderate/vigorous physical activity, exposure to junk food advertising, sleep displacement, vision problems and musculoskeletal problems, (4) given that children are physically, cognitively, and socially different from adults, they use computers differently from adults and the designs of computer workstations for children in the classroom and at home must address these differences, and (5) because children are undergoing dramatic developmental change and maturation, their interaction with computers has the potential to impact later adult risk of experiencing physical, cognitive, and social problems.

Based on these conclusions, a set of guideline principles for wise use of computers by children are presented, covering computer literacy, technology safety, child safety and privacy, and appropriate social, cognitive, and physical development. A selective elaboration of these guidelines, specifically aimed at supporting the appropriate physical developmental health and wellbeing of children interacting with computers, subsequently is set forth in the Straker et al. 2010 paper that delineates a total of 45 guidelines categorized into five major topic areas.

Perspective. Collective evidence reviewed by Straker et al. (2009, 2010) points to both positive and negative consequences of recurrent use of computers by children, with the range and severity of putative negative effects somewhat outweighing possible positive outcomes. This work points to the need for careful management of child–computer interaction, and for using the ergonomic principles and guidelines for such management set forth in this work in a systematic manner to educate parents and teachers, as well as equipment designers and suppliers, about this need, and to also teach children themselves about the wise use of computers.

Bailey and Puckett (2011) offer strong support for the adoption of technology by K-12 schools, based on the argument that the classroom that does not embrace technology is becoming progressively out of touch with the way America's children learn and interact with technology at home and away from school. Their perspective underscores the provocative point that introduction of laptop computers and mobile devices into K-12 classrooms represents a newer and surging wave of school technology. Experience on the part of selected school districts in the Twin Cities (Minnesota) metropolitan area can serve as an example.

- One school district offered free laptops to all students in its junior high schools (at a cost of $1.84 million over six years), but later canceled the program because of cost. It is reported that laptop access improved teaching, learning, and student engagement, but it could not be determined if the program improved student achievement on standardized tests (Boldt, 2010; Patterson, 2009a). Three other metropolitan area school districts reportedly are planning laptop distribution for selected district schools.
- In documentary reports of investment in laptop access in metropolitan area schools, no mention is made of the earlier investigations of Straker and colleagues (Harris & Straker, 2000; Straker et al., 1997) on possible musculoskeletal problems in children associated with classroom use of laptop computers.
- Six metropolitan area districts report the introduction of Apple iPads or iPods for regular classroom use in their schools (Estrada, 2013; Koumpilova, 2012b; Magan, 2013; Ojeda-Zapata, 2009; South Metro School Briefs, 2012).
- Not atypically, K-12 teachers often are less familiar than their students about how to use new classroom technology. For example, a university instructor of secondary school teachers reports that her students in some cases are terrified of classroom technology (Ojeda-Zapata, 2009). The metropolitan area district offering free laptops (first bullet) emphasizes intense teacher training in use of the technology (Patterson, 2009a).
- Sensing a ripe opportunity, Apple has announced a program offering inexpensive digital textbooks on iPads for schools (Newton, 2013).

In a pattern comparable to other examples of new technologies, the introduction of mobile communication devices into K-12 classrooms has been taking place with essentially no analytical input from the HF/E community. The systematic investigations by Straker et al. (2009, 2010) of the effects of classroom computer use on K-12 student performance supports the tentative conclusion that effects of classroom use of mobile communication devices on student performance also are likely to be mixed. Support for this conclusion is provided by a recent quote from the Director of Media and Technology for Edina, Minnesota Independent School District 273, who openly acknowledges that, "Technology has changed everything. We can't ignore it because we don't yet have concrete data on a link between technology and grades" (Koumpilova, 2012a). Given the wholesale investment that school districts across the United States are making in technology, a more prudent approach to technology policy that schools arguably should consider is to defer costly investment in new technology until this link has been more clearly defined.

What is undeniable is that this domain of educational ergonomics represents both a need and a highly promising opportunity for HF/E research and analysis.

4.6.1.3 Online Learning Environments

Because of the distinctive nature of the technology, the possible impact of online learning environments on K-12 student performance is considered here as a separate classroom technology issue. Over the past decade, provision of online courses has become a regular and widely accepted mode of instruction in the world of higher education—for example, online enrollments have continued to grow at rates far in excess of the total higher education student population, such that about one in three higher education students now take at least one course online (Allen & Seaman, 2014). Indeed, a recent discussion of what are termed massive open online courses (MOOCs) offered by some of the most elite universities in the United States cites a widely held view that MOOCs represent the future of U.S. higher education (Heller, 2013). Nevertheless, based on 2012 survey data, only 12 percent of U.S. institutions of higher education offer or are planning to offer MOOCs, whereas 55.4 percent are undecided and 32.4 percent have no plans for MOOCs (Allen & Seaman, 2014).

Not surprisingly, higher education based on MOOCs has attracted strident opposition, linked to concerns about the likely negative impact of MOOCs on academic quality, student performance, and a tradition of face-to-face interpersonal and intellectual interactions of students and teachers in the physical classroom that dates back to the ancient Greeks. Importantly from a HF/E perspective, the survey notes that data about how well MOOCs work are diffuse and scant.

Implementation of online learning environments in K-12 schools has remained comparatively modest, despite calls for much wider application of this technology plus strong endorsement from the Gates Foundation (Mangu-Ward, 2010). Based on a 2007 Sloan Consortium report (Picciano & Seaman, 2007), out of an entire population of 48,000,000 U.S. public school students, an estimated 600,000 (1.3 percent) to 700,000 (1.5 percent) K-12 public school students were enrolled in online learning for 2005–2006.

Reports of the impact of interaction with online learning environments on K-12 student performance are equivocal and nonsystematic. The Sloan report (Picciano & Seaman, 2007, pp. 14–17) presents a balanced mix of both positive comments and expressions of concern about the benefits of this technology. Student enrollees in the Florida Virtual School are reported to score higher on advanced placement tests than regular public school students (Mangu-Ward, 2010). In contrast, the superintendent of a Pennsylvania school district reports that its regular school students outperform cyberschool enrollees in almost every regard based on achievement data (Jubera et al., 2010). Similarly, online tutoring in mathematics for Minnesota students in third through eighth grade is reported to substantially improve student achievement on the Minnesota standardized math tests in 2012 (Magan & Webster, 2012).

Perspective. Survey data suggest that online learning has been growing in K-12 schools, and that this growth will continue for the foreseeable future (Picciano & Seaman, 2007). If K-12 trends follow the more established patterns observed in institutions of higher education, it is possible that online learning will emerge as a substantial contributor to the education of K-12 students, especially at the secondary level. Blended learning environments, combining student enrollment in both fully online and blended (combination of online and regular classroom instruction) courses, represent one particularly likely manifestation of this trend. Nevertheless, the impact of this technology on the academic performance K-12 students has yet to be systematically explored, and unless or until this shortcoming is addressed, the putative benefits of online learning remain uncertain.

4.6.2 IMPACT OF EDUCATIONAL SYSTEM DESIGN FACTORS ON STUDENT PERFORMANCE AND LEARNING

A total of seven educational system design factors are specified in Table 4.2. Evidence regarding what is known about the impact of each of these factors on student performance is addressed below, followed by a perspective on the broader implications of this evidence.

4.6.2.1 Smaller Class Size

Class size constitutes a perennial preoccupation of most if not all school districts in the United States, a typical view being that reducing class size (more accurately, the classroom student/teacher ratio) represents the holy grail for bolstering student achievement. For example, an entire website devoted to this topic (http://www.class sizematters.org/) lists over 55 publications dealing with class size, most of them preaching that smaller is better. On the other hand, Greene and Winters (2005) offer evidence that reductions in class size have had little impact on student achievement. A more systematic, and nuanced, analysis is provided by Biddle and Berliner (2002), who review over 55 (mostly research) studies dealing with class size, and conclude that (1) long-term exposure to small classes generates substantial advantages for U.S. students enrolled in early elementary school grades, and these extra gains are greater the longer students are exposed to such classes, but (2) evidence for the possible advantages of small classes in the upper grades and high school is so far inconclusive. Given that the second author of this study is the senior editor of the 1996 *Handbook of Educational Psychology* (Berliner & Calfee, 1996), this analysis must be viewed as one of the most definitive to date regarding the impact of small class sizes on student achievement.

4.6.2.2 Longer Exposure to Learning

The evidence is persuasive that longer exposure to learning, by extending the length of the school day and school year, is highly likely to positively impact student performance. For example, studies of charter schools that adhere to extended schedules advocated by programs such as Knowledge is Power or Harlem Children's Zone indicate that compared with children in typical public schools, those who are required to spend a greater number of hours per year in the classroom achieve notable gains in learning performance (Brooks, 2009; Gladwell, 2008; Lexington, 2009; Patterson, 2009b; Tuttle et al., 2013). In general, experience with extended school schedule designs, documented in these reports, has been shown to overcome gaps in learning performance between black versus white students and economically more disadvantaged versus less disadvantaged students. President Obama has provided federal endorsement of this design factor with a call for a longer school year (Werner, 2010). This call apparently has attracted attention; it was announced recently that five U.S. states (Colorado, Connecticut, Massachusetts, New York, and Tennessee) have agreed to participate in a pilot program, starting in 2014, that will add 300 hours to the school year in some state schools, affecting nearly 20,000 students in 40 schools (Lederman, 2012).

4.6.2.3   School Choice

School reform initiatives in both the United States and elsewhere over the past decade have provided parents and students with a greater latitude for choosing a school outside the particular locale in which they live. In the United Kingdom, recent policies to improve public schools have included a major emphasis on school choice. It is reported that this reform has had minimal impact on student achievement (The Economist, 2009).

In contrast, this same *Economist* report notes that new schools developed under the "free school" reform in Sweden have captured a tenth of the market in the past decade. The likely reason is that student performance results from this program show that free schools not only are good for their own pupils, but also forced ordinary state schools to shape up (Böhlmark & Lindahl, 2008). The availability of school choice also is cited as one of the major keys to the remarkable gains in student achievement observed among K-12 students in Florida over the past decade (Kersten, 2011). In Minnesota, it is reported that as a result of open enrollment (another type of school choice), schools in the state get better (Boldt, 2011). Collectively, the Sweden, Florida, and Minnesota results offer persuasive evidence for school choice success.

In the United States, the vanguard of the school choice movement has been the advent of charter schools. Over the past decade, the state of Minnesota has been a national leader in approving the installation of new charter schools. For 3rd through 8th and 10th and 11th grade students enrolled in either public schools or charter schools in Minnesota, results (http://education.state.mn.us/mde/index.html, 2006, 2009, 2012), for state standardized math and reading tests administered between 2003 and 2009 (averaged across all seven years for all students taking the tests in the eight different grades), percentages of students statewide achieving proficiency in the two disciplines are 63.2 percent for math (N = 1,402,983 students achieving math proficiency) and 70.9 percent for reading (N = 1,604,126 students achieving reading proficiency) for public school students (enrolled in 366 public school districts), and 43.9 percent for math (N = 23,188 students achieving math proficiency) and 53.4 percent for reading (N = 30,233 students achieving math proficiency) for charter school students (enrolled in 366 charter schools). That is, relative to public school students, percent proficiency achievement for charter school students in Minnesota is about 20 percent less for math and 17.5 percent less for reading for the years 2003 through 2009. These results indicate that the availability of school choice in Minnesota, in the form of charter schools, has not benefitted student achievement.

Experience with school choice in Wisconsin recapitulates that in Minnesota. Across all grades, low-income students who received vouchers to attend private or religious schools in Milwaukee performed worse on statewide reading and math tests than their low-income counterparts in Milwaukee public schools, according to standardized test scores in math and reading for students in these schools released for the year 2010 (Bauer, 2011).

The foregoing summary of available evidence regarding school choice—again nonsystematic—suggests that this design factor has yielded equivocal results in terms of student academic achievement.

4.6.2.4 School Funding

In 2005, Standard and Poor's (2005) evaluated the relationship between operating expenditures allocated to instruction and student achievement in reading and math for students enrolled in school districts in nine different U.S. states for the years 2003 and 2004. The observed r^2 for the correlation between these two measures ranges from a low of 0.0008 for Minnesota, to a high of 0.17 for Florida. For the remaining seven states, r^2 values are between 0.01 and 0.10. In other words, this analysis indicates that, except for Florida, the level of school expenditure devoted to instruction accounts for under 0.1 percent to 10 percent of observed variability in student reading and math proficiency.

On a broader scale, an analysis for 11 different countries of the relationship between public spending per student and student performance for the years 2006/2007 shows a similar relative lack of predictability between the two measures (The Economist, 2010b). The conclusion from these analyses is clear—the level of school funding has essentially no relationship to student performance (Beamish, 2011).

4.6.2.5 School Size

A 2002 analysis by Abt Associates (Page et al., 2002) of the impact of school size on student academic achievement concludes that smaller schools are associated with greater student learning, greater social equity of educational achievement, and greater engagement in school than larger schools. Smaller schools also have higher attendance rates and graduation rates, lower dropout rates, and have been found to be more cost-effective, especially when measured on a per-graduate basis rather than a per-pupil basis.

On the other hand, the school size analysis of Wainer (2007) is less definitive. For 1662 Pennsylvania schools, no relationship is observed between enrollment and 5th grade math scores. In contrast, the data for 11th graders in Pennsylvania schools show a *positive* relationship between school size and math scores. In summary, small schools likely represent an enticing target for parents, students, and perhaps even school districts, but investment in this educational system design factor merits careful cost-benefit analysis.

4.6.2.6   School Start Times

One of the more systematic studies of this design factor is that of Wahlstrom (2002), who carried out a longitudinal study of the impact of changing start times for seven high schools in the Minneapolis (Minnesota) Public School District from 7:15 AM to 8:40 AM. This change was shown to yield significant benefits, including improved attendance and enrollment rates, less sleeping in class, and less student-reported depression. Similar findings from a smaller sample also have been reported (Owens et al., 2010). No impact on academic achievement is reported, one of a number of shortcomings of changing to later high school start times that skeptics point to as reasons for questioning its merits. Wahlstrom acknowledges the controversy by pointing out that this represents a highly charged issue in school districts across the United States.

4.6.2.7   Levels of Teacher Training and Teacher Pay

A 1997 study by Goldhaber and Brewer of the effect of teacher degree level on educational performance found that students of teachers with master's degrees show no better progress in student achievement than students taught by teachers without advanced degrees. Fast forward 13 years and little has changed, despite the fact that U.S. schools pay more than $8.6 billion in bonuses to teachers with master's degrees (Blankinship, 2010). Tellingly, this article also notes that Bill Gates, cochair of the Bill and Melinda Gates Foundation, one of the major private funding sources for educational research, has publically stated that this type of expense represents an example of spending money on something that does not work. Klein (2011) also indicates that the level of teacher pay appears to have no relationship to student achievement.

Perspective. Of the seven educational system design factors specified in Table 4.1, only longer exposure to learning has been shown unequivocally to yield student achievement benefits. Levels of school funding and teacher training show no relationship with student achievement. Smaller classes benefit the performance of lower grade, but not upper grade, students. Thus, smaller classes attract much favorable support, but a Pennsylvania school study shows no achievement benefits for fifth grade students. School choice has had positive effects in Sweden and Florida but has not benefitted student performance in the United Kingdom and in two other U.S. states. Later school start times for adolescents is related to improved sleep quality and other nonacademic benefits, but the impact on academic achievement has yet to be demonstrated.

The benefits of longer exposure to learning may be readily understood from the perspective of the law of training (Schmidt & Lee, 1999; Schneider, 1985), one of the most robust and reproducible of the behavioral laws in psychology. Studies of the law of training across a broad range of cognitive, psychomotor, and motor tasks show that incremental improvements in learning continue to occur with the number of task repetitions (Chapter 3, Section 3.4.4). This law thus appears to apply to classroom student performance, if repetition is taken to refer to the number of school day hours and/or the number of school year days that a student spends in class.

4.6.3 IMPACT OF LEARNING STRATEGY DESIGN FACTORS ON STUDENT PERFORMANCE AND LEARNING

Findings regarding the student performance impact of a total of four learning strategy design factors specified in Table 4.1 are each addressed below, followed by a perspective on the implications of these findings.

4.6.3.1 Cooperative Learning

Cooperative learning environments have been shown to yield robust student learning benefits. As conclusively shown by the Johnson brothers at the University of Minnesota, class designs built around cooperative learning groups or teams reliably yield superior learning outcomes compared with classes based on individualized learning or competitive learning group designs (Johnson & Johnson, 1987, 1989; Johnson et al., 1981).

4.6.3.2 Early Childhood Education

With the possible exception of the socioeconomic status of school communities (Section 4.6.6.2), no other design factor has been shown to exert a greater influence on school performance, and lifelong success generally, than early childhood education (Barnett, 2011). Evidence for this conclusion is robust and diverse. To illustrate the scope of these effects, the example of the High/Scope Perry preschool project can be cited (Parks, 2000), in which 123 low-income 3- and 4-year-olds in Ypsilanti, Michigan were enrolled in a half-day preschool program and followed longitudinally in a project that apparently continues to the present day (making it the longest-running study of this design factor).

Relative to a control group of preschoolers not so enrolled, positive effects documented for the enrollees fall into three categories, namely (1) scholastic success benefits: higher grade point averages and higher graduation rates, higher standardized test scores, fewer instances of placement in special education classes, more time spent on homework, and more positive attitudes toward school; (2) social responsibility benefits: lower overall scores for adolescent misconduct, lower incidence of adolescent violent behavior, property damage, and police contacts, fewer adult arrests and births out of wedlock, and greater family stability; and (3) socioeconomic success: higher levels of monthly earnings, and home and automobile ownership.

Similar results in a more urban setting have been reported for the Chicago Child Centers study (Dickman & Kovach, 2008), prompting a statement by the authors of the study (p. 23) that, "no other social program has the evidence to show this level of savings to society." These dramatic results have prompted studies of the economic benefits of early childhood education (Beckstrom, 2003; Heckman, 2008; D. Smith, 2009). For example, an analysis by Art Rolnick, recent vice president of the Minneapolis Federal Reserve Bank, documents a return on investment of 7 to 16 percent a year from early childhood learning programs for at-risk kids, results that have prompted him to start the Human Capital Research Collaborative at the University of Minnesota Humphrey Institute to champion and support such programs (D. Smith, 2009).

Enthusiasm about the putative benefits of early childhood education has its detractors. Finn (2009) points out that the programs cited above involved relatively small numbers of children, were highly intensive, and had high per pupil costs. He also points out that the long-term positive impacts of these programs for low-income enrollees may not be as significant as claimed. Nevertheless, he goes on to acknowledge the value of early childhood education for youth development as long as the potential limitations of such programs are recognized.

Another cautionary note addressing a different focus of concern is offered by Rosin (2013). She points out that (1) young children—even toddlers—are spending more and more time with digital technology (such as iPads and mobile phones), (2) thousands of applications for these devices, appealing to these kids, are now released every year, (3) in 2006, 90 percent of parents said that their children younger than age 2 consumed some form of electronic media, (4) in the face of this statistic, the American Academy of Pediatrics in 2011 released its latest policy for very young children and media that nevertheless uniformly discouraged passive media use on any type of screen for these kids, and (5) as technology becomes ubiquitous in our lives, American parents are becoming more, not less, wary of what it might be doing to their children.

Based on this analysis, Rosin raises the obvious question as to what effect interaction with these new modes of digital technology has on brain development in very young children. It seems reasonable to suggest that expansion of early childhood education programs inevitably will be accompanied by inclusion of digital technology, based in part on mobile devices, in the educational toolkit. The observations of Rosin thus call attention to a potential concern that merits further investigation, and perhaps caution, on the part of advocates of early childhood education.

4.6.3.3   Amount of Homework

The prospect of the ordeal of homework as a rite of passage for K-12 students represents a widespread expectation on the part of teachers, students, and parents alike. The question is, do academic achievement benefits arise from the application of this design factor? On one side of this question are true believers who have little doubt that hours and hours of homework represent the key to student success in school and in life (Chua, 2011). On the other side are skeptics who cite research indicating that homework does not foster better study habits or, for K-third-grade students, improve standardized test scores (Bennett & Kalish, 2007; Kohn, 2006).

A recent (occasionally tongue-in-cheek) commentary by Menand (2012) offers the following interesting observations about homework:

- François Hollande, the French president, has announced his intent to abolish homework for primary and middle school students in France based on the argument that homework gives an advantage to more affluent students (whose parents are likely to provide homework assistance) over less affluent students (whose parents are less likely to help).

- Contrary to President Hollande's assumption, it is likely that the people most hostile to homework are affluent parents who want their children to spend their afterschool time engaged in more enriching extracurricular activities. Less affluent parents are likely to prefer more homework as a way of keeping their kids off the streets.
- Opponents of homework argue (the reasons vary) that homework is mindless, unrelated to academic achievement, negatively related to academic achievement, and a major contributor to stress in children, teachers, and parents alike.
- Homework is an institution roundly disliked by all who participate in it. Children hate it for obvious reasons; parents hate it because it makes their children unhappy; and teachers hate it because they have to grade it. Grading homework is teachers' never-ending homework.
- Two counts in the standard argument against homework don't appear to stand up. The first is that homework is busywork, with no effect on academic achievement. According to the leading authority in the field, Harris Cooper of Duke University, homework correlates positively—although the effect is not large—with success in school. This is more true in middle school and high school than in primary school, since younger children get distracted more easily. He also thinks that there is such a thing as homework overload—he recommends no more than 10 minutes per grade a night. But his conclusion that homework matters is based on a synthesis of 40 years' worth of research.
- The other unsubstantiated complaint about homework is that it is increasing. After the publication of "A Nation at Risk," by the U.S. Department of Education (1983), which prescribed more homework, the amount of time American students spend on homework has not changed since the 1940s. And that amount isn't much. A majority of students, including high school seniors, spend less than an hour a day during the five-day school week doing homework. Recent data confirm that this is still the case.
- The country with one of the most successful educational systems, according to a recent analysis by Allan (2012, p. 86), is Finland. Students there are assigned virtually no homework; they don't start school until age seven; and the school day is short.
- On the other hand, another highly successful country in terms of student academic performance is South Korea, whose schools are notorious for their backbreaking rigidity. Ninety percent of primary school students in South Korea study with private tutors after school, and South Korean teenagers are reported to be the unhappiest in the developed world.

In summary, evidence for the impact of homework on student performance is equivocal.

4.6.3.4 Teaching Quality

In a 2004 report (p. 4), Goldhaber and Anthony assert that, "A growing body of research shows that the quality of the teacher in the classroom is the most important schooling factor predicting student outcomes." Chieppo (2012) claims that, "There is no factor within the four walls of a school that affects student performance more than teacher quality." An article in *The Economist* (2011a, p. 26) recapitulates this claim by noting that, "Budget, curriculum, class size—none has a greater effect on a student than his or her teacher." Analyses above support the first part of this latter claim insofar as budget and class size are concerned (Sections 4.6.2.1 and 4.6.2.4). What about the larger conclusion that teachers matter most when it comes to student achievement?

There is broad support for this conclusion, such as (1) the U.S. federal No Child Left Behind and Race to the Top initiatives of the Bush and Obama administrations, respectively, both emphasize teacher accountability as the key to improving schools; (2) the basic assumption of the Bill and Melinda Gates Foundation (over $4 billion in education funding) is that provision of high-quality teaching will ensure student success (Bill and Melinda Gates Foundation, 2010); (3) the agenda of the National Commission on Teaching and America's Future rests on what this entity considers the single most important strategy for achieving America's educational goals, namely recruiting, preparing, and supporting excellent teachers (National Commission on Teaching and America's Future, 1996), starting with three basic premises—(a) what teachers know and can do is the most important influence on what students learn, (b) recruiting, preparing, and retaining good teachers is the central strategy for improving our schools, and (c) school reform cannot succeed unless it focuses on creating the conditions in which teachers can teach well; and (4) the international supremacy of Finnish schools (Finnish 15-year-olds have the highest worldwide rankings on math and reading standardized tests) is attributed to how they prepare and manage their teaching corps—Finnish teachers enjoy a high level of support and delegated responsibility, with a strong emphasis on professional training, no streaming, no selection, no magnet schools, no national curriculum, and few national exams (Charlemagne, 2006; Kaiser, 2005).

Perspective. The efficacy of cooperative relative to individual or competitive learning designs in promoting student learning may be understood in the context of social cybernetic theory. Social cybernetics is founded on the broader field of behavioral cybernetics, which assumes that human behavior and learning are controlled as closed-loop or cybernetic processes (K.U. Smith & Smith, 1966). As delineated by K.U. Smith and colleagues over the past four decades, the science of social cybernetics is directed at delineating the reciprocal sensory feedback control relationships and closed-loop behavioral-physiological manifestations and properties of interpersonal and group interactions (K.U. Smith & Kao, 1971; K.U. Smith & Smith, 1973; T.J. Smith et al., 1994a). This topic is addressed in Chapter 8 as part of an analysis of sources of variability in social and team performance.

From a social cybernetic perspective, each individual in a social context, in a process termed *social tracking*, must control the sensory feedback generated not only by his or her own behavioral movements and functioning, but also the sensory feedback created by interacting with one or more social participants. Social tracking effectiveness enables the exercise of group intelligence, shared knowledge, reciprocal peer support, and social learning strategies in cooperative learning contexts (Johnson & Johnson, 1987, 1989; Johnson et al., 1981), strategies unavailable to individual or competitive learning designs. Although it has been shown that social tracking skills start to emerge early in infancy (Thomsen et al., 2011), full mastery of these skills may not be achieved until late adolescence or early adulthood.

The well-documented strongly positive impact of early childhood education on subsequent student performance may be understood from the perspective that learning is largely experiential in nature; that is, as indicated at the outset of this chapter, the HF/E design of the learning environment represents a prominent influence on observed variability in learning performance (K.U. Smith & Smith, 1966; T.J. Smith, 2007; T.J. Smith et al., 1994a). Thus, the earlier that interaction with learning environments occurs, the earlier that context-specific (i.e., experiential) changes to brain function and structure take place, with positive consequences for subsequent cognitive development of the child (Bransford et al., 2000; Heckman, 2008). Indeed, studies on infants show that children who receive strong mental stimulation beginning in the first few months of life are likely to develop normal intelligence (as measured by IQ tests), no matter how poor their home environment (Dawson, 1996).

The possible value of homework attracts passionate debate that lends itself to no straightforward interpretation. Arguments favoring regular assignment of homework rest first on many decades of educational tradition and second on the idea that homework means longer exposure to learning materials outside the classroom, exposure that (according to the law of training) should bolster learning (Section 4.6.2.2). Arguments against assigning homework rest first on research showing that homework requirements are not associated with improved student achievement and second on the idea that completing homework requirements detracts from student interaction with informal learning environments and conditions that arguably are more important for student learning than formal interaction with classrooms (Section 4.6.6.1).

The above analysis of teaching quality appears to support the definitive conclusion that, when it comes to student achievement, teachers matter most. Yet among the design factors identified in Table 4.2, with strongly positive, equivocal, or varied impacts on student performance, only use of classroom technology and cooperative learning strategies, assignment of homework, and participation in teacher training programs arguably are under direct control of the teacher. That is, holding teachers accountable for mediating the impact on student learning of a series of design factors that they cannot control may be deemed at best ineffective and more pointedly a recipe for failure on the part of teachers, students, and the educational system generally. Furthermore, it is now widely recognized that managing order and discipline in classrooms rather than nurturing student learning represents a major challenge for many U.S. teachers, yet it is reported that preparing teachers to meet this challenge (especially new teachers) represents a neglected component of teacher training (Ferdig, 2013; Hanson, 2001; Wingert, 2010). These considerations support a balanced conclusion that teaching quality represents a necessary, but not sufficient, contributor to student achievement.

4.6.4 Impact of Student Character and Emotional Status on Student Performance and Learning

This section summarizes evidence regarding the impact on student performance of student character and emotional status followed by a perspective on the implications of the findings.

4.6.4.1   Student Character

Based on their review of what can be broadly termed student character, and what the authors specifically refer to as student self-confidence, competence, and control beliefs (i.e., student belief in their competence for learning and in their level of self-control over the context in which learning occurs), Schunk and Zimmerman (2006) draw two key conclusions: (1) competence beliefs will not serve to motivate learning unless students also perceive themselves to be in control of the learning context, and (2) research on this topic shows that the nature and extent of these beliefs are predictive of diverse behavioral outcomes such as learning, motivation, self-regulation, metacognition, and volition. Independent findings by other investigators tend to support these conclusions (Miyake et al., 2010; Norman & Spohrer, 1996; Ramirez & Beilock, 2011; Steele, 2010).

Tough (2012) has devoted an entire book to exploring the issue of the influence of student character on academic success. His basic premise is that noncognitive skills, broadly related to student character—such as perseverance, curiosity, conscientiousness, optimism, ability to focus, and self-control—are more predictive of future student success than cognitive skills such as those measured by IQ. He offers a series of personal stories in support of this idea and references neurobiological evidence based on brain studies of children growing up in abusive or dysfunctional environments, indicating that these children find it harder to concentrate, sit still, and/or rebound from setbacks. Functional changes in the prefrontal cortex, linked to early exposure to environmental stress and often associated with disadvantaged socioeconomic conditions, are implicated in these effects. Tough's analysis thus relates the emergence of a seemingly innate factor—student character—to the design of the environment to which a student may be exposed during development. This linkage may account, at least in part, for the strong association between school district socioeconomic status and student standardized test performance in a series of Minnesota school districts, which is addressed in Section 4.6.6.2.

Findings calling into question the significance of student character as an influence on student success are based on the performance of U.S. students on standardized international math, reading, and science tests, administered to students in 75 countries (Organization for Economic Cooperation and Development, 2010), compared with their self-reports as to how self-confident they are about doing well in these subjects (Loveless, 2006). Specifically, a far higher percentage of U.S. students express self-confidence about their abilities in these subjects relative to the percentage who achieve advanced status on the actual tests.

4.6.4.2   Emotional Status

Exposure of children to environmental stressors, such as negative childhood experiences (poverty, abuse, dysfunctional family and/or community conditions) (Tough, 2011), frequent transfers between schools (Mitchell, 2011b), and/or parental divorce (Olson, 2011), adversely impacts both the emotional well-being as well as the academic performance of students. The analysis by Tough (2012), cited in the previous section, introduces evidence that implicates stress effects on the prefrontal cortex as the likely neurobehavioral basis of these outcomes.

Perspective. With reference to this impact category (Table 4.2), the first obvious question is why student character and/or emotional status, with behavioral dimensions such as competence and control beliefs, levels of student self-assurance and self-affirmation, and/or emotional well-being, are treated as a design factors. The perspective offered here is that these behavioral dimensions do not emerge in isolation. Rather, they are established through a given student's social intercourse with family, peers, superiors, and so forth, as an outcome of social tracking, as well as interaction with environmental stressors such as abuse or poverty (Tough, 2012). In other words, whatever contribution innate biological factors may make, variations in the level of character of a student also are feedback-related to the design of the social and environmental context with which the student interacts.

As for the conclusion that the level of motivation of a student for learning is linked to self-perception of the degree of control of the learning environment (Schunk & Zimmerman, 2006), this idea was elaborated in more detail four decades prior to the 2006 *Handbook* chapter by K.U. Smith and Smith (1966) in a cybernetic theory of motivation. These authors point out that (1) the most general trend in human motivational patterns is defined by the nearly universal drive among humans to control their environment, (2) cybernetic learning theory maintains that motivated acts of creation, of discovery, of seeking, and/or of volitional behavior have their genesis in the persisting efforts of young children (even infants [Dawson, 1996; Gopnik, 2012; Thomsen et al., 2011]) to exercise direct control over their own sources of stimulation, (3) the degree to which a student can gain definitive control over the many sources of stimulus feedback provided by learning environments varies with the student's behavioral and social capacities, (4) treating students as control systems implies that they self-govern all modes and levels of learning, the basic premise behind learner-centered education strategies (Brooks, 2009), and (5) the impetus of humans to control their environment has its phylogenetic origins in animals, who manifest a similar drive.

As summarized above, empirical results from different sources reveal that student character in most cases predicts, but in some cases overestimates, actual academic achievement, pointing to largely but not exclusively positive effects of this design factor on student performance. On the other hand, evidence linking emotional well-being and student academic performance appears to be consistently positive across a number of different observations.

4.6.5 Impact of Student Health on Student Performance and Learning

This section summarizes evidence regarding the impact on student performance of two design factors related to student health—namely nutritional adequacy and physical activity—followed by a perspective on the implications of the findings.

4.6.5.1 Nutritional Adequacy

Although there is extensive evidence that adequate nutrition is essential for children's health and well-being, evidence that adequate nutrition is linked to student academic performance is less definitive. Swingle (1997) cites evidence compiled through the mid-1990s pointing to a positive link between adequate nutrition and both cognitive development and learning. One year after enrollment in a free school breakfast program authorized by the state of Minnesota, one school showed improvement in reading and language arts test scores for second and fourth graders; math scores for second graders increased, but those for fourth graders declined, over this period (Draper, 1996).

A longitudinal study of approximately 21,000 nationally representative children from kindergarten through third grade, divided into food insecure and food secure groups, found that food insecurity was associated with impaired social skills development and reading performance for girls but not for boys (Jyoti et al., 2005). In their review of the effects of food insecurity on school performance of first, third, fifth, and eighth graders, Mykerezi and Temple (2009) conclude that food insecurity does *not* appear to have a robust effect on cognitive outcomes.

4.6.5.2 Good Physical Fitness Levels and Participation in Physical Activity

Unlike the somewhat varied results regarding the student performance impact of nutritional adequacy (above), the case for the academic benefits of good physical fitness levels and regular participation in physical activity (PA) among children is stronger. A series of recent reports provide what essentially amounts to consensus agreement on this conclusion. Key findings may be summarized as follows: (1) 11 of 14 studies find benefits of PA participation, and three find benefits of good physical fitness for student academic achievement (Active Living Research, 2007), (2) for kindergarten through fifth grade students, higher participation in PA benefitted math and reading performance for girls but not for boys (Carlson et al., 2008), (3) for fourth, fifth, seventh, and eighth graders, a statistically significant association between physical fitness levels and academic achievement in math and English is observed (Chomitz et al., 2009), and (4) a review of 43 separate studies finds substantial evidence that physical activity can help improve academic achievement, including grades and standardized test scores (Centers for Disease Control and Prevention, 2010).

The foregoing evidence is derived from student PA programs conducted outside the academic classroom. To reduce student restlessness and disruptiveness and greater attentiveness, a number of strategies have been described for promoting student movement patterns other than traditional sitting within the classroom, such as (1) sitting on inflated exercise ball seats (Belden, 2006; Matheson, 2013), (2) using adjustable sit-stand desks (Saulny, 2009), (3) practicing yoga exercises (Miranda, 2009; Moran, 2009), and (4) mimicking hand and arm movements shown on a video while keeping rhythm to accompanying music (Pfitzinger, 2010).

Perspective. There is growing evidence that regular participation in physical activity not only boosts muscle power and fitness levels, it also bolsters brain function. For example, Wilson and Kuhn (2010) point out that both nutritional adequacy and physical activity are essential for nurturing and improving brain function in children. A popular review of this topic (Carmichael, 2007) summarizes research evidence showing that regular exercise is accompanied by neurophysiological changes critical for learning, memory, and other brain processes, including formation of new nerve cells and increased blood supply in selected brain regions and enhanced production of several chemicals that nurture brain structure and function.

These findings appear to account for the consensus findings cited above that, with few exceptions, good fitness levels and participation in regular physical activity have a positive impact on student performance. Why is the evidence less definitive for the impact of nutritional adequacy? One possible interpretation is that, unlike levels of overt malnutrition that have well-documented, dramatically adverse effects on the physical and cognitive development of children, food insecurity in U.S. children (the focus of the nutritional studies cited above) has less dramatic developmental effects and the impact on student academic performance therefore is more subtle.

4.6.6 IMPACT OF COMMUNITY SYSTEM DESIGN FACTORS ON STUDENT PERFORMANCE AND LEARNING

This section summarizes evidence regarding the impact on student performance of two community system design factors—namely, informal learning and the socio-economic status of, as well as the degree of integration with, communities and the schools and/or school districts residing in these communities—followed by a per-spective on the implications of the findings.

4.6.6.1 Informal Learning
The significance of informal learning—defined as that which occurs outside formal classroom environments—is nicely articulated by Mark Twain, who observed, "I never let my schooling interfere with my education." Informal learning is considered in this section because it occurs in the community beyond the boundaries of the classroom or school (in which formal learning is assumed to occur). Nearly a century ago, Dewey (1916) called attention to the significance of informal learning in the acquisition of knowledge by children, as follows:

> From the standpoint of the child, the great waste in school comes from his inability to utilize the experience he gets outside…while on the other hand, he is unable to apply in daily life what he is learning in school. That is the isolation of the school—its isola-tion from life.

A psychological learning theory perspective on informal learning is provided by Bransford et al. (2006), who point out that (1) 79 percent of a child's waking activi-ties during the school-age years are spent in nonschool pursuits, (2) over the human life span, the portion of time spent outside of school (and therefore a potential source of informal learning) is over 90 percent, (3) a great deal of what an adult learns in a lifetime is not covered in school, (4) it is an open question as to ways in which formal learning transfers to life skills and activities outside of school, (5) one view is that informal learning broadens knowledge and understanding on the part of the learner, (6) a contrasting view is that informal learning tends to support undisciplined think-ing and uninformed views, and (7) informal learning is less studied compared with formal learning.

Support for the likely importance of informal learning also is provided by Lexington (2009), who observes that the typical U.S. child averages only 6.5 hours in class and one hour of homework per weekday—this means that a child spends about two-thirds of waking hours interacting with learning environments, largely defined by community design factors, outside the classroom. Falk and Dierking (2010) review a limited body of evidence indicating that informal learning can benefit student academic performance in the domain of science. Key findings cited by these authors are: (1) informal learning contributes to people's understanding of and interest in science, (2) childhood informal learning experiences contribute significantly to adult science knowledge, (3) nonacademically oriented afterschool programs resulted in gains in children's academic grades, school attendance, and graduation rates, and (4) informal learning opportunities are major predictors of children's development, learning, and educational achievement.

Other observers citing the close link between informal learning and knowledge acquisition by children include Bransford et al. (2000), Hartnett (2012), and Mayer (1997).

4.6.6.2 Community Socioeconomic Status and School-Community Integration

This brings us to what I submit represents a highly compelling example of an intimate link between design and student performance; namely, the relationship between the socioeconomic status of a community and the performance of students attending schools in that community (T.J. Smith, 2012). As used here, the term "socioeconomic status" refers to the average levels of both economic well-being and social cohesion and interaction in a given community. This essentially represents a profile of community ecology, characterized by such indicators as life expectancy and other measures of community health, literacy, and educational achievement, crime levels, and levels of employment and welfare dependency, along with less-tangible factors such as personal dignity and safety, and the extent of participation in civil society.

I use the term *community ergonomics* to refer both to the collective set of system design factors that define the socioeconomics of a community as well as to the study of these factors (Cohen & Smith, 1994; Newman & Carayon, 1994; J.H. Smith & Smith, 1994; M.J. Smith et al., 1994).

Evidence implicating the critical influence of community design on educational performance has emerged with results from standardized tests of mathematical ability and reading comprehension administered to eighth graders (age 13 years) in public school districts in the State of Minnesota. The findings show a high correlation between test performance and percentage of low-income students in different districts. Representative results for 48–50 Minnesota urban and suburban public school districts in the Minnesota Twin Cities (i.e., Minneapolis and St. Paul) metropolitan area, for the years 1996, 2002, 2006, 2009, and 2012 are summarized in Figure 4.2.

The choice of this particular set of data for the analysis is based on the following considerations: (1) the state of Minnesota started administering statewide basic skills tests in math and reading in 1996, and then only to eighth graders—up to the present, eighth grade test results thus span the longest period of time for which district-by-district comparison of student performance is possible, and (2) the Twin Cities metropolitan area features school districts with a broad range of low-income eighth grade student percentages, with high percentages for strongly urban districts such as Minneapolis (65.4 percent in 2012) and St. Paul (73.2 percent in 2012) versus low percentages for wealthier suburban districts such as Orono (7.9 percent in 2012) or Edina (9.0 percent in 2012).

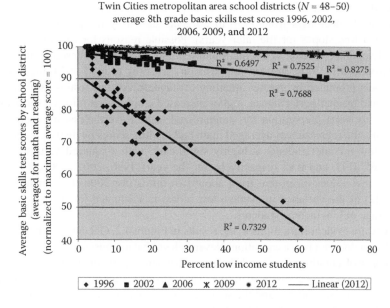

Twin Cities metropolitan area school districts (N = 48–50) average 8th grade basic skills test scores 1996, 2002, 2006, 2009, and 2012

FIGURE 4.2 Basic skills test scores averaged for both math and reading for Minnesota eighth graders by school district for the years 1996, 2002, 2006, 2009, and 2012 in relation to percent low-income students, for different districts in the Minneapolis/St. Paul (Minnesota) metropolitan area based on publicly reported data. (From Draper, N., Math, reading scores up statewide. Schools still plagued by racial gap in performance. In *Minneapolis Star Tribune* [April 19, 2002], pp. A1, A20, A21; Minnesota Department of Education, *Minnesota 2006, 2009 and 2012 Grade 8 comprehensive math and reading assessment test results.* St. Paul, MN: Minnesota Department of Education. Downloaded from http://education.state.mn.us/mde/index.html, 2006, 2009, 2012; O'Conner, D., Spending isn't key to success. In *St. Paul Pioneer Press* [June 13, 1996], pp. 1A, 10A; Smetanka, M.K., & Hotakainen, R., Thousands fail math, reading. In *Minneapolis Star Tribune* [May 29, 1996], pp. A1, A8.)

Figure 4.2 plots the linear regressions of test scores for eighth graders in 48–50 Twin Cities metropolitan area public school districts (average of mathematics and reading scores combined) as a function of the percentage of low income students (the percentage of eighth grade students in different districts receiving free or low-priced lunches), for the years 1996 (diamond symbols), 2002 (square symbols), 2006 (triangle symbols), 2009 (star symbols), and 2012 (round symbols) based on publicly reported data (Draper, 2002; Minnesota Department of Education, 2006, 2009, 2012; O'Conner, 1996; Smetanka & Hotakainen, 1996). Each symbol in the figure represents results for one district. For districts with eighth grades in more than one school, both test score results and low income percentages are averaged across all of the eighth grades in the district. To enable comparison of results for different years (and for revised tests in 2006, 2009, and 2012), the average test scores by district are normalized for each year, with the highest average score assigned a value of 100. Note that the regression lines for the 2006, 2009, and 2012 results are virtually indistinguishable.

Results in Figure 4.2 indicate that variance in average test scores across districts progressively decreases from 1996 to 2012 (as shown by the progressively lower regression line slopes for the later years). Possible reasons for this effect include revised versions of basic skills math and reading tests for Minnesota eighth graders introduced in 2006, greater emphasis on "teaching to test" in Minnesota public schools prompted by greater attention to standardized test results as a gauge of school and teacher performance, and/or increased emphasis on tutoring services for students to assist preparation for standardized tests. Nevertheless, for each of the five years, the regression of averaged math and reading scores by district on percentage of low income students is highly significant, with r-squared values of 0.73, 0.77, 0.65, 0.75, and 0.83 respectively, for the years 1996, 2002, 2006, 2009, and 2012. These values mean that from about two-thirds (for 2006) to over four-fifths (for 2012) of the variance in average test scores is accounted for by regression on percentage of low-income students.

There are evident limitations to the results in Figure 4.2. One obvious point is that correlation does not prove causation. Nevertheless, that a robust relationship between test scores of eighth graders and the socioeconomic status of their school district communities persists across a span of 17 years strongly suggests that the school performance of these students and the socioeconomic conditions in the communities in which they reside are inextricably intertwined in some fashion. Another limitation is that these are data from one grade for a select number of school districts in only one metropolitan area of one U.S. state. It is by no means certain that similar patterns would be observed elsewhere. Also, the data pertain to test results for only two subjects, namely math and reading. Finally, implicit in this analysis is the assumption that test performance is equated with learning proficiency.

Perspective. Despite the limitations cited above, let us take the data in Figure 4.2 at face value and assume that they would be replicated in other areas of the United States (and possibly in other countries as well) that encompass school districts whose communities embody a reasonable distribution of socioeconomic wellbeing. In support of this assumption, a recent report (Swanson, 2009) cites large gaps in high school graduation rates between U.S. cities and suburbs—the average urban rate in the nation's largest cities was 53 percent compared with 71 percent in the suburbs. Based on the foregoing assumption, the findings in Figure 4.2 support a number of compelling conclusions. First, across a 17-year period, the strong correlation between average eighth grade math/reading test scores by school district and the percentage of low-income eighth grade students in different districts remains highly consistent and reproducible. Second, the percentage of low-income eighth grade students in a given school district arguably represents a socioeconomic design factor varying in a manner that is largely independent of biological factors such as gender and IQ. This conclusion is bolstered by findings from a recent study showing that, in healthy students, there is little difference in IQ and mathematics test scores between socioeconomically advantaged and disadvantaged students of different ages (Waber et al., 2007).

Third, it is rare in educational psychology to be able to account for two-thirds to over four-fifths of the variance in a dependent variable of learning performance with a single independent variable. No other design factor specified in Table 4.2

approaches this effect. That such a high and reproducible degree of dependence of test performance on percentage of low-income students is in fact observed implies that remedial strategies unrelated to underlying prevailing socioeconomic conditions in a school district will at best address only about one-fifth to one-third of the variance in math and reading test scores for Twin Cities metropolitan area eighth graders.

What specific socioeconomic design factors in a community tend to either nurture or undermine learning performance of students who attend schools in that community? M.J. Smith et al. (1994) point out that high parental involvement in schools and formation of school-community alliances are linked to good educational performance, whereas low socioeconomic and educational status of parents and poor nutrition of children are linked to poor performance. Another important obstacle to student achievement is the number of immigrant children who enter the U.S. K-12 system with limited or no mastery of English (Mitchell, 2011a); unlike private schools, public schools in Minnesota are obligated to accept such students.

The importance of parental involvement is reiterated by Falk and Dierking (2010) and by Hartnett (2012), both of whom point to research showing that one of the best predictors of student success in school is family life. The former authors indicate that a major way in which high parental involvement benefits student performance is through involving their children in informal learning experiences. One plausible hypothesis is that the results in Figure 4.2 reflect a distinctly greater level of attention by socioeconomically advantaged parents to the performance of their students relative to attention paid by less socioeconomically advantaged parents. However, Mayer (1997) reports findings from her research showing that there is essentially no relationship between family income and a child's life outcomes such as dropping out of high school or teen pregnancy.

Other design factors that may account for the link between community design and student learning outcomes have been identified by different observers. In addition to family interaction and support, noted above, these also include development and refinement of socioemotional skills and nurturing of personal character qualities of self-confidence, dedication to hard work, and trustworthiness as Tough (2012), cited above, has emphasized.

Brooks (2005) frames these points in terms of the significance of introducing and improving elements of cultural, social, moral, cognitive self-awareness, and aspirational capital in children in order to promote achievement both inside and outside of school. In the same vein, Heckman (2008) makes the following points about key socioeconomic and community design factors that critically influence both the educational and societal trajectories of children:

1. Compared to 50 years ago, relatively more American children are being born into disadvantaged families where investments in children are smaller than in advantaged families.
2. Many major economic and social problems such as crime, teenage pregnancy, dropping out of high school, and adverse health conditions are linked to low levels of skill and ability in society.

3. In analyzing policies that foster skills and abilities, society should recognize the multiplicity of human abilities.
4. Currently, public policy in the United States focuses on promoting and measuring cognitive ability through IQ and achievement tests. The accountability standards in the No Child Left Behind (and more recently the Race to the Top) programs concentrate attention on achievement test scores and do not evaluate important noncognitive factors that promote success in school and life.
5. Socioemotional as well as cognitive skills, physical and mental health, perseverance, attention, motivation, and self-confidence contribute to socioeconomic success and to performance in society at large and even help determine scores on the very tests that are commonly used to measure cognitive achievement.
6. Family environments of young children are major predictors of cognitive and socioemotional abilities as well as a variety of outcomes such as learning, crime, and health.
7. The absence of supportive family environments harms child outcomes. Early societal interventions directed at disadvantaged children promote schooling, reduce crime, foster workforce productivity, and reduce teenage pregnancy. These interventions are estimated to have higher benefit-cost ratios and rates of return than later interventions such as reduced pupil-teacher ratios, public job training, convict rehabilitation programs, adult literacy programs, tuition subsidies, or expenditure on police.
8. Life-cycle skill formation is dynamic in nature. Skill begets skill; motivation begets motivation. Motivation cross-fosters skill and skill cross-fosters motivation. If a child is not motivated to learn and engage early on in life, the more likely it is that when the child becomes an adult, he or she will fail in social and economic life.
9. A major refocus of policy is required to capitalize on knowledge about the life cycle of skill and health formation and the importance of the early years in creating inequality in America and in producing skills for the workforce.

The foregoing considerations argue for the idea that schools and school districts should pay closer attention to socioeconomic conditions in their surrounding communities, an interaction that I term school-community integration. Examples in Minnesota of such an emphasis include the recently announced federal Promise Neighborhood initiative (Havens & Patterson, 2010), aimed at promoting community ownership of education, a parent outreach program in the St. Paul public school district in which teachers visit students' homes to learn more about their lives and family (Koumpilova, 2011), and a program at a Minneapolis (Minnesota) school that teaches parents how to help build the reading skills of their children at home; this program reportedly has boosted reading performance of the children of parents involved in the program (Brandt, 2013). If these school-community integration efforts achieve the goal of improving student performance, feedback effects on communities may result in a virtuous circle, given strong evidence that educational achievement reduces urban poverty (Swanson, 2009).

A leading champion of school-community integration, in the context of education in the United Kingdom, is Kathryn Riley (2007, 2009). In order to serve as an effective leader of an urban school, Prof. Riley identifies four challenges dealing with school-community integration that such a leader must meet and master, as follows:

- Challenge I: Making sense of the big picture in our cities—their changes and complexity
- Challenge II: Understanding more about local communities and about children's lives and experiences as influenced by these communities
- Challenge III: Fostering trusting relationships between schools and communities
- Challenge IV: Understanding the role of their school and their own role as a school leader in relation to the surrounding community

Whatever the case may be in the United Kingdom, the limited attention paid in the United States to the implications for student academic performance of the linkage between school system and community system functioning suggests that few educational system leaders and policymakers in the United States are aware of the significance of this relationship as a key influence on student learning.

The basic message offered by this perspective is that the aims and objectives of designing educational systems (educational ergonomics) and those of designing community systems (a branch of HF/E science termed community ergonomics (Cohen & Smith, 1994; Newman & Carayon, 1994; J.H. Smith & Smith, 1994; M.J. Smith et al., 1994) are intimately coupled. In particular, the data in Figure 4.2 suggest that community ergonomic interventions that yield improved quality in community socioeconomic conditions therefore also should yield improved learning performance in community schools. This prediction has prompted T.J. Smith (2007, p. 1542) to suggest that school districts should adopt the principle that, "the boundaries of a school are the boundaries of its community."

4.7 LIMITATIONS

There are evident limitations to the analysis in this chapter, encompassing errors of both omission and commission. One possible criticism is that the issues addressed in this chapter are well recognized—that is, the importance of design for nurturing student achievement and the functioning of educational systems is widely accepted. As another example, the possible influence on student learning of well-recognized factors unrelated to design—such as differences in learning ability (addressed for example by gifted student programs)—are ignored. Thirdly, in terms of putative effects on student learning, some important design factors may not be considered. Each of these possible limitations is addressed briefly below.

4.7.1 INFLUENCE OF NONDESIGN-RELATED FACTORS ON STUDENT LEARNING

Reviewed in Chapter 3 is work summarizing extensive evidence that the preponderance of variance observed in learning performance is attributable to the design (the context) of the task being learned. As for the significance of generalized intelligence or cognitive functioning for learning ability (the "g" factor) (Ackerman & Lohman, 2006; Bransford et al., 2006; Gustafsson & Undheim, 2006), the scientific validity of this idea has been critically evaluated—and called into question—by both Bredo (2006) and K.U. Smith and Smith (1966). Admittedly, support for context specificity in learning is based on evidence compiled primarily from young adults, not K-12 children. However, those advocates of generalized intelligence theory must not only counter this evidence, they also must explain why children might be exempt from the applicability of this evidence.

4.7.2 Analysis Offers Nothing New

The ideas and evidence set forth here regarding the primacy of design as an influence on student learning may not be new to the HF/E community, but most assuredly the concept remains alien to the educational community. As noted in Section 4.2, no reference is found to HF/E in the 2006 edition of the *Handbook of Educational Psychology* (Alexander & Winne, 2006) (a major influence on what teachers are taught about how student learning occurs), and as noted in Section 4.3, only a subset of design factors specified in Table 4.2 are addressed in the *Handbook* as possibly influencing student learning (de Jong & Pieters, 2006; O'Donnell, 2006; Schunk & Zimmerman, 2006).

4.7.3 Some Design Factors Not Considered

The critique that not all design factors have been considered almost certainly is true. However, my objective in this report is to address design factors whose impact on student learning has been demonstrated with reasonable certainty based on available evidence, with a particular emphasis on factors with strongly positive effects.

4.8 CONCLUSIONS

This chapter has extended the emphasis in Chapter 3 on context specificity as a major source of variability in cognitive performance and learning to the particular case of learning in children. Unlike the evidence reviewed in Chapter 3—based largely on performance evaluations conducted under controlled experimental conditions—the analysis in this chapter is based almost exclusively on studies of context-specific effects on student learning carried out in largely uncontrolled environments such as the classroom or the community. As such, the evidence collected thereby must be considered less scientifically definitive and subject to a broader range of possible interpretations than that compiled under more controlled circumstances. Nevertheless, considering the range of design factors whose possible influence on student learning has been explored in a reasonably systematic manner (Table 4.2), the weight of evidence reviewed in this chapter strongly supports the conclusion that the influence of context specificity on variability in learning of K-12 students is as prominent as that demonstrated for older individuals.

Specifically, the review in this chapter supports three broad conclusions: (1) student learning outcomes, and more broadly the edifice of K-12 education itself, are largely defined and structured in terms of an extensive system of design factors and conditions, (2) the time is long overdue for the educational system to acknowledge the central role of HF/E design as a major influence on student performance and learning, and (3) K-12 educators and administrators should emphasize allocation of resources to design factors listed in Table 4.2 for which a positive impact has been reliably demonstrated; namely, environmental design of classroom and building facilities, longer exposure to learning, cooperative learning designs, early childhood education, teaching quality, participation in physical activity and physical fitness, and school-community integration—but should treat expenditure on factors listed in the table with equivocal, varied, or weak influence on student performance with more caution and/or skepticism.

Let me close by reiterating a theme alluded to in Section 1.3.3 of Chapter 1 and in the first section of this chapter regarding the relevance of this book to the most ubiquitous manifestation of human performance variability; namely, human work. Work is the engine of the human condition (T.J. Smith, 2001). The children of today are the workers of tomorrow. Making informed decisions today about how best to support student learning—and this report argues that these decisions must focus on the design of learning environments—will help ensure a productive, effective, and competitive workforce of tomorrow.

5 Variability in Human Performance under Displaced Sensory Feedback

Thomas J. Smith

5.1 OVERVIEW

One of the most dramatic manifestations of variability in human performance is that which occurs when the feedback that individuals receive of their own movement is displaced in some manner. Research studies on displaced feedback, dating back almost 16 decades, have been carried out in the context of a more general interest in psychological science regarding the nature of space and time integration of efferent and afferent neural mechanisms in the control of overt behavior. The term "displaced sensory feedback" thus refers to the introduction of either spatial transformations (spatially displaced feedback) or temporal delays (delayed feedback) in the relationships between motor (efferent) behavior and the sensory (afferent) feedback generated by that behavior. For conditions in which sensory feedback is distorted in some manner, but not spatially or temporally displaced, the term "perturbed sensory feedback" is applied.

Figure 5.1 schematically illustrates the implications of these two modes of displaced feedback for variability in human performance using teleoperation (Section 5.4) as an example. The figure suggests that effective control of a telemanipulator during teleoperation depends upon space and time compliance between operational feedback from the telerobot slave sector and motor behavioral control of this sensory feedback by the operator. If human factor design problems cause spatial or temporal displacements in sensory feedback, then compliance between sensory feedback and its control is compromised and performance suffers (discussed in more detail in Section 5.4).

Most displaced feedback studies have focused on displacements in visual feedback. However, there also is an appreciable literature on the analysis of the behavioral effects of delay on either auditory or force feedback. The advent and refinement of technology that enables operator control of remote devices and vehicles also has been accompanied by a distinct line of research dealing with displaced feedback effects on what is called *teleoperation* of these systems.

The focus of this chapter is on the *degree* of behavioral variability that ensues when an individual is asked to perform under conditions of displaced sensory

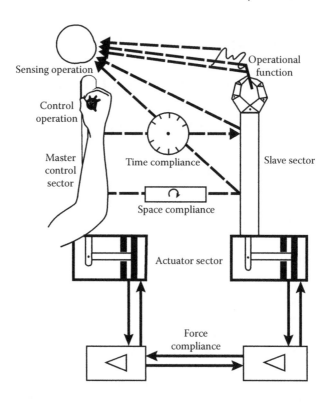

FIGURE 5.1 Schematic illustration of spatial and temporal feedback relationships in tele-operation. Compliance in spatial and temporal feedback between movements of the operator and those of the slave is essential for optimal performance of the telemanipulator. (Reproduced from Smith, T.J. et al., *The behavioral cybernetics of telescience—an analysis of performance impairment during remote work* [SAE Technical Paper Number 94138]. Warrendale, PA: Society of Automotive Engineers International, Figure 2, 1994b.)

feedback, as well as the amount of behavioral adaptation that may take place (with a possible reduction in such variability) under training or learning trials with these conditions. I will anticipate this discussion by noting that when naïve individuals are exposed to displaced feedback conditions, the effects are immediate and profound, with pronounced decrements in behavioral control capabilities. Displaced feedback conditions represent a design factor. The dramatic disruptions in behavior that these conditions cause thus constitute a highly appropriate line of inquiry for understanding the nature and extent of human performance variability.

It is a relatively straightforward proposition to test these effects—for spatially displaced feedback—on yourself. One simple test is to rotate your computer mouse by 90 degrees and try to accurately control the trajectory of the mouse cursor on the computer screen. You will find performance of this task difficult and disconcerting. This is an example of the behavioral effects of performance with a 90-degree angular displacement in visual feedback. Another test (if you have access to the equipment) is to stand behind an overhead projector with a transparency containing a tracking path (such as a star) placed on top of the large Fresnel lens of the projector. The test consists of two trials, each of which should be timed. First, use a marking pen to trace the tracking path on the transparency while looking directly at the transparency. This is the control condition. Then, repeat the tracing task, but this time watch the projection of the movements of your marking pen while tracing the tracking path on a display screen placed in front of the .projector. This is the experimental condition.

You will find performance of this latter condition difficult and disconcerting. This is an example of the behavioral effects of performance under reversed (operational visual feedback (Figure 1.9 generated by movements of the marking pen to the right indicates that the pen is moving to the left, and vice versa) plus inverted (operational visual feedback generated by upward movements of the marking pen indicates that the pen is moving downward, and vice versa) displacement of visual feedback. I have had students perform this test in my university classes over the past decade, and increases in elapsed task times from the control to the experimental conditions typically range from 3- to 11-fold. Evident degradation in behavioral control is instantaneous as soon as a student is asked to start spatially displaced visual feedback tracking. Verbal comments by students after a trial indicate a concomitant loss of body sense plus emotional effects. Specifically, loss of a sense of the position of the upper limb used for tracking is reported—these observations recapitulate those experienced by subjects participating in the earlier spatially displaced visual feedback studies of Smith (K.U. Smith & Smith, 1962). Students also report emotional consequences related to increased stress, again recapitulating earlier findings (cited in Section 5.2).

To experience self-demonstration of feedback delay, it is likely that some users interacting with a computer mouse may have encountered incidents of short, but frustrating, delays between movements of the mouse and movements of the mouse cursor on the computer screen. Another example of feedback delay that you occasionally may encounter because of transmission lags is a temporal lag between auditory feedback from a speaker (say on a TV) and the lip movements of that speaker. Don't watch the lip movements while listening to the auditory feedback under these conditions—the effect is disconcerting and somewhat confusing.

As for the question of *why* performance under displaced feedback conditions causes such pronounced effects on behavioral variability, a number of alternative theories have been advanced. This chapter will not attempt an objective review of all of these alternatives. Instead, interpretation of the source of displaced feedback effects will primarily reference behavioral cybernetic theory (Section 1.3.3), given that the inspiration for development of this theory by K.U. Smith, and for a career devoted to researching these effects, was based on evaluation during WWII of the effects of feedback delay on the operation of an early servo-controlled tracking system (K.U. Smith, 1945, 1962a; K.U. Smith & Smith, 1962).

The remainder of this chapter will provide a historical perspective (Section 5.2), a behavioral control systems analysis of displaced visual feedback effects (Section 5.3), a compilation of performance effects of spatially displaced and delayed sensory feedback (Section 5.4), a summary of displaced sensory feedback effects on teleoperation (Section 5.5), and conclusions (Section 5.6).

5.2   HISTORICAL PERSPECTIVE

5.2.1   Early History of Spatially Displaced Visual Feedback Research

Scientific interest in spatially displaced visual feedback dates back to the middle of the nineteenth century to the early work of Helmholtz (1856–1866), who studied the effects of and adjustment to spatial displacements in visual feedback. Subsequently, over the ensuing century, interest in the behavioral effects of this condition continued to attract research attention based on the use of mirrors or specialized glasses to study effects of inverted vision (Ewert, 1930; Kohler, 1951a, 1951b, 1953, 1955; Siipola, 1935; Snyder & Pronko, 1952; Stratton, 1896, 1897, 1899).

In the past six decades, the effects on human performance of spatial displacements in visual feedback have been the subject of numerous investigations whose findings are summarized in a series of reviews (Harris, 1965; K.U. Smith, 1972; K.U. Smith & Smith, 1962; T.J. Smith & Smith, 1985, 1987a, 1988a, 1990; W.M. Smith & Bowen, 1980; Welch, 1978). Effects on use of visual feedback to guide and control motor behavior has been the exclusive focus of spatially displaced feedback research, whereas effects of delayed feedback have been evaluated in the context of both visual and force feedback studies.

There are a number of general features of behavioral disturbance evoked by spatial displacements in visual feedback that appear consistently from study to study. Oscillatory instability in movement control becomes more pronounced, accompanied by increased variability and extremes in movement velocities and accelerations. As a consequence, the accuracy of movement guidance and tracking accuracy suffers. Perception is degraded and may disappear altogether; learning concomitantly is impaired. When asked, subjects report feeling confused, uncomfortable, or uncertain about their own behavior. Skilled performers are particularly sensitive to these effects. A group of talented graphic artists asked to execute a drawing under conditions of size distortion of visual feedback displayed a distinct lack of motivation for continuing after a few sessions (K.U. Smith & Smith, 1962, Chapter 11).

5.2.2   Early History of Delayed Feedback Research

Two distinct lines of inquiry mark the early history of delayed feedback research. One deals with studies of performance effects of delays in visual feedback, the other with delays in auditory feedback.

5.2.2.1   Delayed Visual Feedback

Challenges associated with performing under delayed visual feedback emerged during WWII with the introduction of new, servo-controlled aided tracking systems with delay inherent in their operation. This was one of the problem areas associated with the advent of new military technology that prompted the U.S. Army to form a team of civilian psychological scientists to investigate how design problems inherent to these new systems could be modified to better meet the behavioral control capabilities and limitations of military personnel assigned to operate these systems. It is recognized that human factors science in the United States originated during WWII with this effort (K.U. Smith, 1987). Thus, it is fair to say that the previously unappreciated problem of delayed visual feedback made a key contribution to the emergence of this new science.

One of the members of the Army team of civilian scientists was K.U. Smith. His encounter with feedback delay problems inherent to servo-controlled aided tracking systems (preceding paragraph) represented a key inspiration for the rest of his scientific career. His description (K.U. Smith, 1962a, p. 19) of what it is like to operate under delayed visual feedback represents a classic account of the difficulty an operator faces in controlling such a system.

> In a velocity tracking system...[with delay]...the novice operator feels lost, because his control movements seem loosely connected with the gun or cursor. One of the rare experiences in World War II was to climb into the Sperry belly turret of the B-17 and try to aim the turret with its two .50-caliber guns by means of the velocity tracking system with which it was equipped. It was as good as a carnival ride. No matter what the novice gunner did, the turret and guns gave him the wrong answer about his movements. If he thoughtlessly tried to direct the guns at the target in a hurry, he might find himself overshooting in a wild spin. The basic trouble was that the operator was receiving delayed visual feedback of his motions, occasioned by a lag of some 0.5 to 1 second between the execution of control movement of the sight and the effect of this movement on the action of the turret.

In the two decades following WWII, a number of studies of delayed visual feedback were carried out using various experimental designs and delay intervals. In 1945, the Foxboro Company reported findings from a series of aided tracking studies, with a lag of 0.1 second in transmitting an error display to the tracker imposed with each study (K.U. Smith, 1962a, p. 31). Slight decreases in tracking accuracy were observed in every case.

Warrick (1949) imposed transmission lags in visual feedback ranging from 0 to 320 milliseconds in a series of compensatory (error) tracking tasks. A linear relationship between reductions in tracking accuracy and feedback delay interval was predicted with these studies, but not in fact observed. In contrast, Levine (1953) did observe a linear relationship between feedback delay interval and loss of compensatory tracking accuracy for delay intervals ranging from 0 to 2.7 seconds.

Conklin (1957, 1959) carried out two extensive studies of the effects of visual feedback delay on tracking accuracy. The first study, involving compensatory tracking, imposed partial lags in tracking error indication ranging from 0 to 16 seconds. When the target course had a coherent pattern, tracking accuracy dropped sharply with the shorter lags, and then tended to level off at a chance level of performance with delays of several seconds. When the target course was random, however, accuracy reduced to chance (and below!) with delays no longer than 1 second.

The aim of the second study was assess to what extent different control lags between 0 and 1 second, introduced in steps of 0.2 second, affected direct pursuit and compensatory tracking accuracy. Target courses of different complexity were used. Results showed that the function relating tracking accuracy and delay in error feedback was essentially linear in the range studied. Comparing the relative effects of the lag on the two types of tracking, Conklin found that subjects learned to reduce error more with pursuit tracking than with compensatory. In fact, pursuit tracking with a 1-second display lag was better than compensatory tracking without any error delay. These results suggest that the deficiencies of compensatory tracking may result from intrinsic delayed sensory feedback factors in the compensatory display.

The general conclusion to be drawn from these early studies is that delayed visual feedback of motion is invariably detrimental to tracking performance. This is strong evidence for the assumption that the basic defect in velocity and aided tracking systems is the partial or selective feedback delay inherent in their operation.

5.2.2.2 Delayed Auditory Feedback

Although delayed feedback phenomena were first studied in relation to delayed visual feedback (above), the effects of delayed auditory feedback established initial broad scientific interest in the problem. Not long after the war, Lee (1950a, 1950b, 1951) extended the exploration of delayed visual feedback to time displacement of the sounds of oral speech. Lee used a magnetic audiotape recorder to delay the sound return that a person gets of his or her own speaking. These studies showed that the ability of most persons to adapt to such delayed hearing was extremely limited.

Very shortly after Lee (1950a, 1950b, 1951) completed his primary work, study of delayed speech was extended by many others. The general procedure of such experiments was to record speech sounds of a reader or speaker on magnetic tape, store these sounds for a variable period by means of a tape loop, and then feed back the sounds to headphones worn by the subject with the delay period determined by the length of the tape loop.

Without exception, these studies of delayed auditory feedback reported severe disturbances in speech when the sounds of talking or reading are delayed by even small fractions of a second. Subjects stutter, make errors in talking, reduce the rate of talk, change the pitch and nature of the voice, and in some cases block altogether. In talking spontaneously, some subjects can avoid these effects by not attending to the sounds of the delayed voice, but most subjects cannot do this. The effects of the delay are greatest with lag intervals of about 0.2 second, and at this interval many subjects block after one to two minutes of attempting to talk against the delayed side tone, as it is sometimes called (Lee, 1950a). When subjects are required to read they seem to have greater difficulty in avoiding the effects of the delay.

The main *qualitative* effect of the delayed auditory feedback is what Lee (1951) designated as "artificial stutter." The stutter thus produced is very similar to real stutter in that repetitive utterance of a constricted syllable sound may go on for several repetitions of the sound. A limited number of observations have been made on delayed auditory feedback of movements other than speech. Limited effects of the time lag in hearing related to control of such movements also have been observed.

The most generally observed *quantitative* effect of delayed hearing, a slowing down of speech, was reported in the first studies by Lee (1950a, 1950b, 1951) and Black (1951, 1955) with delays of 0.03 to 0.18 second. Rawnsley and Harris (1954) used sound spectrograms for observation and reported that phrases spoken with delayed auditory feedback were invariably spoken more slowly than under normal conditions. Fairbanks (1955) observed that the delay period producing the greatest decrease in reading rate was about 0.2 second.

Fairbanks (1955) and Fairbanks and Guttman (1958) carried out systematic studies of the rate of speech as a function of delay interval. The latter authors evaluated the different types of errors produced by delayed auditory feedback of different magnitudes. Several measures of speech, articulatory accuracy, correct word rate in words per second, reading duration, and number of errors varied systematically with the magnitude of delayed feedback. Peak disturbances were found at 0.2 second. Longer hearing lags were associated with relatively fewer errors than the number occurring at this peak lag value. The function relating different types of error to magnitude of hearing lag were all much the same.

Lee (1950a) made the first observations on the effects of delayed hearing on sound-controlled manual activities. He observed that a skilled wireless telegraph operator was incapable of rhythmic tapping with delayed auditory feedback. Kalmus et al. (1955) also observed disturbances of tapping or clapping with delayed hearing.

Besides artificial stutter, there are many other special movement effects of hearing lag in speech. Fairbanks and Guttman (1958) found that the repetitive movements of artificial stutter made up about 70 percent of all additive errors, which were the most prominent of all errors observed. Chase (1958) also attempted to analyze the nature of artificial stutter. Using himself as subject, he found that although he did not stutter with delay, he had difficulty, once started, in stopping the repetition of syllables. In other observations on 20 subjects, he found that 17 of them tended to repeat a single syllable sound faster under delay conditions than they repeated this same sound normally. The likely basis of this effect is that under delayed auditory feedback, the individual shifts the nature of control of syllable sounds.

The experience of K.U. Smith (T.J. Smith & Smith, 1988a, p. 6) also points to the possibility of persistent, disruptive after-effects on speech control following exposure to feedback delays in hearing one's own speech. Smith established what he called the Behavioral Cybernetics Laboratory at the University of Wisconsin in the early 1960s, centered around the pioneering use of a computer system to study effects of spatial and temporal displacements in sensory feedback on behavioral control of different types and patterns of movements (T.J. Smith & Smith, 1987a, 1988a). In one of the early applications of this system, effects of delayed auditory feedback were studied by using an analog-to-digital converter to convert analog speech signals recorded from a subject to digital form, storing the digitized speech data in the computer's memory buffer, and playing the sounds back to the subject using a digital-to-analog converter after a short delay (limited by the 8-kbyte size of the computer memory) enabled by the buffered data. This was the first instance in which a computer system (as opposed to tape loops or transmission lags) was used to impose delayed auditory feedback under controlled experimental conditions (K.U. Smith et al., 1963a, 1963b). The original digital computer used in the laboratory, a Control Data Corporation 160-A (the second one produced), is now on display in the Cray Computer Museum in Chippewa Falls, Wisconsin, with a short description of its role in pioneering computer-mediated human factors research on human performance and cognition (T.J. Smith & Smith, 1988a, p. 15).

The first subject to test this new computerized capability for experimental control of delayed auditory feedback was Smith himself. In the course of serving as a subject in a series of delayed auditory feedback experiments dealing with speech control over a period of weeks, he noted the appearance of stuttering and speech irregularity problems that persisted for months after the experiments themselves had been terminated (T.J. Smith & Smith, 1988a, p. 6).

Emotional behavior, too, is affected by hearing lags in speech. Lee's (1950a) original studies indicated that with continued effort to speak against the lagged tone, marked emotional tension, sweating, frustration, fatigue, and reddening of the face occurred. Hanley et al. (1958) recorded the galvanic skin response in 50 subjects during speaking with a hearing lag. As the sound intensity of the lagged hearing was increased, the latency of the skin resistance change decreased, and the magnitude of the effect increased. A skilled musical quartet, asked to perform under delayed auditory feedback (Ansell & Smith, 1966), absolutely refused to continue the task after a few minutes, claiming that further exposure to the condition would irreparably damage their performance skill.

In general, the movement and emotional studies of delayed side-tone indicate that such lags have the same motivating and disorganizing effects as space or time displaced sensory lags in visual behavior. However, in the sound-delayed feedback in speech, the ability to cope with the delay is far less in most people than it is in adapting to space-displaced visual feedback. Moreover, there is fairly good evidence that the sound delay affects most decisively the difficult variable movements of speech that require a high degree of sensory control and affect less the well-established discrete units of talking.

5.2.3 SUMMARY

Three general conclusions can be drawn from the observations summarized above regarding displaced sensory feedback. First, no behavioral activity or process so far examined is immune from the adverse effects of displacements in sensory feedback. This conclusion applies to the performance as well as the learning of the behavior. Second, the learning curves for task performance under displaced sensory feedback conditions show training effects over time, but control levels of performance are not achieved even after training periods lasting as long as 20 days (K.U. Smith & Smith, 1962; Welch, 1978, Chapter 13, pp. 275–279). Finally, results from different studies indicate that the particular effects of a given displacement condition are task-(context-) specific, dependent on the spatiotemporal characteristics and integrative properties of the motor control patterns involved.

5.3 BEHAVIORAL CONTROL SYSTEMS ANALYSIS OF DISPLACED VISUAL FEEDBACK EFFECTS

This section introduces specific examples of decremental changes in tracking performance under conditions of both spatially displaced, and delayed, visual feedback. The purpose of this analysis is twofold; namely, to illustrate with these examples: (1) the degree of change in performance that occurs under these two modes of displaced visual feedback in anticipation of a more comprehensive survey of these effects in the next section, and (2) study designs that enable a behavioral control systems interpretation of performance variability inherent to these effects.

The examples described here are based on two of my own studies (T.J. Smith et al., 1994b, 1998). The displaced visual feedback effects observed with these studies are illustrative but also aligned with a large body of evidence compiled with previous research (next section). On the other hand, the behavioral control systems analytical approach adopted with these studies is scientifically distinctive.

5.3.1 Examples of Spatially Displaced Visual Feedback

Examples from two studies of T.J. Smith et al. (1994b, 1998) are summarized to illustrate performance effects of spatially displaced visual feedback. The first example is from the 1994 study and is illustrated in Figure 5.2. The arrows on the left side of the figure graphically depict the spatial transformations that occur when visual feedback from a target is televised, reversed (R), inverted-reversed (IR), or inverted (I) before being viewed by an operator. The histograms on the right side of the figure illustrate the performance decrements that occur (as multiples of task performance levels under normal televised viewing conditions) for a series of hands-on viewing tasks (reported in K.U. Smith & Smith, 1962), and for teleoperation in which subjects were asked to view a display screen to operate a telerobot in order to execute two remote telemanipulation tasks. The original studies in which results for these tasks are cited is in the legend to Figure 5.2; the figure also lists the numbers of studies for each condition. The teleoperation results anticipate further discussion in Section 5.5 of teleoperation performance under spatially displaced visual feedback.

The comparisons in Figure 5.2 are based on a series of assumptions. First, only data for untrained performance are considered. Second, contrasts between telerobot

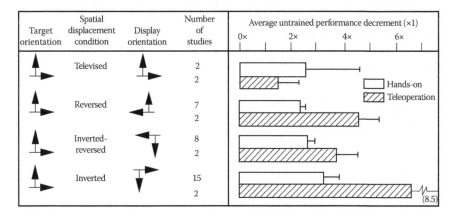

FIGURE 5.2   Average values from a series of hands-on and teleoperation studies for performance decrements under televised viewing (top) and reversed, inverted-reversed, and inverted visual feedback conditions. The effects of each displacement condition on display relative to target orientation are shown by the arrows. Mean values for performance decrements are plotted as histograms on the right, based on findings from the number of studies listed in the fourth column. Error bars are ±1 standard deviation. For the spatial displacement conditions, performance decrements are calculated as (performance level with televised perturbed visual feedback)/(performance level with televised normal visual feedback). For the televised viewing condition, performance decrements are calculated as (performance level with televised normal visual feedback)/(performance level with direct normal visual feedback). For further information on the spatial displacement conditions, televised viewing conditions, and references to the studies and tasks, please see the 'Note' on page 321. (Reproduced from Smith, T.J. et al., *The behavioral cybernetics of telescience—an analysis of performance impairment during remote work* [SAE Technical Paper Number 94138]. Warrendale, PA: Society of Automotive Engineers International, Figure 4, 1994b.)

versus hands-on performance are based on studies of different specific tasks in different laboratories at different times by different investigators, making the validity of any comparison problematic. For this reason, only visual-manual performance tasks are considered, and only normalized performance changes for each task are presented, calculated as the ratio of the level of task performance for a given spatially displaced visual feedback condition relative to that for the normal televised viewing condition. The comparisons therefore should be considered as suggestive only.

With these assumptions in mind, the results in Figure 5.2 support the following general conclusions:

- The relatively modest standard deviations for average hands-on performance decrements under the three spatial displacement conditions (R, IR, R) suggest that for each condition, effects are reasonably consistent across different studies, tasks, and performance measures.
- When all three spatial perturbation conditions are examined in the same hands-on study, inversion of visual feedback consistently produces the largest performance decrements and reversal the smallest, with inversion-reversal having intermediate effects. However, differences in performance decrements across these three hands-on displacement conditions are relatively small, and range from about two- to three-fold.
- In the two teleoperation studies, inversion also is most disturbing, but inversion-reversal has the smallest effect.
- The results suggest that teleoperation may be more labile to spatial displacements in visual feedback than hands-on performance.
- Relative to direct viewing, televised viewing during either hands-on or teleoperation tasks degrades performance. However, teleoperation performance does not appear to be degraded more than hands-on performance during televised viewing.

The 1998 study of T.J. Smith et al. featured a compensatory pursuit tracking task. Subjects were asked to move a joystick to align the position of a tracking cursor as closely as possible with that of the target spot. The target spot was displayed on a computer monitor and traversed a randomly varying path in two dimensions. Three X-axis and three Y-axis sinusoidal functions of different amplitudes and frequencies were added to generate the two-axis target path, with a speed of 0.3 Hz and randomly varying trajectory. During a trial, spatial displacement in visual feedback was introduced by displacing the angle of tracking cursor movement relative to that of joystick movement. Thus, at a displacement angle of 180 degrees (reversed visual feedback), left horizontal movement of the joystick would cause right horizontal movement of the tracking cursor. At a displacement angle of 90 degrees (angularly displaced visual feedback), left horizontal movement of the joystick would cause upward vertical movement of the tracking cursor. Ten levels of angularly displaced visual feedback (0, 30, 45, 60, 75, 90, 105, 120, 150, and 180 degrees) were employed, with the level varied randomly from trial to trial.

Dependent variables collected were root mean square (RMS) error of the position of the tracking cursor relative to that of the target and gain (ratio of the amplitude of joystick movements relative to that of the target movements).

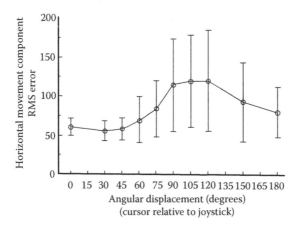

FIGURE 5.3 *X*-axis component error in a pursuit tracking task as a function of angular displacement in visual feedback. (Reproduced from Smith, T.J. et al., SAE Technical Report Number 981701. Warrendale, PA: Society of Automotive Engineers International, Figure 8, 1998.)

Figure 5.3 shows mean *X*-axis component RMS error as a function of degree of angular displacement; error bars are standard deviations. RMS error starts to increase at a displacement angle of about 45 degrees (first shown by Gould & Smith [1963], and termed the *breakdown angle* by these authors), peaks at a displacement angle of about 120 degrees, and moderates at displacement angles beyond 120 degrees. Ellis et al. (1991) report a similar pattern. The variability (standard deviation) in RMS error increases markedly with higher angles of spatial displacement. There is approximately a 2.4-fold increase in RMS error between that observed at a displacement angle of zero degrees (control condition) and that observed at displacement angles of 105 and 120 degrees.

Figure 5.4 is shows mean *X*-axis component gain in tracking movements as a function displacement angle; error bars are standard deviations. At displacement angles up to 75 degrees and above 120 degrees, gain is almost constant at levels close to 1.0. Gain is reduced at displacement angles where error is highest (Figure 5.3). The main effect of angular displacement on gain is significant ($F_{7,63} = 2.241$, $p = .0179$). As with RMS error, variability (standard deviation) in gain increases at higher RMS error levels. To my knowledge, the T.J. Smith et al. (1998) report represents the first in which performance gain relative to displacement angle has been reported in a spatially displaced visual feedback study.

5.3.2 Example of Delayed Visual Feedback

Findings from two different studies of T.J. Smith et al. (1994b, 1998) are summarized to illustrate performance effects of delayed visual feedback. The goal of the 1998 study was to document effects of different fixed delay intervals on accuracy in a compensatory pursuit tracking task. The task involved tracking a target generated as a solid line sweeping right to left on a computer monitor. Three sinusoidal functions

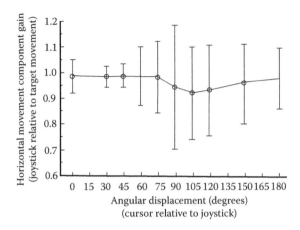

FIGURE 5.4 *X*-axis component gain in pursuit tracking movements as a function of spatial displacement angle. (Reproduced from Smith, T.J. et al., SAE Technical Report Number 981701. Warrendale, PA: Society of Automotive Engineers International, Figure 9, 1998.)

of different amplitudes and frequencies were added to generate a sinusoidally varying target function with a frequency of 0.3 Hz and randomly varying amplitude. During a trial, subjects were asked to move a joystick toward or away from them in a one-dimensional pursuit tracking manner to align the vertical position of the tracking cursor as closely as possible with the shifting position of the target line.

Computer methods were used to impose a series of fixed-delay intervals ranging from 0 to 3.0 seconds between joystick movements and visual feedback from corresponding movements of the displayed tracking cursor. Nine levels of feedback delay (0, 0.2, 0.4, 0.7, 1.0, 1.5, 2.0, 2.5, and 3.0 seconds) were employed. The dependent measure was RMS error in tracking accuracy.

Figure 5.5 shows levels of mean RMS error as a function of visual feedback delay interval for this study; error bars are standard deviations. Mean error increased over eightfold with an increase in delay from 0 and 3.0 seconds. The variability (standard deviation) in RMS error increases markedly with longer display intervals. The increase in error is linear for delay levels from 0 to 1.0 seconds; that is, there appears to be no threshold below which delay intervals greater than 0 do not evoke increased error. This observation of linearity in error increase for delay between 0 and 1.0 seconds recapitulates earlier observations of Levine (1953) and Conklin (1959), which were reviewed in Section 5.2. The rate of error increase tapers off somewhat at delay levels above 1.0 second, but error does not level off at 3.0 second delay. The main effect of delay is highly significant ($F_{8,64} = 211.5, p < .0001$).

The goal of the second T.J. Smith et al. (1994b) study was to apply a distinctive experimental design based on a behavioral control systems analytical paradigm (Chapter 1, Section 1.3), to document performance effects of delayed visual feedback under conditions in which delay is continuously varied. The motivation for this approach is the analysis introduced in this study that teleoperation (Section 5.5) may expose the operator to variable feedback delay conditions.

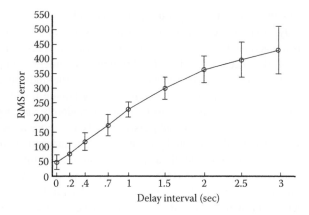

FIGURE 5.5 Tracking error in a pursuit tracking task as a function of visual feedback delay interval. (Reproduced from Smith, T.J. et al., SAE Technical Report Number 981701. Warrendale, PA: Society of Automotive Engineers International, Figure 2, 1998.)

The experimental challenge in this study, therefore, was to introduce more than one level of delayed visual feedback as the dependent variable in a given trial. For this purpose, a methodology based on a well-established experimental paradigm in engineering and physiology was implemented, involving imposition of delayed visual feedback as a continuously varying forcing function (Jagacinski & Flach, 2003, pp. 137–157). The second advantage offered by this approach is that it provides insight into the dynamic control system properties of behavioral performance under displaced sensory feedback conditions.

The study therefore entailed evaluation of the control system response characteristics of tracking performance under variable feedback delay imposed at different sinusoidal frequencies. Figure 5.6 illustrates this approach, showing the imposition of delay

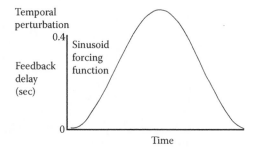

FIGURE 5.6 Conceptual illustration of the use of a sinusoidal forcing function to assess the ability of subjects to directly perceive the effects of delayed visual feedback. Feedback delay, in the range between 0 and 0.4 seconds (ordinate), was imposed in a continuously varying manner at different temporal frequencies (abscissa) to yield the results shown in Figures 5.7 and 5.8. (Reproduced from Smith, T.J. et al., *The behavioral cybernetics of telescience—an analysis of performance impairment during remote work* [SAE Technical Paper Number 94138]. Warrendale, PA: Society of Automotive Engineers International, Figure 6, 1994b.)

in visual feedback that is continuously varied in a sinusoidal manner in a given trial. In the actual study, delay was continuously varied in a sinusoidal manner between 0 and 0.4 seconds in each trial (Figure 5.6), and seven different sinusoidal frequencies were employed (.05, 0.1, 0.2, 0.3, 0.5, 1, and 2 Hz) in random order by subject and trial along with the control condition of a 0-second delay. As with the study described above (T.J. Smith et al., 1998), the task involved compensatory pursuit tracking of a target generated as a solid line sweeping right to left, with a randomly varying amplitude, on a computer monitor. During a trial, subjects were asked to move a joystick toward or away from them in a one-dimensional pursuit tracking manner to align the vertical position of the tracking cursor as closely as possible with the shifting position of the target line. The dependent measure is RMS error in tracking accuracy.

The histogram in Figure 5.7 shows aggregate results for RMS tracking error (Y-axis) at 0 delay and at the 7 variable delay frequencies (X-axis), with error at each delay condition averaged over all subjects, all trials, and all test days. Analysis of variance (ANOVA) results show a significant main effect of delay condition ($F_{7,696} = 19.2, p < .0000$). Post hoc analysis indicates that this effect occurs because of two significant differences ($p < .05$): (1) average RMS error observed at 0-second delay is significantly less than that observed at each of the variable delay conditions, and (2) error for 0.2-Hz variable delay frequency is significantly greater than that for 1-Hz variable delay frequency. This latter observation recapitulates the earlier observations of Lee (1950a), Fairbanks (1955), and Fairbanks and Guttman (1958) that peak disruptive effects of delayed feedback occur at a delay interval of about 0.2 seconds (Section 5.2), observations subsequently also confirmed in the delayed feedback studies of K.U. Smith (1962a).

Figure 5.8 shows a Bode plot (Chapter 2, Section 2.3.2; Jagacinski & Flach, 2003, pp. 137–157) of the performance gain observed (Y-axis) at each of the 7 variable delay frequencies (X-axis), with gain values averaged over all subjects, all trials,

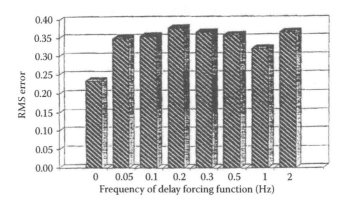

FIGURE 5.7 Tracking error (ordinate) observed with no delay in visual feedback and with feedback delay (between 0 and 0.4 seconds) imposed as a sinusoidal forcing function (Figure 5.6) at different frequencies (abscissa). (Reproduced from Smith, T.J. et al., *The behavioral cybernetics of telescience—an analysis of performance impairment during remote work* [SAE Technical Paper Number 94138]. Warrendale, PA: Society of Automotive Engineers International, Figure 13, 1994b.)

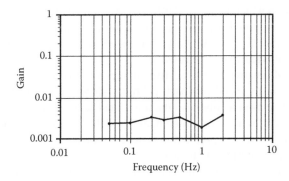

FIGURE 5.8   Bode plot (Chapter 1, Section 1.3) of performance gain as a function of variable visual feedback delay frequency. (Reproduced from Smith, T.J. et al., *The behavioral cybernetics of telescience—an analysis of performance impairment during remote work* [SAE Technical Paper Number 94138]. Warrendale, PA: Society of Automotive Engineers International, Figure 14, 1994b.)

and all test days. ANOVA results show a significant main effect of variable delay frequency ($F_{6,693} = 9.0$, $p < .0000$) on gain. Post hoc analysis indicates that the effect occurs because of the following significant differences ($p < .05$): (1) the three highest gain levels (at variable delay frequencies of 0.2, 0.5, and 2 Hz) each are significantly greater than the lowest gain at the variable delay frequency of 1 Hz; and (2) gain at the variable delay frequency of 2 Hz is significantly higher than that at the variable delay frequency of 0.05 Hz.

Perhaps the most intriguing finding from this study is the Bode plot illustrating the performance gain-variable delay frequency relationship in Figure 5.8. As with other control systems research, the forcing function paradigm used here was developed with the aim of assessing the frequency response characteristics of behavioral control under variable delayed feedback conditions. In cases where a control system is sensitive to imposed perturbation, the Bode plot of gain versus forcing function frequency has a typical pattern: performance gain shows little change over a particular range of frequencies (stable control range), but then declines steadily at frequencies above that range (Jagacinski & Flach, 2003, pp. 137–157).

For example, in a study of eye movement tracking of an external visual target presented as a sinusoidal forcing function at different frequencies, Vercher et al. (1993) found that performance gain was almost invariant between target frequencies of 0.1 and 0.5 Hz, but then declined to less than 10 percent of the stable level as target frequencies were increased from 0.5 to 2 Hz. The Bode plot in Figure 5.8 does not show this pattern. It indicates instead that with variable delay imposed as a sinusoidal forcing function between frequencies of 0.05 and 2 Hz, performance gain remains relatively steady, with no obvious drop in gain at the higher frequencies. Indeed, the highest gain observed is at the highest frequency of 2 Hz.

There are two alternative interpretations of this finding: (1) the control system is stable over the range of variable delay frequencies employed, but gain will start to drop at some point as variable delay frequency is increased above 2 Hz, or (2) performance gain is independent of variable delay frequency. Given that this study is the first in which behavioral effects were evaluated in the face of displaced sensory feedback imposed as a variable forcing function, and that (to my knowledge) this experimental design has yet to be replicated in subsequent sensory feedback studies, neither alternative can be ruled out.

The second alternative is provocative. It suggests that even though behavioral control is sensitive to visual feedback delay (as shown in Figure 5.5), the system does not control delayed feedback in the traditional sense. This conclusion supports the thesis advanced by K.U. Smith (1962a) that delay effects are insidious because subjects perceive delay as a problem in spatial displacement rather than timing. To my knowledge, this interpretation also would provide the first example of a biological control system sensitive to a particular mode of sensory feedback but incapable of controlling it. Finally, the second alternative has sobering implications for telescience in that it suggests that there is inherent behavioral incapacity for effective control of concurrent delayed visual feedback.

5.4 COMPILATION OF PERFORMANCE EFFECTS OF SPATIALLY DISPLACED AND DELAYED SENSORY FEEDBACK

Section 5.1 noted that the focus of this chapter is on the *degree* of behavioral variability that ensues when an individual is asked to perform under conditions of displaced sensory feedback. This section surveys findings bearing on this objective in order to document the consistent and substantial degradation in performance that occurs under such conditions. In the concluding section of this chapter (Section 5.6), a behavioral cybernetic interpretation of these findings is offered that is in line with the emphasis in this book on *sources* of performance variability.

The findings reviewed here are derived from a sustained program of displaced sensory feedback research carried out in the Behavioral Cybernetics Laboratory of K.U. Smith at the University of Wisconsin during the 1960s and 1970s. There are shortcomings and advantages to this approach. One obvious shortcoming is that this is old news. Why revisit results that are decades old and already well documented? A second shortcoming is that the analysis ignores the large body of research on displaced feedback carried out by other laboratories.

My view is that the advantages outweigh the shortcomings. First, the investigations carried out by this program into different types of movements and behaviors in relation to different modes and degrees of feedback displacement, and to different training/learning adaptation regimens under displacement conditions are unmatched by any other displaced sensory feedback research program. As such, the program featured a reasonably consistent analytical methodology inspired by a coherent theory of behavioral control, also distinctive for this research area. However, program findings are dispersed in a diverse set of publications; this chapter offers the opportunity to pull the material together in a coherent manner. Finally, despite the mature nature of this material, it is likely that it will be unfamiliar to many readers.

Since conclusion of the major experimental effort with this program in the 1970s, most empirical research on displaced sensory feedback has focused on performance effects during teleoperation. Key findings from this research, in relation to their relevance to performance variability, are summarized in Section 5.5.

A review of the Wisconsin program begins with Table 5.1, which lists a series of studies carried out in the University of Wisconsin Behavioral Cybernetics Laboratory (Section 5.2) of different types of cognitive behavioral performance under displaced sensory feedback conditions. The table lists the type of performance decrement observed (third column of table) for 19 different cognitive tasks (first column of table), as a result of imposition of spatial and/or temporal displacements in sensory feedback (second column of table). Specific tasks studied in this research program include visual-manual tracking (both pursuit and compensatory), assembly, manipulation, handwriting, graphic drawing, reading, speech, musical performance, steering performance, typing/keyboard entry, social tracking, eye movement control of retinal feedback, visual perception, memorization, head and postural movement tracking, and ventilatory movement tracking. Numbers in the last column of the table refer to references (listed at the bottom of the table) to different published studies in which these tasks were evaluated.

Collectively, the tasks listed in Table 5.1 are representative of nine different classes of cognitive behavior addressed in the course of this displaced sensory feedback research program (T.J. Smith & Smith, 1988a); namely, speech, music, handwriting, learning and memory, machine performance, behavioral-physiological self-regulation, active vision, social interaction, and movement integration, many of which are widely cited as among the most advanced modes of human cognitive skill and proficiency. Under the displacement conditions listed in Table 5.1 (second column), cognitive performance in all cases, and learning of the task in many cases, are compromised by the imposition of these conditions.

Among the tasks listed in Table 5.1, cognitive proficiency among subjects under compliant feedback conditions already was highly developed for some tasks (i.e., speech, musical performance, writing, and reading), whereas other tasks (i.e., some of the tracking tasks) were relatively unfamiliar to subjects. Motor behavior is required for the execution of all of the tasks listed in Table 5.1. Performance of some tasks (i.e., tapping, tracing, assembly, various tracking tasks) is primarily dependent on visual-motor tracking skills. Other tasks (i.e., handwriting, reading, speech, musical performance, memory) require a combination of psychomotor skills and previously developed knowledge and memory capabilities.

There are a number of general features of cognitive behavioral disturbance evoked by sensory feedback displacement that appear consistently from study to study. Oscillatory instability in movement control becomes more pronounced, accompanied by increased variability and extremes in movement velocities and accelerations. As a consequence, the accuracy of movement guidance and tracking accuracy suffers. Perception is degraded and may disappear altogether; learning concomitantly is impaired.

Tables 5.2 and 5.3 summarize findings from a series of studies that illustrate the decrements in task performance that occur as a consequence of these displaced sensory feedback effects for both spatial (Table 5.2) and temporal (Table 5.3) displacements in visual feedback. For the *spatial displacement studies,* subjects viewed a televised display receiving input from a camera viewing the task platform; camera orientation was adjusted to implement the different spatial displacement conditions. Table 5.2 lists the type of spatial displacement condition (first column of table); in the second column, a number referring to the reference for the study (this number refers to references cited by number in the footnote to Table 5.1), plus the number of

TABLE 5.1

Different Tasks Evaluated under Displaced Sensory Feedback Conditions, Based on Studies Carried Out at the University of Wisconsin Behavioral Cybernetics Laboratory

Task	Perturbed Sensory Feedback Condition	Performance Decrement	Reference
Handwriting	Inversion; reversal; inversion-reversal; angular displacement; feedback delay	Contact and travel times accuracy; learning	1–3,6,7,24
Symbolic drawing	Inversion; reversal; inversion-reversal; angular displacement; feedback delay	Contact and travel times accuracy; learning	1–3,6
Target tapping	Inversion; reversal; inversion-reversal; angular displacement; feedback delay	Contact and travel times; learning	1,2,4
Maze/star tracing	Inversion; reversal; inversion-reversal; angular displacement; feedback delay	Accuracy; learning	1,2,4,5
Pin-assembly	Inversion; reversal; inversion-reversal; angular displacement; feedback delay	Movement duration; learning	1,2,4
Graphic drawing	Size distortion	Accuracy; learning	1
Reading	Inversion; reversal; inversion-reversal; angular displacement	Reading times and errors	8
Speech	Feedback delay	Stuttering; slowing and errors of speech; long-term speech disturbances	9
Musical performance	Feedback delay	Performance time	10
Jaw movements during speech	Feedback delay	Duration of jaw activity	11,28
Postural tracking of a visual target	Inversion; reversal; inversion-reversal	Tracking accuracy	12
Steering performance in a driving task	Angular displacement; feedback delay	Steering accuracy	13,23
Retinal projection of eye movement-controlled visual image	Reversal	Skittered vision; severely impaired vision pain and discomfort	14,15

(continued)

TABLE 5.1 (Continued)
Different Tasks Evaluated under Displaced Sensory Feedback Conditions, Based on Studies Carried Out at the University of Wisconsin Behavioral Cybernetics Laboratory

Task	Perturbed Sensory Feedback Condition	Performance Decrement	Reference
Eye movement tracking of visual target	Reversal; feedback delay	Tracking accuracy; learning	14–17
Memory of a visual image	Feedback delay; feedback intermittency	Memory error	18,19
Breath pressure tracking of visual image	Feedback delay	Tracking accuracy; learning	20
Head movement tracking of visual image	Reversal; angular displacement; feedback delay	Tracking accuracy; learning	21
Electromyographic tracking of visual image	Feedback delay	Tracking accuracy; learning	22
Social tracking	Provision of visual feedback division of feedback control	Tracking accuracy; learning	25–27,29

Note: Numbers denote the following references: K. U. Smith and Smith (1962); 2, K. U. Smith (1962a); 3, McDermid and Smith (1964); 4, K. U. Smith et al. (1963c); 5, Gould and Smith (1963); 6, K. U. Smith and Greene (1963); 7, K. U. Smith and Murphy (1963); 8, K. U. Smith et al. (1964); 9, K. U. Smith et al. (1963b); 10, Ansell and Smith (1966); 11, Sussman and Smith (1971); 12, K. U. Smith and Arndt (1970); 13, K. U. Smith et al. (1970); 14, K. U. Smith and Molitor (1969); 15, K. U. Smith et al. (1969); 16, K. U. Smith (1970); 17, K. U. Smith and Putz (1970); 18, K. U. Smith and Sussman (1969); 19, Sussman and Smith (1969); 20, Henry et al. (1967); 21, K. U. Smith and Ramana (1969); 22, Rubow and Smith (1971); 23, Kao and Smith (1969); 24, K. U. Smith and Schappe (1970); 25, Sauter and Smith (1971); 26, K. U. Smith and Kao (1971); 27, Ting et al. (1972); 28, Abbs and Smith (1970); 29, Rothe (1973).

subjects for the study, in brackets; the task (third column of the table); the dependent measure (fourth column of table); the days of training for the training studies, with the number of training trials in parentheses (fifth column of table); and the magnitude of performance decrements observed before and after training (rightmost two columns of table) relative to performance levels under nondisplaced visual feedback. The asterisks next to the performance decrements in the rightmost two table columns indicate that the decrements observed are statistically significant ($p < .05$).

In summary, the spatial displacement studies summarized in Table 5.2 encompass nine different types of tasks under five different conditions of spatial displacement in visual feedback: reversal, inversion, inversion-reversal, size reduction, and angular displacement of the visual image.

For every study cited in Table 5.2, except for angular displacement of 40 degrees or less (last study listed in table), untrained performance with spatially displaced

TABLE 5.2

Decremental Changes in Task Performance under Spatially Displaced Visual Feedback Conditions Before and After Training

Spatial Perturbation Condition	Reference – [Subjects][a]	Task	Measure	Days of Training (Trials)[b]	Performance Change (x1)[c]	
					Before Training	After Training
Reversal	1 – [24]	Writing	Contact time	2 (4)	2.8×*	2.5×*
			Travel time	2 (4)	2.5×*	2.6×*
	1 – [24]	Drawing	Contact time	2 (4)	–	1.8×*
			Travel time	2 (4)	–	1.7×*
	1 – [24]	Maze tracing	% Time off path	3 (9)	2.3×*	1.7×*
	7 – [12]	Writing	Contact time	2 (2)	–	1.8×*
		tracing	Contact time	2 (2)	–	2.5×*
	8 – [4]	Reading	Time × error	5 (5)	3.1×*	1.1×
Inversion	1 – [24]	Writing	Contact time	2 (4)	4.1×*	3.9×*
			Travel time	2 (4)	3.8×*	3.4×*
	1 – [24]	Drawing	Contact time	2 (4)	–	2.7×*
			Travel time	2 (4)	–	2.4×*
	1 – [24]	Maze tracing	% Time off path	3 (9)	3.9×*	2.6×*
	1 – [24]	Dial setting	Manipulation time	7 (21)	3.7×*	2.2×*
			Travel time	7 (21)	3.6×*	1.3×*
	1 – [24]	Knob turning	Manipulation time	7 (21)	2.9×*	1.8×*
			Travel time	7 (21)	3.1×*	1.4×*
	1 – [24]	Button pushing	Manipulation time	7 (21)	3.0×*	2.0×*
			Travel time	7 (21)	3.2×*	1.4×*
	1 – [24]	Switch pressing	Manipulation time	7 (21)	3.5×*	2.2×*
			Travel time	7 (21)	3.3×*	1.4×*
	7 – [12]	Writing	Contact time	20 (20)	4.0×*	1.4×*
			Travel time	20 (20)	5.1×*	1.1×
	7 – [12]	Drawing	Contact time	20 (20)	5.7×*	1.7×*
			Travel time	20 (20)	6.8×*	1.3×
	7 – [12]	Writing	Contact time	2 (2)	–	2.5×*
	7 – [12]	Tracing	Contact time	2 (2)	–	3.8×*
	8 – [4]	Reading	Time × error	5 (5)	3.5×*	1.2×

(continued)

TABLE 5.2 (Continued)

Decremental Changes in Task Performance under Spatially Displaced Visual Feedback Conditions Before and After Training

Spatial Perturbation Condition	Reference – [Subjects][a]	Task	Measure	Days of Training (Trials)[b]	Performance Change (x1)[c]	
					Before Training	**After Training**
Inversion-reversal	1 – [24]	Writing	Contact time	2 (4)	3.0×*	3.5×*
			Travel time	2 (4)	2.6×*	3.0×*
	1 – [24]	Drawing	Contact time	2 (4)	–	2.1×*
			Travel time	2 (4)	–	1.9×*
	1 – [24]	Maze tracing	% Time off path	3 (9)	2.0×*	2.1×*
	1 – [10]	Assembly	Duration	4 (4)	1.9×*	1.6×*
	7 – [12]	Writing	Contact time	2 (2)	–	2.0×*
	7 – [12]	Tracing	Contact time	2 (2)	–	3.5×*
Size reduction (to 1/3 size)	1 – [18]	Panel control	Contact time	5 (40)	2.1×*	2.0×*
			Travel time	5 (40)	3.3×*	3.7×*
Angular diplacement	5 – [24]	Maze tracing	Errors	10 (10)		
10 degrees					–	1.0×
20 degrees					1.1×	0.9×
30 degrees					–	1.0×
40 degrees					1.1×	1.1×
50 degrees					–	1.6×*
60 degrees					2.2×*	2.5×*
70 degrees					–	3.9×*

[a] Number of subjects in brackets, preceded by reference number. Reference number refers to references cited by number in footnote to Table 5.1.

[b] Number of training trials in parentheses.

[c] Performance change calculated as (performance level with televised displaced visual feedback/performance level with televised nondisplaced visual feedback).

* Change from nondisplaced visual feedback performance level is statistically significant ($p < .05$).

visual feedback is significantly impaired relative to that with nondisplaced feedback. Statistically significant performance decrements before training range from 2.1-fold to 4.1-fold. These significant decrements persist after training with four exceptions; namely, reading under reversed and inverted visual feedback and writing and drawing travel time under inverted visual feedback. Statistically significant performance decrements after training range from 1.3-fold to 3.9-fold. For the 16 studies listed in Table 5.2 for which both before- and after-training observations were made for the same task, the after-training performance decrement is smaller in all but three cases, suggesting that learning under spatially displaced visual feedback typically can occur.

TABLE 5.3

Decremental Changes in Task Performance under Delayed Visual Feedback Conditions[a]

Reference – [Subjects][b]	Task	Measure	Performance Change (x1) at Specified Delay[c]		
			0.2 sec	0.4 sec	Max (sec delay)[d]
2 – [2]	Writing letters	Contact time	–	–	2.4× (.52)*
2 – [2]	Drawing	Contact time	–	–	1.1× (.52)
2 – [2]	Star tracing	Contact time	–	–	4.6× (.52)*
2 – [2]	Maze tracing	Contact time	–	–	5.0× (.52)*
21 – [8]	Head movement tracking of visual target	Tracking error	1.3×*	1.6×*	2.3× (0.8)*
28 – [8]	{auditory delay interval[e]:		0.1 sec	0.2 sec}	
	Speech	Speech errors	4.1×*	11.1×*	14.0× (0.3)*
17 – [12]	Tracking of visual target:	Tracking error	1.2×*	1.2×*	2.0× (1.6)*
	Eye movement	Tracking error	1.2×*	1.2×*	5.0× (1.6)*
	Head movement Head-eye movement	Tracking error	0×	1.2×*	1.9× (1.6)*
19 – [8]	Memory of visual image	Recall error	1.5×*	1.9×*	3.3× (0.8)*
29 – [2]	Visual-manual tracking:				
	Individual	Tracking error	1.1×	1.5×*	2.6× (1.5)*
	Social	Tracking error	1.5×*	1.8×*	3.0× (1.5)*

[a] All studies examined visual feedback delays, except for Abbs and Smith (1970, reference number 28), who examined auditory feedback delay.

[b] Number of subjects in brackets, preceded by reference number. Reference number refers to references cited by number in footnote to Table 5.1.

[c] Performance change in multiples of one (indicated by small ×), observed at delay intervals of 0.2 sec, 0.4 sec, and maximum delay interval examined[d]. Performance change is calculated as (performance level with feedback delay)/(performance level with no feedback delay).

[d] Performance change in multiples of one[c] at maximum delay feedback delay interval examined in study. The maximum delay interval examined, in seconds, is indicated in parentheses.

[e] Delayed auditory feedback delay intervals of 0.1, 0.2, and 0.3 sec examined in this study.

* Change from nondisplaced visual feedback performance level is statistically significant ($p < .05$)

In showing that humans can partially adapt to spatial displacements in visual feedback, this research tends to confirm and extend the original thesis of Helmholtz (1856–1866) that visual space perception is learned rather than innate in nature. However, in the studies summarized in Table 5.2, the relationship between eye movements and projection of the visual image on the retina remained normal even though the image itself was spatially displaced. In order to investigate how the spatial organization of image projection by movements of the eye itself onto the retina affects visual perception, a scleral contact lens with a built-in dove prism was devised that could reverse eye-movement projection of the visual image onto the retina (K.U. Smith, 1970; K.U. Smith & Molitor, 1969). The single subject for this experiment was K.U. Smith, who reported (1970, p. 180) the following experimental results:

> This scleral-lens and dove prism was worn by me for about forty-five minutes, which was as long as the skittered vision produced by the reversing prism could be tolerated. When first put on, the prism caused almost complete blindness in the affected eye. After some ten or fifteen minutes, it was possible to try to fixate distant objects but this could be done only by looking through the reversing prism. Moreover, the experience was extremely upsetting and painful even though the scleral lens itself caused no difficulty or pain. It was necessary to cover the affected eye periodically to get some relief from the unsettling effect of the vision. Vision itself was blurred for the most part. The impression at the end of the experimental session was that it would be impossible to wear such prisms on both eyes for any period of time and that the effects were too unsettling and painful to try on subjects other than the experimenter himself.

In 1978, Welch (p. 124) claimed that this was only study of its sort conducted. As far as I know, this claim remains true. Keeping the limitations of the results in mind (one study with a single subject), the implications are that partial adaptation (learning) under spatially displaced feedback can occur (compare before training and after training results in Figure 5.2), as long as the relationship between eye movements and projection of the visual image on the retina remains normal. However, if eye-movement projection of the visual image onto the retina is spatially displaced, the study described above suggests that adaptation will no longer occur.

Table 5.3 summarizes performance decrements observed for 10 different types of tasks under *temporal delay of either visual or auditory feedback*, with delay intervals ranging from 0.1 to 1.6 seconds. Table 5.3 lists in the first column, a number referring to the reference for the study (this number refers to references cited by number in the footnote to Table 5.1), plus the number of subjects for the study, in brackets; the task (second column of the table); the dependent measure (third column of table); and in the rightmost three columns of the table, performance decrements at delay intervals of 0.2 and 0.4 seconds, and the maximum delay (in parentheses, range .52 to 1.6 seconds) used in the study. The asterisks next to the performance decrements in the rightmost three table columns indicate that the decrements observed are statistically significant ($p < .05$).

Across the different studies, performance is degraded significantly under all but three of the feedback delay conditions examined. For the delayed visual feedback studies, statistically significant performance decrements range from: 1.2-fold to 1.5-fold for 0.2-second delay; 1.2-fold to 1.9-fold for 0.4-second delay; and 1.9-fold to 5.0-fold for the maximum delay intervals. Also for statistically significant decrements with the delayed visual feedback studies, with an increase from 0.2- to 0.4-second delay, the performance decrement is higher for the longer delay for three out of five studies; and with an increase from 0.4 second to maximum delay, the performance decrement is higher for all seven studies. The latter finding is aligned with the example presented in Section 5.3 that performance is increasingly degraded with longer delay intervals.

Results for the third study listed in Table 5.3 (Abbs & Smith, 1970, reference 28) show that performance decrements for delayed auditory feedback are notably higher than those for delayed visual feedback, at comparable delay intervals.

Of the studies cited in Table 5.3, only that of Sussman and Smith (1969, reference 19) examined training effects on performance under feedback delay. The findings indicate no training effect over five trials for delay intervals of 0.4 second and higher and a small effect at 0.2-second delay. A comparable finding was made by K.U. Smith and Molitor (1969) in a study of eye movement tracking under feedback delay. This study found no training effect at 0.2-second delay and a small effect at 0.1-second delay after 50 trials.

5.4.1 INTERPRETATION

There are a series of possible objections to the interpretation provided above to the findings summarized in Tables 5.1 through 5.3. One objection is that these findings deal only with psychomotor tasks and therefore do not encompass knowledge-based cognitive abilities and skills. In response, it first should be noted that early research supporting the role of task-specific variance in cognition dealt primarily with knowledge-based tasks (Chapter 3). However, it may be argued that knowledge abilities play a critical role in the cognitive behaviors of reading, handwriting, musical performance, and memory recall, whose performance is markedly degraded by both spatial and temporal displacements in visual feedback.

It also may be argued that, for the tasks enumerated in Tables 5.1 through 5.3, individual differences make a comparable or greater contribution to performance variability than that observed under displaced feedback conditions. This in essence is the position of Harris (1965), who maintains that loss of innate position sense accounts for the performance decrements observed under displaced visual feedback. For the tasks addressed by the Wisconsin program, this possibility cannot definitively be ruled out, given that the referenced studies were not explicitly designed to compare and contrast individual differences versus task specificity related to displaced feedback effects as sources of performance variability.

However, it is germane to cite Salvendy (1985), who notes that across a range of occupations, a between-operator performance range of 2 to 1 encompasses 95 percent of the working population and that within-operator variability is substantially less. Similarly, both Nof (1985) and Bullinger et al. (1987) indicate that 100–150 percent variation in human individual differences may be expected for task performance. The higher individual difference performance ranges reported by these authors (1.5-fold to twofold) are less than all but two of the 27 statistically significant before-training performance decrements found for spatially displaced visual feedback (Table 5.2), and also lower for all but two of the 10 statistically significant performance decrements found for delayed visual feedback (Table 5.3). These observations suggest that task specificity related to displaced feedback has a greater influence on performance variability than individual differences.

However, evidence contrary to this conclusion is summarized in the review of individual differences by Egan (1988), who cites a series of studies indicating that between-operator performance differences ranging between 2 to 1 and 7 to 1 have been observed in five studies of text editing, and between 3 to 1 and 10 to 1 in three studies of information search. Based on additional evidence, this author goes on to conclude (p. 551) that, "When user differences are assessed in the same experiment as design differences, user differences usually account for much more of the variability of performance."

Based on the following points, it may be argued that in drawing this conclusion, Egan (1988) did not in fact critically assess the possible role of task-specific variance in performance variability. Specifically, the studies cited by this author did not:

- Carefully assess the sources of variability for a range of different performance measures
- Carefully assess the influence of a range of design alternatives on performance variability
- Adhere to an analysis of variance protocol in which measures of task performance following practice are correlated with initial scores and with scores from performance on non-practiced tasks
- Alter the spatiotemporal properties of sensory feedback provided to subjects in any systematic way to ascertain whether such design-related displacements substantially influence variability in task performance in the same manner as has been shown with other studies (Tables 5.2 and 5.3).

5.5   DISPLACED SENSORY FEEDBACK EFFECTS ON TELEOPERATION

The practical implications of the displaced sensory feedback effects documented in the preceding section have attracted growing recognition and importance with the advent and steady expansion of teleoperated systems.

In 1952, Goertz and Thompson (1954) developed the original teleoperator to handle radioactive materials in a nuclear facility. Since then, five basic themes have marked the emergence and elaboration of the field of telerobotics and teleoperation, namely: (1) efforts to improve behavioral motorsensory compliance between master and slave, with the aligned objectives of improving teleoperation performance as well as imbuing the operator with a sense of presence in the remote but real environment in which the telerobot actually is operating, a phenomenon termed *telepresence* (Draper et al., 1998; Sheridan, 1992a, 1999; T.J. Smith & Smith, 1990) (NASA [1985] defines telepresence as, "a teleoperation situation in which the operator has sufficient cues to simulate sensations that would be experienced in direct manual performance of the operations"); (2) a sustained struggle to understand and delineate the different patterns and sources of displacements in sensory feedback inherent to teleoperated systems; (3) development of mathematical engineering models and strategies aimed at stabilizing and improving systems control properties of teleoperated systems in the face of sensory feedback displacements in systems control; (4) investigation into the behavioral consequences of displacements in sensory feedback for teleoperator performance; and (5) an impressive expansion in the application of teleoperated systems across diverse domains, encompassing such areas as high-hazard waste handling, space telerobotics, laparoscopic surgery and telesurgery, and remote vehicle operation on land, underwater, and in the air, for both military and civilian applications.

This section will focus on the fourth of these themes in relation to performance variability during teleoperation, and also on themes 1 and 2 that pertain to the sources of this variability.

Reviews have been provided by various authors both of the operational effects of selected types of displacement and of selected applications of teleoperated systems. Bejczy (2002), Sheridan (1989, 1992b), and Vertut and Coiffet (1985) review the history of telerobotics and teleoperation generally. Held and Durlach (1991) and Sheridan (1992b) review the effects of delayed feedback on teleoperation; the latter review includes a 30-year history of research on the effects of time delay on teleoperation in space. Draper et al. (1998) and Sheridan (1992a, 1999) review the background, principles, and operational requirements of telepresence. Ballantyne (2002) and Cao and Rogers (2007) review the background and development of robotic surgery. Spong and Hokayem (2006) provide a 50-year history of bilateral teleoperation. T.J. Smith et al. (1989, 1990, 1994b, 1996, 1998), T.J. Smith and Smith (1990), and T.J. Smith and Stuart (1990) review the influence of displaced sensory feedback factors on the fidelity of telepresence and on performance variability with teleoperation.

5.5.1 HUMAN FACTORS ISSUES WITH WORKSTATION TELEPRESENCE

During human performance with a telerobot via a teleoperated workstation, sensory feedback to the operator is generated by two sources: direct interaction with the workstation, and workstation-mediated interaction with the telerobot. A teleoperator may be provided with multiple modalities of sensory feedback via this interface, encompassing visual, force, haptic, and/or perhaps even auditory feedback. Concomitantly, the operator may be confronted with, and required to behaviorally control to the extent possible, one or more modes of displacement of this feedback, encompassing spatial displacement, feedback delay, frictional resistance of telemanipulator effectors, and/or operator misjudgments of telerobot inertial properties under microgravity conditions.

Indeed, T.J. Smith and Smith (1990) estimate that visual-manual performance during teleoperation may require as many as 19 principal motor control factors through use of the teleoperator's motor system to control both the visual system (11 principal visual perception factors) as well as five distinct modes of sensory feedback (K.U. Smith, 1962a, 1972; K.U. Smith & Smith, 1962; T.J. Smith & Smith, 1987a, 1988b). This control process potentially may be influenced by as many as 18 principal telerobotic system design factors, some of which may give rise to displaced sensory feedback conditions. These estimates are derived from basic human factors considerations regarding telerobotic operation (T.J. Smith & Smith, 1988b) plus analyses by various scientific specialists in telerobotic research (Chandlee et al., 1988; Huggins et al., 1973; Krishen, 1988a, 1988b; R.L. Smith, 1988).

To optimize workstation telepresence, the spatial, temporal, and physical properties of workstation and telerobot sensory feedback, as determined by the human factors design of the teleoperation system, must be compliant with the sensory feedback control capabilities of the human operator. However, some appreciation of the scientific challenge confronting systematic investigation of the factors that potentially can influence such compliance may be gained by considering that, based on cross-multiplication of the above estimates (T.J. Smith & Smith, 1990), there are almost 19,000 possible combinations of performance, design, and feedback factors ($19 \times 1 \times 5 \times 18$) that potentially can influence performance variability in visual-manual control of a telerobot. That it probably is not feasible to completely evaluate even a reasonable subset of such possible combinations raises the question of research priorities and emphasis in human factors analysis of performance variability in teleoperation.

In line with the fourth theme cited above, the assumption adopted here is that spatiotemporal displacements in sensory feedback constitute a major source of variability in human performance with telerobotic systems, and that human factors analysis aimed at mitigating displacement effects likely is essential for optimization of workstation telepresence.

One of the first comprehensive conceptual and technical investigations of human factors engineering requirements for effective workstation telepresence was carried out during the fifties and sixties in the context of an extensive General Electric research and development program on cybernetic anthropomorphic (manlike) machines (CAMs), under the direction of Mosher (Mosher, 1964; Mosher & Murphy, 1965). The broad objective of the program was to develop mechanical devices, "which would serve as operator-controlled extensions of the body to expand the strength and endurance capabilities of human performance." The program dealt with four general types of CAMs: arm-claw manipulators, bipedal walking machines, quadripedal walking machines, and ambulatory exoskeletons that enclosed the operator within bodylike frames. These were true telerobotic devices, in that they each comprised master control, actuator, and slave subsystems that provided computer-mediated, servo-controlled spatial, temporal, and force feedback compliant with all modes and patterns of motor-sensory feedback from the human operator's articulated arm, hand-wrist, leg, foot, and torso movements.

In general, Mosher's work defined a series of important human factors issues pertaining to human performance variability with telerobotic systems, including (1) need for compliant sensor, actuator, and effector designs for effective human-machine system integration, (2) guidance of interactive performance via behavioral feedback control mechanisms, (3) critical role of temporal and spatial feedback factors in human-system compliance and integrative control of robot movement, and (4) need for human-computer-robot integration arising out of use of computers to mediate control of telerobots.

In particular, Mosher's work also established that four different dimensions of force feedback are needed to control body movement accurately and should therefore be incorporated into a telerobot control system: (1) force of movement at different levels of exertion, (2) upper limit of force required, (3) prediction of force needed to execute movements of a given velocity and power, and (4) detection of relative displacement between movement and its sensory feedback. A series of system design and operational factors that influence the muscle-tendon-joint sensibility of the operator were identified, including force feedback ratios between operator and robot, drift on bias forces, friction thresholds, nonlinearity of force interactions, saturation factors, and force-signal integrity as influenced by system kinematics. As part of this human factors system analysis, Mosher also demonstrated that force feedback related to the limits of exertion of powered grips, lifts, and pulls must be built into telerobots as a safety precaution to prevent their destructive action (Rahimi, 1987), a finding with continued implications for present-day robotic systems.

Although Mosher's work documented the essential role of force feedback for guidance and control of telerobotic systems, K.U. Smith also established that effective workstation telepresence could not be achieved without also introducing spatial compliance in sensory feedback control of telerobotic action (K.U. Smith, 1965b, 1965c; K.U. Smith & Smith, 1988). Several characteristics of Mosher's original CAMs point up the importance of full spatial compliance for effective telepresence. For example, one problem that emerged in the design of the bipedal walking machines was inability of the operator to determine whether the machine was going down an incline versus falling backward (or going up an incline versus falling forward).

To resolve this problem, at Mosher's invitation to provide a human factors analysis, K.U. Smith (1965b, 1965c) demonstrated first, that the operator's normal posture had to be maintained with respect to leg motion, and second, that the operator had to be able to sense any displacement of either leg of the machine with respect to the position of the machine cab. It was concluded that these cab-leg displacements had to be sensed by the operator via active force and position feedback. After appropriate torso-leg/cab-leg servo feedback mechanisms were incorporated into the design of the walking machine, it was found that even a novice operator could effectively control the dynamic balance of the machine.

A second example of the importance of spatial factors in telepresence emerged with research on manipulator CAMs. Tests showed that the operator of a manipulator experienced extension of his or her dynamic body image as long as just the manipulator was moved, but that this sensation was lost when the manipulator claw was moved. This loss of telepresence was related to the fact that the grasp actions of the claw were not spatially compliant with the operator's hand movements, although the force compliances were adequate.

These two examples cited above demonstrate that body image is defined in large part by the spatial organization of body movement, and that the skill, precision, and safety of telerobotic operation are enhanced when the human operator perceives robot movements as an extension of his or her own movements. The general conclusion is that telerobotic design must incorporate compliances of both spatial and force factors between operator movements and machine movements before effective workstation telepresence can be established.

Although this conclusion is based on CAM research dating back some five decades (Mosher, 1964; Mosher & Murphy, 1965; K.U. Smith, 1965b, 1965c), I am not aware of more recent evidence that contradicts the insights it provides into the human factors of workstation telepresence or that suggests that modern designs of telerobotic systems may have eliminated human factors issues related to displaced sensory feedback. It is noteworthy that, some two decades after publication of this work, a recognized expert in the field referred to Mosher's machines as still among the most sophisticated telerobots yet developed (Raibert, 1987).

The human factors issues raised above, related to teleoperation, were illustrated earlier in Figure 5.1, using the example of a movement-actuated (master–slave) telerobotic manipulator. As suggested in the figure, effective control of the telemanipulator depends on space, time, and force (kinesthetic) compliance between operational feedback from the telerobot slave sector and motor behavioral control of this sensory feedback by the operator. If human factors design problems cause spatiotemporal displacements in sensory feedback, then compliance between sensory feedback and its control is compromised and performance suffers. Section 5.5.2 summarizes findings indicating that performance decrements attributable to spatial or temporal displacements in sensory feedback are a pervasive feature of teleoperated systems.

Figure 5.9 illustrates the concept that workstation telepresence is critically dependent on spatial compliance between sensory feedback and its motor control mechanisms using the same master control-slave telerobot model as in Figure 5.1. Teleoperators normally use the body of the manipulator as a reference point when making control inputs. However, if remote cameras are placed such that the view of a remote work site is not normal with respect to the telemanipulator, visual feedback from the manipulator will be spatially displaced. In Figure 5.9, for example, visual feedback has been spatially displaced by reversing the lateral motion relationships

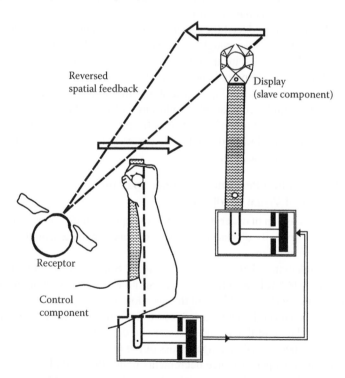

FIGURE 5.9 Example of spatial displacement (reversed visual feedback) in operation of a telemanipulator. (Reproduced from Smith, T.J. et al., *The behavioral cybernetics of telescience—an analysis of performance impairment during remote work* [SAE Technical Paper Number 94138]. Warrendale, PA: Society of Automotive Engineers International, Figure 3, 1994b.)

between arm-controlled movement of the control component and derivative movement of the slave component. Consequently, the operator experiences spatial displacement between visual feedback from telemanipulator movements relative to that from arm movements, telepresence is compromised, and performance suffers.

5.5.2 PERFORMANCE VARIABILITY DURING TELEOPERATION

This section reviews findings regarding the nature and degree of performance variability observed during teleoperation. The purpose and rationale for this review is twofold. First, for selected studies, comparison of performance under displaced feedback conditions for hands-on versus teleoperation conditions is possible, thereby drawing attention to how performance variability under these conditions differs between the two conditions. The second, broader, objective is to underscore the point, in line with the focus of this chapter, that displaced sensory feedback effects are inherent to teleoperation performance. That is, teleoperation arguably represents a distinctive example of a systems area in which operators unavoidably and continuously confront displaced feedback effects during routine operations. The teleoperation performance environment thus offers a test bed, with an ever-growing expansion, for exploring operational performance demands under displaced feedback and their behavioral control consequences.

Figure 5.2 already introduced the difference between hands-on and teleoperation performance variability under roughly comparable spatially displaced visual feedback conditions. The results are intriguing. As noted in Section 5.4 and Table 5.2, hands-on performance is impaired under spatially displaced relative to nondisplaced visual feedback conditions. Yet Figure 5.2 shows that teleoperation performance is impaired even more under comparable conditions. Why should this be the case? The remainder of this section summarizes findings from relevant studies of teleoperation under both spatially displaced and delayed sensory feedback conditions that may provide insight into this question.

5.5.2.1 Teleoperation Performance under Spatially Displaced Visual Feedback

The foregoing discussion of experience with CAM operation (Section 5.5.1) indicates that mitigation of spatially displaced feedback effects may be essential for effective teleoperation performance.

Pioneering telemanipulator research by R.L. Smith, Stuart and colleagues (Chandlee et al., 1988; R.L. Smith, 1988; R.L. Smith & Stuart, 1989; Stuart & Smith, 1989; Stuart et al., 1991) has addressed the question of how spatial displacements in visual feedback produced by camera-induced transformation of the telemanipulator reference plane relative to that of the operator can degrade task performance of the teleoperator and thereby risk damage or loss of the telemanipulator.

Table 5.4 compares performance decrements under three conditions of spatially displaced feedback observed with the telemanipulator research cited above relative to hands-on performance under the same conditions. Although the findings summarized in the table recapitulate in part the graphical results in Figure 5.2, they embody a somewhat different set of both hands-on and teleoperation data. In presenting these

TABLE 5.4

Performance Decrements under Spatially Displaced Visual Feedback: Hands-On Versus Teleoperation Tasks

Spatial Displacement	Performance Condition	No. of Studies	Performance Decrement (x1)[a]			Ref[b]
			Mean	S.D.	Range	
Reversed	Hands-on	4[c]	2.7×	0.4	2.3–3.1	c
	Teleoperation	1[d]	5.0×	–	–	d
		1[e]	4.0×	–	–	e
Inverted-reversed	Hands-on	4[c]	2.4×	0.5	1.9–3.0	c
	Teleoperation	1[d]	4.2×	–	–	d
		1[e]	3.1×	–	–	e
Inverted	Hands-on	16[c]	4.0×	1.1	2.9–6.8	c
	Teleoperation	1[d]	7.9×	–	–	d
		1[e]	5.0×	–	–	e

[a] Performance decrement calculated as (untrained performance level with televised displaced visual feedback)/(untrained performance level with televised normal visual feedback).

[b] Footnote to Table 5.1 contains references (Ref) for hands-on studies.

[c] For each of the three spatially displaced feedback conditions (first column), the numbers of hands-on studies listed for each of these conditions (third column) are identical to the numbers of before-training results listed for the same conditions in Table 5.2 (second-to-last column of Table 5.2). The performance decrement statistics listed for each of the hands-on conditions in the fourth, fifth, and sixth columns) are based on averages for the before-training studies for each of the three spatially displaced feedback conditions listed in Table 5.2. The reference numbers for each of the before-training, hands-on studies are listed in the last column of Table 5.2. These numbers refer to the numbered references in the footnote to Table 5.1[b].

[d] Teleoperation performance decrements refer to changes in task completion times (for spatial displacement conditions relative to times with normal televised viewing) for a telemanipulation task (grasping a moving 6 wooden blocks and dropping them into a box), using a Kraft Telerobotics remote manipulator (R.L. Smith & Stuart, 1989; Stuart & Smith, 1989).

[e] Teleoperation performance decrements refer to changes in task completion times (for spatial displacement conditions relative to times with normal televised viewing) for a telemanipulation task (successively grasping and placing two task pieces into receptacles located on different levels), using a Kraft Telerobotics remote manipulator (Stuart et al., 1991).

data, the dual purposes are to emphasize once again the difference in performance variability between direct viewing and teleoperation under spatially displaced feedback conditions as well as to call attention to the originality of the Smith and Stuart research on this problem area.

As with the results in Figure 5.2, the comparisons in Table 5.4 are based on a series of assumptions. First, only data for untrained performance are considered. Second, contrasts between telerobot versus hands-on performance data are based on studies of different specific tasks in different laboratories at different times by different investigators. For this reason, only visual-manual performance tasks are considered and only normalized performance changes for each task are presented, calculated as the ratio of the level of task performance for a given spatially displaced visual feedback condition relative to that for the normal televised viewing condition.

Results in Table 5.4 are comparable to those in Figure 5.2 in indicating that for all three spatially displaced feedback conditions, average performance decrements for hands-on tasks are consistently less than those for the teleoperation tasks; for reversed and inverted-reversed visual feedback, the performance decrements observed for the hands-on tasks are comparable and lower than decrements observed for inverted visual feedback; and for teleoperation tasks, among the three spatial displacement conditions, performance decrements for the inverted-reversed feedback condition are lowest, intermediate for the reversed feedback condition, and highest for the inverted feedback condition. Thus, spatially inverted visual feedback has the greatest decremental impact on visual-manual performance for both hands-on and teleoperation tasks.

5.5.2.2 Other Sources of Spatially Perturbed Visual Feedback during Teleoperation

In addition to the spatial displacement effects documented in Figure 5.2 and Table 5.4, teleoperation research over the past five decades has identified a number of other spatial perturbations in visual feedback that have the potential to adversely influence teleoperator performance. These include direct versus televised viewing of visual field, angular displacement of the visual image, magnification or minimization of the visual image, a restricted field-of-view, distorted depth cues, and/or distorted viewing attributable to extreme illumination, contrast, glare, and/or reflectance conditions. Findings from studies addressing some of these effects are summarized below.

1. *Direct versus televised viewing of the visual field.* Using a Fitts' placement task, Massimino and Sheridan (1994) compared telemanipulator performance under direct viewing versus televised telemanipulator viewing conditions. Mean task times were consistently higher for televised relative to direct viewing for all index of difficulty (ID) levels. Specifically, the performance decrement for televised viewing was about 1.9-fold at the highest ID level (ID = 6) evaluated, and about 2.2-fold at the lowest ID level (ID = 4) evaluated.

2. *Angular displacement of the visual image.* Two studies report relevant results for this condition. Du and Milgram (2002) evaluated an aiming task, viewed using a head-mounted display, under angular displacements of visual feedback of 0 (control), 90, 135, and 180 (reversal) degrees. Relative to the control, angular displacement induced aiming performance decrements of about 1.8-fold, 2.3-fold, and 1.4-fold, at 90, 135, and 180 degrees displacement, respectively. This pattern recapitulates earlier results reported by T.J. Smith et al. (1998) and shown in Figure 5.3, featuring a peak performance error at a displacement angle in the range 120–135 degrees, and smaller error at displacement angles of 90 and 180 degrees.

   Chadwick and Pazuchanics (2007) evaluated the effects of angular displacement of the visual image on the ability of subjects to effectively control a remote ground vehicle. Angular displacements of 0 (control), 90, and 180 (reversal) degrees were imposed. Relative to the control condition, X-axis guidance performance decrements of about 1.4-fold were observed at 90 degrees displacement, and of about 1.2-fold at 180 degrees displacement.

3. *Magnification/minimization of visual image.* In a study of accuracy of pilots in touching an aircraft down on a target point on a runway, Roscoe et al. (1966) report that relative to nondistorted viewing available to a pilot: (1) under 0.86-fold magnification (i.e., minimization) of the visual image of the runway, pilots consistently landed about 75 feet beyond the target point; that is, pilots perceived the landing target point to be further away than it actually was, (2) under 2.0-fold magnification of the visual image of the runway, pilots consistently landed about 150 feet in front of the target point; that is, pilots perceived the target point to be closer than it actually was, and (3) interestingly, under 1.2-fold magnification of the visual image of the runway, pilots consistently landed on or very near to the target point. To be sure, this was not a teleoperation study. However, it is reasonable to assume that comparable decremental effects would be observed in teleoperator performance under magnification or minimization of visual feedback.

4. *Restricted field-of-view (FOV).* Pazuchanics (2006) compared the ability of teleoperators to execute navigation tasks with an uninhabited ground vehicle (UGV) under both narrow (the customary design condition for UGV guidance) and wide FOV conditions. For measures of three navigation tasks—elapsed time, number of collisions, and number of turn-arounds—performance with the wide FOV was consistently superior to that with the narrow FOV. Relative to wide FOV performance, narrow FOV performance decrements ranged from about 1.25 to about 4.0 for these three tasks.

   In a study of subjects traversing a virtual environment, Toet et al. (2008) report a 1.5-fold increase in elapsed transit times under restricted FOV conditions relative to nonrestricted conditions.

5. *Distorted depth cues.* Moore et al. (2007) evaluated the ability of subjects to judge the passability of a robot through an aperture under both direct viewing and teleoperation viewing conditions. For aperture sizes ranging from 40 to 70 cm, subjects consistently overestimated the aperture gap under the teleoperation relative to the direct viewing condition. That is, for these gap sizes, they judged that the robot could pass through the gap during teleoperation viewing when in fact it couldn't, suggesting a distorted depth cue effect during teleoperation. The degree of teleoperation gap width overestimation was approximately constant at a level of about 5 cm, for the 40- to 70-cm gap width range. For aperture sizes of 35 and 75 cm, the teleoperation gap width overestimation was about half this level. For aperture sizes of 30 and 80 cm, direct and teleoperation viewing gap width estimations were identical.

5.5.2.3   Reversed Visual Feedback in Minimally Invasive Surgery

The last three decades have featured the advent of a revolution in surgery; namely, the use of a laparoscopic probe with both a camera and a surgical effector at its end to carry out minimally invasive surgery. Ballantyne (2002) reviews the history of this technology. As Ben-Porat et al. (2000) point out, this technology offers, "several advantages for the patient over conventional surgery, such as a shorter recovery time, a decreased risk of infection, and less pain and trauma. Health care organizations also tend to favor these techniques because they decrease hospitalization time and costs." However, these authors also point out that, "the physician who performs endoscopic surgery faces some difficulties that are inherent to this technique...[including] impaired depth perception, impaired orientation due to changes in perspective, lack of tactile feedback, and constrained movement of the instruments."

The consequence, as Ballantyne (2002) notes, is that, "Most laparoscopic...operations are difficult operations to learn, master, and perform routinely. Surgeons face a long learning curve." Poignant testimony to this challenge is provided in the following testimony regarding an accomplished surgeon trained on traditional surgical procedures (Holden et al., 1999):

> No one could have predicted that here he would be, 56 years old and at the peak of his surgical prowess, grappling with a new technology that was changing the nature of the field right before his eyes … he was too young to retire, but he felt uneasy being a general surgeon who couldn't remove gallbladders. The question was simple. Was he fit to make it up the slope of his next learning curve?

Among the learning challenges facing a surgeon untrained in laparoscopic surgery, possibly one the most formidable is mastering the skill of operating under conditions of reversed visual feedback. As Ben-Porat et al. (2000) explain, in the case of gastrointestinal laparoscopic surgery, "To move the instrument's tip inside the body in a chosen direction in the horizontal plain, the surgeon's hand has to move in the opposite direction, due to the fulcrum effect at the entry port in the patient's abdominal wall. Practitioners recommend positioning the camera so that it captures an image that resembles the surgeon's natural view of the surgical site...With this camera placement, the movement on the monitor is in the opposite direction to the movement of the hand, reversing the normal compatibility of motor and visual input."

That reversed visual feedback is an inherent feature of at least some forms of laparoscopic surgery has prompted recent research attention to this issue on the part of the laparoscopic surgical community (Ballantyne, 2002; Ben-Porat et al., 2000; Holden et al., 1999; Tessier et al., 2012) (it is worth noting that none of these reports reference the earlier research on reversed visual feedback, either hands-on or teleoperation studies, cited in the footnotes to Figure 5.2 and Tables 5.1 and 5.4). Relative to nondisplaced visual feedback, Ben-Porat et al. (2000) report an untrained performance decrement in trial time of about 2.2-fold for reversed visual feedback in a laboratory study of a manipulation task, with teleoperation viewing conditions meant to resemble those prevailing during actual laparoscopic surgery. This decrement is smaller than those observed in earlier studies of telemanipulation under reversed visual feedback (Figure 5.2 and Table 5.4). The likely reason for this difference is that the Ben-Porat et al. study did not actually employ a telerobot (as the earlier studies did), but instead used a servo-controlled rod to move an object across a smooth task board to different target locations.

5.5.2.4   Interpretation

Findings cited above of studies of other sources of spatially perturbed visual feedback during teleoperation indicate that, except for one of the narrow versus wide FOV tasks, observed teleoperation performance decrements relative to hands-on viewing are distributed over a relative modest range of 1.25-fold to 2.3-fold. These decrements are notably and consistently lower than performance decrements observed for teleoperation for reversed, inverted, and inverted-reversed spatial displacement conditions (Figure 5.2 and Table 5.4). This suggests that, among all of the perturbed visual feedback factors with the potential to degrade teleoperation performance, it is the latter three overt spatial displacement conditions that are likely to have the greatest adverse influence.

It was pointed out above that hands-on performance is impaired under spatially displaced (reversal, inversion, or inversion-reversal) relative to nondisplaced visual feedback conditions, but that teleoperation performance is impaired even more under comparable conditions (Figure 5.2). The question was raised as to why this should be the case?

My view is that the foregoing analysis does not provide a definitive answer to this question. The first thing to keep in mind is that pick-and-place telemanipulation with an actual telerobot, the task evaluated for the teleoperation studies cited in Figure 5.2 and Table 5.4, is inherently difficult even for direct viewing performance, for reasons delineated in the various teleoperation studies cited above. Displaced visual feedback conditions may have magnified this difficulty. Second, other sources of perturbed visual feedback during teleoperation summarized above, in addition to spatial displacement, may have exacerbated the displacement effects. Another factor that may have exacerbated the telemanipulation results under displacement is a small amount of delay between controller actuation and slave response, due either to inertial properties of the telerobot or transmission lags, or both. Whatever the case, feedback delay effects on variability in teleoperation performance have attracted a great deal of research attention, dating back almost to the inception of the technology. These effects are addressed below.

5.5.2.5   Teleoperation Performance under Feedback Delay

Because of the possibility of transmission lags, particular during guidance of teleoperated vehicles at remote distances (Sheridan, 1992b; T.J. Smith et al., 1998, 1989; T.J. Smith & Stuart, 1990), and/or linked to the inertial properties of a telerobot, effects of feedback delay on teleoperation performance have attracted research attention and operational concern since the early days of the technology. Held and Durlach (1991) cite a total of 21 publications dating from the early 1960s to the mid-1980s dealing with research on the effects of time delays on manual tracking and remote manipulation and on methods for mitigating these effects. Jung et al. (2000) cite six subsequent studies in the 1990s showing that latencies in virtual environments disrupt both objective measures of teleoperation performance as well as the subjective sense of telepresence.

The discussion of feedback delay in Section 5.4, along with summaries of representative delay effects in Figure 5.5 and Table 5.3, indicate that performance under delay

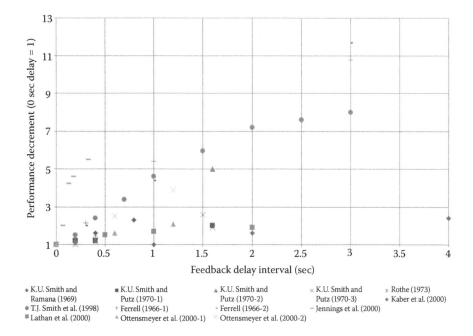

FIGURE 5.10 Performance decrements (ordinate) observed under delayed sensory feedback, for different delay intervals (abscissa) based on results from nine different hands-on and teleoperation studies (listed in the legend), and 13 data sets from these studies. For all studies except that of Ferrell (1966), the delay condition evaluated was delayed visual feedback.

varies in relation to delay magnitude. Figure 5.10 illustrates the nature of this relationship in a more comprehensive manner with a plot of performance decrements under feedback delay observed in 9 different studies and 13 different sets of data, identified in the legend to the figure. The studies whose feedback delay data are plotted in Figure 5.10 were selected based on two criteria: (1) performance effect data for at least two feedback delay levels are described, and (2) data for performance under no delay also are presented.

Four of the studies and six of the data sets whose data are plotted in Figure 5.10 involved evaluation of *hands-on* performance under delayed visual feedback—these studies are Rothe (1973), K.U. Smith and Ramana (1969), K.U. Smith and Putz (1970) (three data sets, labeled 1970-1, 1970-2, and 1970-3 in figure legend), and T.J. Smith et al. (1998). Tasks and dependent measures for the Rothe (1973), K.U. Smith and Ramana (1969), and K.U. Smith and Putz (1970) studies are summarized in Table 5.3, and those for T.J. Smith et al. (1998) are summarized in Section 5.3.

The remaining five studies (and seven data sets) whose data are plotted in Figure 5.10 all involved remote control of a task effector, either under direct or video viewing conditions. Task descriptions for each study are summarized as follows:

- Ferrell (1966) (two data sets, labeled 1966-1 and 1966-2 in the figure legend) evaluated a movement to target task and a contact with target task, both under delayed force feedback. For both tasks, the dependent measure was task completion time. Delay intervals were 0.3, 1.0, and 3.0 seconds.

- Jennings et al. (2000) evaluated control of a simulated helicopter in a virtual environment flight simulator with video viewing. The task involved compensatory hovering control of the aircraft in a fixed position under four visual feedback delay intervals of 0.067, 0.134, 0.184, and 0.334 seconds. The dependent measure was average position error.
- Kaber et al. (2000) studied teleoperation control of a telerobot in a pick-and-place task under virtual reality viewing conditions. The dependent measure was task completion time under three visual feedback delay intervals of 1, 2, and 4 seconds.
- Lathan et al. (2000) evaluated a tracing task with the tracing cursor displayed on a monitor and actuated using a spine biopsy telesurgery simulator. The dependent measure was RMS error under three visual feedback delay intervals of 0.5, 1, and 2 seconds.
- Ottensmeyer et al. (2000) evaluated two tasks, object grasp and transfer, and suture cutting, during teleoperation of a laparoscopic effector with simulated telesurgery conditions. For each task, the dependent measure was task completion time under two visual feedback delay intervals of 0.6 and 1.2 seconds.

In Figure 5.10, the feedback delay interval evaluated in the different studies is plotted on the abscissa and the task performance decrement produced by the delay conditions is plotted on the ordinate. The task performance decrement is calculated as follows: the dependent measure of task performance at 0 seconds delay is defined as 1.0, and the dependent measure of task performance at delay levels greater than 0 seconds is calculated as multiples of 1.0.

Results for the nine studies plotted in Figure 5.10 indicate that, without exception, task performance is adversely affected (task performance decrements observed) under feedback delay. Moreover, higher feedback delay intervals are associated with higher performance decrements. These results recapitulate those summarized in Figure 5.5 and Table 5.3, but they also extend the findings presented earlier in this chapter in documenting a relatively wide distribution of performance decrements at different delay intervals. The distribution of results in Figure 5.10, across the different studies, also indicates that there appears to be no systematic difference in delay effects on hands-on versus teleoperation tasks.

5.6   CONCLUSIONS

The scientific motivation for the focus of this chapter is the insight into the nature of cognition revealed by the findings. Specifically (as noted in previous sections), spatial or temporal displacements in sensory feedback produce profound impairments in both learning and real-time performance for every cognitive task so far evaluated. In addition, in most instances, improvements in performance observed under displaced sensory feedback conditions during learning trials never reach performance levels observed under normal feedback conditions. Displacements in sensory feedback represent a (rather dramatic) example of variability inherent to all modes and patterns of sensory feedback arising from behavioral interaction with environmental design factors (Chapter 3). However, the pronounced cognitive demands that such displacements place on behavioral control, with their consequent pronounced performance decrements, reveal a distinct modality of performance-design interaction (Chapter 1, Section 1.1.3, Principle 1) unlike interactive modes associated with behavioral control of nondisplaced feedback. The empirical observations of this research program therefore point to the conclusion that displacements in sensory feedback, produced by task-specific ergonomic design features of the performance-design interface, make a substantive contribution to context-specific variance in cognitive performance.

This conclusion must be tempered by findings by a number of investigators that the spatial ability of an operator predicts performance ability during teleoperation. In these studies, spatial ability is assessed with standard tests. Representative findings may be summarized as follows:

- Riley and Kaber (1999) report that subjects' spatial ability is correlated with their judgment of of the degree of telepresence while performing in a virtual environment. The authors attribute this observation to a link between spatial ability and subject proficiency in performing mental rotations and orientation to understand features of virtual objects and remain on selected routes during navigation in virtual environments.
- Chen (2008) and Chen and Joyner (2006) report that spatial ability predicts performance in a simulated multitasking environment combining gunnery and teleoperation of a robotic vehicle.
- Chen and Barnes (2010) and Chien et al. (2011) report that spatial ability predicts subject performance in human-robot teams that require supervisory control of multiple robots.

Given that teleoperation performance typically features spatially displaced visual feedback demands (Section 5.5.2), the implication of these findings is that spatial ability may be associated with improved proficiency in performance under spatially displaced visual feedback conditions (to my knowledge, a similar relationship has not been observed for performance under delayed visual feedback). The larger implication is that performance under displaced sensory feedback is not entirely attributable to context-specific effects, but rather represents both empiricist (task specificity) and nativist (spatial ability) contributions to performance variability under displaced feedback (terms introduced in Chapter 1).

Nevertheless, the dramatic nature and extent of performance decrements observed under displaced sensory feedback strongly suggest that task specificity is by far the more prominent contributor to displaced sensory feedback performance variability. The only caveat regarding this conclusion is that, unlike the analysis of cognitive performance variability presented in Chapter 3, the displaced feedback observations are not underpinned by definitive scientific evidence regarding the role of context specificity in displaced feedback performance variability.

A behavioral cybernetic interpretation (Chapter 1, Sections 1.3.3 and 1.3.4) is offered here to explain the basis of performance variability observed under displaced feedback. Behavior cybernetic theory assumes that behavior is organized in terms of closed-loop links between sensory feedback (generated by environmental design factors) and sensory feedback control (mediated by motor behavioral mechanisms) (Figure 1.8). As findings presented in this chapter suggest, some of the strongest support for the theory is derived from studies of the degree to which hands-on or teleoperation task performance is compromised by lack of spatial or temporal compliance between the control capabilities of behavior and sensory feedback from task design factors (Figures 5.1 and 5.9) (K.U. Smith, 1962a, 1972; K.U. Smith & Smith, 1962; T.J. Smith & Smith, 1987a, 1990; T.J. Smith, 1993). In particular, the assumption is that motor and receptor systems of the body have evolved as spatially specialized systems that function through variable spatial and temporal functional feedback tracking between muscle activities and sensory input inherent to receptor and effector activation (Figure 5.1). Imposition of spatial or temporal displacements from external environmental sources—such as feedback delay or spatially displaced visual feedback—disrupts the ability of the brain to effectively control these intrinsic body tracking patterns, with behavioral performance decrements being the consequent result.

6 Human Error and Performance Variability

Thomas J. Smith

6.1 INTRODUCTION

It might seem as though human errors—"actions other than those desired" (Konz, 1995, p. 138)—represent the dark side of human performance variability. This chapter will argue otherwise. Specifically, sections below note that analyses of human error embody both empiricist and nativist interpretations, in line with the dichotomy introduced in Chapter 1 (Goldhaber, 2012) (Section 6.3); from the empiricist perspective, most if not all incidents of human error feature some degree of context specificity (Section 6.3); consequently, human error represents an inherent attribute of human performance variability (Section 6.4); human error can be interpreted from a control systems perspective (Section 6.5); scientifically, there is a remarkable inconsistency and discordance in evaluations of the nature and sources of human error on the part of different human factors and safety professionals—arguably, this calls into question the validity of the term "human error" as a meaningful scientific and operational concept (Section 6.6); and a focus on management and control of system hazards, rather than on analysis of human error, offers a more promising opportunity for reducing risks contributing to, and for mitigating the consequences of, adverse system events and failures (portions of this chapter adapted from T.J. Smith [2002]).

It is not my intention here to provide a comprehensive review of the diverse conceptual and analytical perspectives contained in the massive literature on human error; there are numerous publications that serve this purpose. Key reviews of the topic over the past three decades include Bennett (2001), Dekker (2004, 2006, 2011), Glendon et al. (2006), Hallinan (2009), Hollnagel (2002), Hollnagel et al. (2006), Kirwan (1992a, b, 2005), Leplat and Rasmussen (1984), Meister, (1982, 1984), Miller (1982), Miller and Swain (1987), Morrow et al. (2005), Norman (1981), Park (1987), Perrow (1999), Peters and Peters (2006, 2013), Pope (1990), Rasmussen (1982, 1990), Reason (1990, 1997, 2000, 2013), Sharit (1999, 2006), Sheridan (2008), T.J. Smith (2002), Strauch (2004), Wallace and Ross (2006), Whittingham (2004), and Woods et al. (2010).

6.2 HISTORICAL PERSPECTIVE

Awareness of and consequent attempts to mitigate human error undoubtedly have prevailed throughout human evolution. A substantial portion of what we know today as *law* has been developed in an effort to standardize or constrain behavior to reduce the likelihood of error. Of the 282 extant provisions of the Code of Hammurabi promulgated some 3700 years ago, 100 are directed at discouraging erroneous behavior that results in harm to property or to other people. Following are six selected examples from the code (www.commonlaw.com/Hammurabi.html):

218. If a surgeon has operated with the bronze lancet on a patrician for a serious injury, and has caused his death, or has removed a cataract for a patrician, with the bronze lancet, and has made him lose his eye, his hands shall be cut off.

219. If the surgeon has treated a serious injury of a plebeian's slave, with the bronze lancet, and has caused his death, he shall render slave for slave.

220. If he has removed a cataract with the bronze lancet, and made the slave lose his eye, he shall pay half his value.

225. If he [a veterinary surgeon] has treated an ox, or an ass, for a severe injury, and caused it to die, he shall pay one-quarter of its value to the owner of the ox, or the ass.

229. If a builder has built a house for a man, and has not made his work sound, and the house he built has fallen, and caused the death of its owner, that builder shall be put to death.

233. If a builder has built a house for a man, and has not keyed his work, and the wall has fallen, that builder shall make that wall firm at his own expense.

The sixth and eighth commandments of the Hebrew bible, proscribing, respectively, killing and stealing and extant in written form for over 2600 years (http://en.wikipedia.org/wiki/10_commandments), have the same purpose. The Hebrew bible also describes (http://en.wikipedia.org/wiki/Halakha) two categories of sin, intentional and unintentional, that are identical to categories of human error suggested by contemporary observers (Reason, 1990, p. 8).

In his seminal text on mining work in Northern Europe (the first scientific treatise on human work), Agricola (1556) took note of a series of design factors that were introduced in underground metal mines of his time to benefit the safety of mineworkers by reducing the likelihood of error. Examples are limiting workday lengths to two shifts to reduce the chance of sleep deprivation and exhaustion, use of bellows for air circulation to prevent adverse behavioral effects of air contaminants in the mine, and design of a haulage machine (for hauling buckets of ore up vertical shafts from underground) that makes use of an arm-plus-leg-powered pulley system to reduce physical demands on the operators. This author also notes that mineworker error, manifest in incompetence or negligence in job performance, represents grounds for dismissal of the worker by the mine foreman, and mineworker injuries or fatalities often are attributed to error on the part of the foreman, through negligence in assuring safety conditions in a mine.

A search of Google books indicates that the year 1603 featured the earliest explicit appearance of the term "human error" in a published book (in Latin, dealing with theology).

The foregoing synopsis suggests that societal recognition of the significant role of error in human performance variability has an ancient history. In the modern era, understanding human error represents a highly active area of inquiry. A July 2013 search on Google recorded over six million hits for the term "human error." The comprehensive bibliography of Pope (1990, pp. 71–75), my compilation of works published from 1990 to 2004 (T.J. Smith, 2002; T.J. Smith & Mehri, 2006), and a Google search for books dealing explicitly with human error published between 2005 and 2012 records approximately 300 texts and monographs on safety and human error published between 1899 and 2012, an average of about 2.6 per year.

6.3   CONTEXT SPECIFICITY IN HUMAN ERROR

In the modern era, the debate between empiricist versus nativist views (Goldhaber, 2012) of the sources of human error can be said to originate with the unsafe acts versus unsafe conditions accident causation dichotomy introduced by Heinrich (1941). This author states (p. 268) that, "psychology lies at root of sequence of accident causes," and goes on to conclude (based on an analysis of 75,000 industrial accident cases) that 88% of the incidents reviewed were attributable to what he calls "unsafe acts of persons" (p. 20). One can mount a strong argument for the influence of this perspective on contemporary infatuation with the role of psychology in human error, and on contemporary assertions that human error is the root cause of 70–90 percent of occupational, driving, and aviation accidents (Woods et al., 2010, p. 3). Some observers attribute unsafe acts to human fallibility (Sharit, 2006, p. 713) linked to such nativist qualities as information processing, long-term memory, decision-making, personality, situation awareness, and/or emotional control capabilities or limitations. Similarly, Swain's widely cited *performance shaping factors* (PSFs) (Meister, 1982; Miller & Swain, 1987), assumed to predispose to human error, include a set of nativist idiosyncratic factors such as personality, intelligence, motivation, and attitudes.

However, what contemporary observers almost universally fail to note is that also on page 20, Heinrich (1941) cites findings from two other studies showing that, when multiple causes contributing to the same accident are considered, unsafe acts and unsafe conditions contribute almost equally to industrial accident causation. Identification of the contribution of unsafe conditions to human error represents an empiricist interpretation of error causation. An example of lack of recognition of this point is the opening sentence in the report by Salminen and Tallberg (1996): "It is generally accepted that 80–90% of accidents are due to human error." Another example is the analysis of Reason (1990, Chapter 2, pp. 19–52), who provides a systematic review of psychological and cognitive science studies of human error in individual performance dating back to the nineteenth century but ignores an equally compelling body of research with an equally prestigious pedigree, indicating that human error in individual performance is highly context-specific in its origins.

Following is a representative sample of the context-specific perspective on human error based on a series of contemporary observations that emphasize a prominent role for the influence of design factors on human error.

- Miller (1982) identifies a series of design factors influencing human error, asserts that work design factors contribute more to the incidence of human error than is necessary, points out that blaming the worker for error is becoming an unpopular defense and a poor excuse, and suggests that management often resists approaching engineering safety problems rationally, but accepts "human error" as unpreventable because human behavior is believed to be unchangeable.

- A series of performance-shaping situational and task and equipment design factors are identified by Meister (1982, p. 6.2.4) and by Miller and Swain (1987, p. 223) that influence human reliability, including the work environment, cleanliness, staffing, work hours and breaks, work space and work layout, availability and suitability of supplies, training, job instructions and procedures, task complexity, frequency and repetitiveness, team structure, interface design factors, and supervisory, incentive and organizational influences.
- Pulat (1997, p. 318) asserts that, "the most effective method of controlling human errors is the ergonomic design of work."
- With regard to accidents arising out of operation of high risk technologies, Perrow (1999, p. 7) maintains that, "the cause of an accident is the complexity of the system."
- With regard to aviation safety, Maurino (2000, p. 956) suggests that, "a contemporary safety paradigm should therefore consider errors as symptoms rather than causes of safety breakdowns, because error-inducing factors are latent in the context…"
- Hollnagel (2002) suggests that human action outcomes differ from intended or required outcomes due to variability in context and conditions rather than action failures.
- Sharit (2006, p. 718) points out that, "Human actions are embedded in contexts and can only be described meaningfully in reference to the details of the context that accompanied and produced them."
- Sheridan (2008) identifies defects in the design of procedures, training, maintenance, and management as factors contributing to error causation.

These observations are supported by empirical evidence from a few studies of factors contributing to the occurrence of accidents on-the-job. For example, the results both of Adams et al. (1981) and Albin (1988) indicate that task-specific factors linked to the ergonomics of work design are directly involved in about 50 percent of all occupational accidents. In the latter study of antecedent events for slip and fall accidents in U.S. surface mines for the years 1985–87, work design factors were implicated directly (18%) or by interaction (31%) in 49% of these accidents, whereas only 11% of the accidents were linked to behavior alone.

More recently, Antonovsky et al. (2014) used a guided interview approach to analyze maintenance operations at a major petroleum production facility. Thirty-eight experienced maintenance personnel were asked to consider a maintenance activity in which they were involved and which did not produce the expected outcome, and across a selection of 27 factors grouped into three different categories (action errors, situation awareness, and organizational threats), identify which factors, in their judgment, contributed to the adverse event. In other words, the analysis did not focus on predetermined events and the researchers did not provide any guidance as to which events to analyze; instead, event selection was up to the workers. Four factors—two organizational threats, a situation awareness factor, and an action error factor—each were associated with over half of the event cases. Factors associated with the 14 highest percentages of event cases featured eight organizational threats, three situation awareness factors, and three action error factors. Factors associated with the 13 lowest percentages of event cases featured five organizational threats, three situation awareness factors, and three action error factors. In summary, asked to identify factors that made a likely contribution to self-selected adverse maintenance events, workers themselves more frequently identify organizational design factors relative to situation awareness or action error factors associated with innate behavior. This study suggests that workers may have an acute awareness of the importance of design—context—in error causation.

However, Sanders and McCormick (1993) advocate caution in any attempt to arrive at conclusions about causes of accidents. These authors cite findings (p. 668) from a single-cause Japanese study of chemical industry accidents indicating that 41 percent were directly attributable to microergonomic or macroergonomic design flaws. Yet they point out that there are inherent limitations to accident causation analysis because every occupational accident likely has multiple antecedent factors, accident causation is context-specific—results typically are not generalizable beyond the specific area or context in which the incident occurred, and information about the accident necessary to infer causation typically is not available.

6.3.1 Management Responsibility for Human Error

There is an extensive body of literature devoted to the critical role of management in reducing workplace error. Over 450 years ago, Agricola (1556) cited management responsibility for preventing shift-related fatigue among mineworkers. In the modern era, dozens of texts have been published over the past century that emphasize the importance of management in accident prevention (Pope, 1990, pp. 71–75). In his seminal text on industrial accident prevention, Heinrich (1941, p. 44) notes that

> The immediate and proximate causes of industrial accidents...are known to lie in two general groups, namely, mechanical or physical and personal. Both causes are controllable by management, and in the case of both, management has an unexcelled opportunity to exercise remedial action.

Recent observers share this perspective. Petersen's (1971, p. 19) first principle of industrial safety is, "An unsafe act, an unsafe condition, an accident: all these are symptoms of something wrong in the management system." Turner (1978) maintains that "man-made disasters" are a result of everyday organizational decision-making. Pope (1990) and Reason (1997) dedicate entire texts to the key role of organizational management in ensuring workplace safety. Maurino (2000), Miller (1982), Morrow et al. (2005), and Perrow (1999) emphasize that safety of contemporary technological systems and management of contemporary organizational systems share a common feature of complex systems design, and that fidelity of the former is intimately linked to effective design of the latter. I have attempted to pull together the literature on organizational design and management strategies for reducing worker error to enhance workplace safety with a chapter devoted to the macroergonomics of hazard management (T.J. Smith, 2002).

Organizational systems and management structures that operate such systems by their very nature represent design factors. Thus, the observers cited above, and many more not cited, who acknowledge management as a key influence on human error also acknowledge by inference that context specificity likely represents a pervasive source of human error.

6.4 HUMAN ERROR AS AN INHERENT ATTRIBUTE OF HUMAN PERFORMANCE VARIABILITY

Some analyses of human error focus on unsafe acts, unsafe conditions, management, and/or other conditions as contributory factors without considering error from the broader perspective of human performance variability. Other observers, however, point out that human error is inherent to such variability. Skinner (1953) offered an early perspective on this conclusion in claiming that without error variability inherent in behavior, there would be no learning. Meister (1982) points out that error is a function of human variability and each human action therefore offers a potential opportunity for error. In the title of his report, Hollnagel (2002) explicitly identifies a link between accident causation and performance variability and goes on to observe that there always is variability in sociotechnical systems; the challenge is to detect uncontrollable variability, performance variability is necessary for human participants in such systems to learn and for the systems to develop, human actions are variable and performance variability is the central issue in understanding human error, and therefore human action outcomes differ from intended or required outcomes due to variability of context and conditions rather than action failures. Dekker (2007) points out that, "The dominant safety paradigm in human factors has long been based on searching for ways to limit human variability in otherwise safe systems."

Arguably, the most definitive support for the idea that variability is inherent to error occurrence is provided by the field of statistics. Statistical analyses, involving tests such as ANOVA, the *t*-test, regression analysis and so forth, are based on the premise that sample size and/or sample variability should be targeted at levels appropriate to achieve statistical significance in the analytical outcomes. The variability that inevitably persists is termed random error or experimental error and is calculated as standard error (SE) of the mean. Statistical confidence intervals, calculated using the SE, are termed error bars. In other words, as far as statisticians are concerned, the idea that variability and error are synonymous was settled some 350 years ago in the seventeenth century with emergence of mathematical origins of the field.

What are the implications of this idea? Aspects of human performance variability discussed in the foregoing chapters suggest the following likely answers to this question:

A. *Variability in human error emerges as a consequence of variability in movement behavior.* Ultimately, human error outcomes at whatever level are generated as a result of movement behavior. The following observations of Harbourne and Stergiou (2009) and Stergiou et al. (2006) regarding the nature and sources of variability in movement behavior (Chapter 2) thus may be assumed to apply, to some degree, to variability in human error.

- Chaotic temporal patterns of movement variability are associated with healthy movement behaviors.
- In contrast, reduction or deterioration of such patterns is associated with a decline in healthy flexibility of movement behaviors coupled with the onset of behavioral rigidity and inability to adapt.
- There thus appears to be an optimal amount of temporal variability, characterized by a complex chaotic structure, associated with healthy movement behaviors. A reduction in such variability equates with increased behavioral rigidity. An inappropriate increase in such variability equates with an increase in noise and instability in movement behavior.

These observations appear to be aligned with the assumption of Dekker (2007) that safety, once established, can be maintained by requiring human performance to stay within prescribed boundaries of variability.

B. *Human error emerges as a consequence of human system complexity.* In the domain of human movement variability (Chapter 2), Harbourne and Stergiou (2009) point out that:

- Complexity may be hidden within the time series of a movement sequence as it emerges over time.
- A nonlinear, deterministic temporal pattern of movement variability is generated when a movement that occurs at one moment affects, and in turn is affected by, movements that occur either before or after the given movement.
- Nonlinear dynamics methods (Chapter 2, Section 2.3.2) based on chaos theory can be deployed to analyze this complexity.

These observations appear to be aligned with the assumption of Hollnagel (2002) that systems performance is an emergent phenomenon and with the observation of Karwowski (2012) that

Complex human work systems often exhibit emergent global behaviors that are not predictable from their local properties. In the context of HFE problems, global behavior cannot be explicitly described by the behavior of the component systems, and therefore, it may be unpredictable and remain unexpected to designers, system users, or operators, leading to unrecognizable states that are often defined as system errors or aberrations that lead to degraded human performance, industrial accidents, system failures, and catastrophic losses.

C. *Human error is inherently unpredictable.* Two themes bear on this conclusion. The first dates back to Locke (1632–1704) (1974) (Chapter 3, Section 3.2), who maintained that all knowledge derives from experience, that experience by its very nature is variable, and that knowledge, and performance upon which knowledge is based, therefore is probabilistic and inherently unpredictable in nature. We may extend with some confidence this line of reasoning to error performance.

The second theme derives from evidence that most if not all error performance features some degree of context specificity (Section 6.3). As noted in Section 1.1.2, the basic idea of context specificity is that design factors in the performance environment make some contribution to observed variability in performance. Thus, if context specificity makes a contribution to variability in error performance, the implication is that interaction of an operator or worker with a new design introduces new modes and patterns of error performance variability that cannot necessarily be predicted in advance. The quote from Karwowski (2012) above endorses this idea of unpredictability in error performance. Hollnagel (2002) also notes that human action outcomes may differ from intended or required outcomes due to variability of context and conditions rather than to action failures.

It is worth recalling the observations of Silver (2012) (Chapter 1, Section 1.2.2) regarding the difficulty in making accurate predictions about human performance and about the erroneous consequences of inaccurate predictions. This author advocates (in Chapter 8 of his book) the application of Bayesian statistics to improve the likelihood of making correct predictions. To my knowledge, this approach has yet to be employed in human error prediction. It might prove worthwhile for human error analysts to investigate Silver's proposition.

D. *As a behavioral process, human error can be understood from a control systems perspective.* The background to this idea is the observation introduced in Section 1.3 that a broad range of human performance systems featuring dramatically different levels and modes of biological and sociotechnical complexity exhibit adaptive behavior. A control systems interpretation of such adaptability, grounded in behavioral cybernetics and applicable across all levels of performance complexity, was offered in Section 1.3.3. The rationale for the assumption that a control systems framework also applies to error performance may be summarized as follows: behavioral cybernetic theory assumes that behavior as a control systems process reciprocally links sensation, perception, and cognition to motor behavior through motor control of sensory feedback (Figure 1.8; K.U. Smith, 1972, p. 52); variability in motor behavior therefore likely serves as the substrate for variability observed in different modes of performance, including error performance (point A above); performance variability facilitates behavioral choices among options, selection of behavioral strategies, and behavioral flexibility to adapt to variations in the environment, and in so doing, enables individuals or groups to be more successful (Harbourne & Stergiou, 2009); and Dekker (2007) points out that most stories of error indicate that failures represent breakdowns in behavioral adaptation associated with attempts to cope with complexity.

The next section addresses in more detail a control systems interpretation of error performance.

6.5   A CONTROL SYSTEMS PERSPECTIVE ON ERROR PERFORMANCE

A popular target of the application of control systems analysis to human error is tracking performance (Gerisch et al., 2013; Jagacinski & Flach, 2003, Chapter 10; K.U. Smith & Smith, 1962, p. 52). More generally, Woods et al. (2010, pp. 75–77) discuss human error from a control theory perspective. Sheridan (2004) describes a practical application of this approach with a closed-loop model of driver distraction. In his analysis, Sheridan makes three assumptions about the driving control system. The first is that vehicular guidance during driving is based on a combination of feedback and feedforward control (Figure 1.7). The second is that distraction of the driver (that may occur for a variety of reasons) has the potential to disrupt two key inputs into the driving control system; namely, the projective (that is, goal-setting) behavior of driving (mediated by the forward model, see Chapter 1, Section 1.3.4), as well as sensory perception of the driving environment in real time (mediated primarily by visual and auditory feedback). The third driving control systems assumption is that the disruptions introduce error signals in these inputs that cannot be effectively feedback-controlled by output of the system; namely, guidance of the vehicle by the driver. In summary, the analysis of Sheridan (2004) maintains that error in driving performance emerges as a consequence of system variability attributable to environmental distraction that compromises the driving control capabilities of the driver.

The foregoing analysis can be extended to a more generalized control systems interpretation of error performance based on the two key principles described below:

1. *Feedforward control of error by means of a forward model.* In cybernetic terms, what different manifestations of cognition have in common is predictive activity. In other words, for effective guidance of behavior you have to be able to predict the sensory and perceptual consequences of your actions (Blakemore et al., 2001; Hawkins & Blakeslee, 2004; Miall et al., 1993). Cognitive demands arise from the need to control future behavior, a process termed feedforward control (Chapter 1, Section 1.3.2). Feedforward control thus involves cognitive projection of past memories and associations to enable anticipation of future events and the behavioral requirements for their control, so that control actions can occur to prevent behavioral errors from occurring when these events transpire. Various observers have noted that error is likely to occur when an operator's internal cognitive model of how a system should function is not aligned with how the system actually functions (Dekker, 2007; Hollnagel, 2002; Leplat & Rasmussen, 1984; Peacock, 2002; Rasmussen, 1982; Reason, 1990). The observation of Reason (1990, p. 5) is that, "the notions of intention and error are inseparable." The behavioral cybernetic perspective assumes that intention is observable (as a manifestation of motor behavior), and error may occur when sensory feedback that is actually generated by a given behavior is not aligned with predictions of the forward model regarding anticipated sensory feedback from the behavior.

   Recently, neurophysiological evidence has been reported documenting neuronal populations in both the cerebellum and the medial frontal cortex whose cognitive control functioning specifically subserves learned error prediction (Brown & Braver, 2005; Miall & Reckess, 2002; Miall et al., 1993). The evidence suggests that memory of sensory feedback from past action (the predictive model) is referenced against real-time sensory feedback from current action (perception) and the model updated (learning) based on any detected discrepancy.

2. *Integration of cognitive and motor behavior.* The second control systems principle of human error assumes that cognition emerges as a consequence of active motor control over sensory feedback, a concept that has been delineated from both behavioral and neurobiological perspectives (Jackson & Decety, 2004; K.U. Smith, 1972; T.J. Smith & Henning, 2005; T.J. Smith & Smith, 1987a, b). This principle leads to the concept that variability in overall systems behavior, and thus the propensity for human error, is closely referenced to the properties of motor behavioral control, in that such control is what an operator employs to interact with environmental design factors.

With regard to error performance, one of the key implications of motor behavioral variability is the imperfect nature of motor control. This phenomenon (Jagacinski & Flach, 1993, p. 222; Schmidt & Lee, 1999) is based on evidence showing that when an operator interacts with a target in a repetitive fashion, successive movement patterns may be very similar but never are exactly identical. This is a feature of all effector movements. Such lack of fidelity of movement patterning is exacerbated under a variety of design conditions, such as stimulus-response incompatibility, displaced sensory feedback, fatigue, or job stress. This means that when different operators interact with common tasks with the same design, both movement patterns and cognitive behaviors may vary among operators in both space and time.

In terms of understanding of human error, the above considerations collectively point to the conclusion that cognitive demands as they are commonly experienced can be understood as resulting from challenges to closed-loop control of motor behavior. In such cases, the behavioral consequences are increased performance variability, possibly accompanied by compromised ability for projective movement guidance of behavior, with a concomitant elevation of risk for error.

6.6   THERE ARE EVIDENT LIMITATIONS TO THE TERM "HUMAN ERROR" AS A MEANINGFUL SCIENTIFIC AND OPERATIONAL CONCEPT

The largely unimpeded ascendancy of the unsafe acts–unsafe conditions paradigm of human error in the decades following its introduction by Heinrich (1941) has recently received more critical attention. As noted in Section 6.3, Miller (1982) indicates that blaming the worker is becoming an unpopular defense and a poor excuse. Miller and Swain (1987) point out that too often the worker is blamed, when in fact the work situation is poorly designed and predisposes the worker to error. Hollnagel (2002) observes that humans no longer should be seen as the primary cause of accidents, and that human actions should not be described in binary correct–incorrect terms. However, he also notes that humans play an important role in how accidents occur. Woods et al. (2010) question the legitimacy of the concept of human error. Sheridan (2008) cites the assertions of prior observers that controversy exists about the concept of human error, and more pointedly, human error is an outmoded idea. Yet this author also notes that the idea of human error remains in popular use.

The following scientific and operational limitations to the concept of human error may be summarized in support of the skepticism cited above:

- *Unsafe acts or unsafe conditions.* The conventional unsafe acts versus unsafe conditions either-or dichotomy of human error is called into question by the context specificity perspective. Evidence regarding context specificity in human error (Section 6.3) indicates that, across different performance domains, design factors in the performance environment have a documented influence on observed variability in error performance. That is, unsafe acts and unsafe conditions are reciprocally linked, and error performance should be viewed as an outcome of the systems integration of these two effects. The idea that human error emerges as a consequence of human system complexity (Section 6.4, point B) is aligned with this view.

- *Sharp end versus blunt end.* A popular notion among human error analysts is the distinction between performance defects at the "sharp end" (i.e., the user or operator—pilot, driver, surgeon, soldier, or machine operator— immediately exposed to and involved with conditions or circumstances in which error may occur) versus those at the "blunt end" (design, procedures, training, maintenance, management, etc., somewhat removed from these circumstances), in contributing to error causation (Sheridan, 2008; Woods et al., 2010, pp. 8–10). Yet as documented in Section 6.3, there is widespread agreement that management is ultimately responsible for many, if not all (Petersen, 1971), occurrences of human error. From this perspective, performance variability in one or more managers, as well as the management system under their control and that of workers embedded in this system, are inextricability linked insofar as both acceptable and unacceptable modes of performance are concerned. So which party deserves the sharp end versus the blunt end label? Once again, as with the previous bullet, a view of error performance as a closed-loop process, in this case involving intimate feedback links between different human participants in a performing system, calls into question a paradigm of error causation built upon a dichotomous foundation.

- *A profusion of perceptions.* Error means different things to different observers. I am reminded of the fable of the blind men and the elephant. Even within the HF/E community of human error analysts, an animated debate persists about the meaning, the causes, and the appropriate models for human error. Yet varying perspectives beyond this community on the nature and meaning of human error appreciably expand this debate.

For example, as noted in Section 6.4, the assumption of statistics is that cohort variability due to individual differences represents error that should be minimized to achieve rigor in statistical analysis.

Another example is the treatment, across the broad range of human and biological science, of results obtained from dependent measures of performance in experimental research studies. Specifically, if a subject's performance on a given measure deviates from expected or desired accuracy, this deviation often is termed error. To illustrate this point, I have tabulated the percentage of articles in selected issues of the journal *Human Factors* that refer to deviation in the accuracy of dependent measure performance as "error." Results are shown in Figure 6.1.

Data in Figure 6.1 are from 10 selected issues of *Human Factors* (indicated in the legend to the figure) dating from August 1970, to June 2013. I have selected one issue in the first year of each decade from 1970 through 2010; additionally, one issue each in 2011 and 2012, and three issues in 2013, also are selected. In each selected issue, I counted the number of articles that referred to inaccuracies in dependent measure performance as "error" and calculated the percentage of this number relative to the total number of articles published in the issue. This percentage is plotted on the ordinate of the figure.

For the selected *Human Factors* issues, results in Figure 6.1 indicate that the percentages of articles that refer to deviation in the accuracy of dependent measure performance as "error" range from a low of about 17 percent (Summer 2000 issue) to a high of about 77 percent (June 2013 issue). There is a considerable variability in these percentage levels, but at least two articles (in the Summer 2000 issue), and a maximum of about three-fourths of the articles (in the June 2013 issue) label

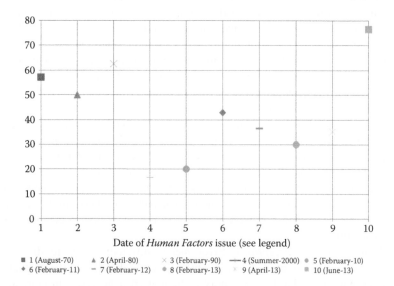

FIGURE 6.1 Percent of articles in selected issues of *Human Factors* that refer to inaccuracies in dependent measure performance as "error."

inaccuracy in dependent measure performance as "error." The *Human Factors* issues selected clearly do not comprise a representative sample, yet the results suggest that, over five decades of publication of the journal, the term "error" is recurrently applied to observations of deviations in dependent measure performance from hypothesized performance.

This analysis raises the following questions. Does the term error as understood by statisticians and experimentalists represent human error? Is such error caused by unsafe acts, unsafe conditions, or both? Does such error represent the sharp end or the blunt end? Is management ultimately responsible for such error? Does such error emerge as a consequence of system complexity? I hope that my sarcasm is evident.

The serious point of this analysis is to suggest that ubiquitous reliance on the term error, in many different contexts, with ongoing debate about definitions, categories, and sources, points to evident limitations to the term "human error" as a meaningful scientific and operational concept.

- *Error management or hazard management.* An idea that appears to have emerged relatively recently in the HF/E community of error analysts is the assumption that error can be managed (Kearns & Sutton, 2013; Loh et al., 2013; Nikolic & Sarter, 2003, 2004; Shappell, 2011; Wiegmann, 2010). Nikolic and Sarter define error management as the process of detection, localization, and recovery (or correction) of error. What these authors do not seem to realize is that when an error occurs, its detection and localization may be possible, but it certainly is too late to be effectively managed. What can be effectively managed is the *risk* of error (Cacciabue, 2005); that is, hazards predisposing to error—these are the actual targets for correction. The next section addresses this concept in more depth.

6.7 PRINCIPLES OF HAZARD MANAGEMENT

The control of hazards by biological systems to avoid danger and secure survival is as old as life itself. Emergence of organized systems of work and increasingly complex technology and sociotechnical systems has prompted human appreciation of hazard control as a key to safety, security, and productivity (T.J. Smith, 2002). This suggests that management of hazards should represent an integral aspect of programs directed at analysis and control of human error and safety. Nevertheless, the recognition of safety as a legitimate management function is relatively recent, emerging in the United States gradually only over the past five decades or so. Further, there is by no means a consensus within the safety community regarding what it is about safety that should be managed. Indeed, "loss control," "accident prevention," and "safety" itself are cited far more often than hazards as the proper focus of management.

Accordingly, definitions, scope, rationale, and background for hazard management first are introduced. Subsequently, a behavioral cybernetic model of hazard management is briefly reviewed and the key principles of hazard management embodied in the model then are summarized.

6.7.1 Definitions, Scope, and Historical Perspective

In this section, the terms "safety management" and "safety program" refer to any organizational function or program with a general focus on safety and accident prevention, whereas the term "hazard management" refers to a safety program with a specific focus on detecting, evaluating, and abating hazards. "Hazard" refers, in a general sense, to any work design factor that elevates the risk of detrimental performance by a worker (employee or manager) or an organizational system. "Safety performance," in a general sense, refers to the integrated performance of all organizational and individual entities whose activities affect safety.

Heinrich (1959, Appendices I and II) and Montgomery (1956) provide a chronology and historical background of the safety movement, from the classical to the modern era. Pope (1990) provides an extensive set of safety definitions. "Safety" refers to freedom/security from danger, injury, or damage. "Hazard" is derived from Old French and Middle English terms of the same spelling referring to games of chance—the term thus connotes chance or risk.

Even today, the meaning of the term "safety" is not uniformly understood within the safety community and elsewhere. Pope (1990) asserts that the term lacks absolute definition and is not acceptable for precise administrative language. Petersen (1971, p. 26) suggests that, "safety is not a resource: it is not an influence: it is not a procedure: and it is certainly not a 'program.' Rather, safety is a state of mind, an atmosphere that must become an integral part of each and every procedure that the company has." Zaidel (1991), with reference to driving safety, suggests that, "safety... represents behaviors, situations, or conditions that...are associated with either higher or lower probability of accidents."

The latter definition invokes a long-standing and cherished assumption in the safety field that the primary concern of safety is dealing with accidents. Indeed, the first text of the modern era on occupational safety is entitled *The Prevention of Accidents* (Calder, 1899). Webster's definition (McKechnie, 1983) of "accident" is instructive: "an event that takes place without one's foresight or expectation; an event which proceeds from an unknown cause, or is an unusual effect of a known cause, and therefore not expected; chance." In a similar vein, Pope (1990, p. 107) offers a series of definitions of the term, all of which refer to the unexpected, unplanned, and uncontrolled nature of accidents.

Some in the safety community observe that given their unpredictable and uncontrollable nature, a focus on accidents is incompatible with a philosophy of safety management. Guarnieri (1992) notes that the word "accident" has almost disappeared from safety science and engineering use.

The safety field is huge with an extensive literature—Pope (1990, pp. 71–75) cites some 163 texts and monographs relevant to the field published since 1899, over half published since 1975. The primary purpose of this section therefore is to develop and address major themes, concepts, principles, and practical considerations germane to the management of hazards and to address the relevance of this analysis to variability in performance generally and error performance in particular.

6.7.2 RATIONALE AND BACKGROUND

Accidents by their very nature are inherently unmanageable. "Safety management" has emerged as a major focus of the safety field. However, the imprecise meaning of "safety" leaves unclear what it is precisely that is being managed. Connotations of the term "loss control" may go beyond safety-related loss, and hazard control arguably represents an essential prerequisite for loss control.

The term "hazard management" avoids these shortcomings (T.J. Smith, 2002). The term "management" refers to, "the executive function of planning, organizing, coordinating, directing, controlling and supervising." Hazards represent concrete, recognizable entities whose existence can be characterized and quantified. With adequate data, informed judgments and predictions can be made about possible outcomes associated with the prevalence and persistence of particular hazards. Reduction in risk associated with hazard modification or elimination thus can be estimated. In other words, hazard conditions represent a highly eligible target for management.

Today, hazard control appears to represent the prevailing focus of the safety management profession. Grimaldi and Simonds (1989, p. 16) illustrate this point: "it seems clear that the pursuit of safety must be pointed toward the identification of hazards, determination of their significance, evaluation of the available correctives, and selection of the optimal remedies. When this path is followed, it ends with the control of unwanted events at an irreducible minimum. 'Accidents' in this case are not the target. Instead, it is the *hazard* that causes the unwanted harmful event, which is to be eliminated. This is the primary concern of safety management."

Hazard elimination also has attracted the attention of human error analysts in the HF/E community, although truth be told, many human error publications make no mention of the importance of hazard mitigation in reducing human error. Swain (1973) advocates an error cause removal program for industry, which essentially represents a call for hazard management. Miller (1982) mentions the importance of physical hazard control to the management of occupational safety. Swain introduced the idea of performance-shaping factors (Meister, 1982; Miller & Swain, 1987), that may otherwise be termed hazards, with the potential to influence human error. A more recent concept is the idea of resilience engineering (Hollnagel et al., 2006; Sheridan, 2008; Woods et al., 2010), which presupposes that human errors and machine failures will likely occur and that error control efforts therefore should be placed on means to anticipate disturbances and to be able to recover and restore the system to the original state or, if need be, some acceptable state that is different but still safe. This essentially represents a hazard management idea. A highly explicit treatment of hazard management is provided by M.J. Smith and Beringer (1987), who devote most of their chapter to discussions of the measurement of hazard potential, the implications of this analysis for safety performance, and the application of human factors to the control of workplace hazards.

What kind of hazards are we referring to as the object of hazard management? The prevailing view is that accidents have multiple, complex causes related to both physical-environmental hazards and behavioral variability (Petersen, 1971) (Section 6.4, point B). The behavioral cybernetic perspective on hazard management (below) references this multiple causation idea, in which safety performance of the organizational system is viewed as a process of interaction between the human and design elements of the system, and hazards are viewed as manifestations of flaws in workplace design subject to organizational management and control. More specifically, the nature of risk associated with hazards can be understood from a human factors and safety performance perspective, with the central concept that across many different categories and forms, a hazard represents a job or workplace factor that gives rise to a fundamental mismatch between worker behavior and workplace conditions (K.U. Smith, 1979).

K.U. Smith is one of the pioneers in adoption and development of hazard management as a human factors-based systems theory and process for safety improvement (K.U. Smith, 1975a, b, 1979, 1988, 1990). With a series of hazard management projects and programs in the Behavioral Cybernetics Laboratory at the University of Wisconsin-Madison, Smith elaborated the theory, principles, and applications of hazard management in the 1960s and 1970s as an early formulation of worker-management participation and union-executive cooperation in operational management of safety and production efficiency in the workplace. These initiatives of Smith are among the first specific instances in management science where hazard management principles were applied to safety management of complex systems.

6.7.3 BEHAVIORAL CYBERNETIC MODEL OF HAZARD MANAGEMENT

As a complement to the control systems perspective on human error offered above (Section 6.5), this section describes a control systems model of hazard management that assumes a closed-loop or behavioral cybernetic relationship (Chapter 1, Section 1.3.3) between workplace design conditions (including hazards) and safety performance of organizational systems (behavior) to account for differences observed in the safety record of different systems in different contexts.

The model is illustrated in Figure 6.2. The shaded region of the figure features classic negative feedback (i.e., servomechanism) control elements introduced in Figure 1.6, including a hazard detection system (the sensor), which provides a means of monitoring sensory feedback from workplace hazards (input); a safety reporting system (the controller), which provides a means of referencing actual with desired safety performance; actual safety performance (output) of the organizational system (the effect); and system safety performance targets and goals (the reference). The model assumes that effective management control of safety is mediated by comparison of feedback from actual system safety performance with safety performance targets and goals. Error between actual and desired performance is used to guide management decision-making directed at tightening system control over workplace hazards (hazard feedback control), thereby reducing system performance error and improving accident and injury prevention. Many years ago, Juran (1964) introduced a similar servomechanism model to characterize the management of system quality, a model that will be discussed in Chapter 9.

A further implication of the model in Figure 6.2 is that it points to a role for ergonomics in facilitating breakthrough in hazard management and safety performance. Juran (1964, p. 3) defines managerial breakthrough as "change, a dynamic, decisive

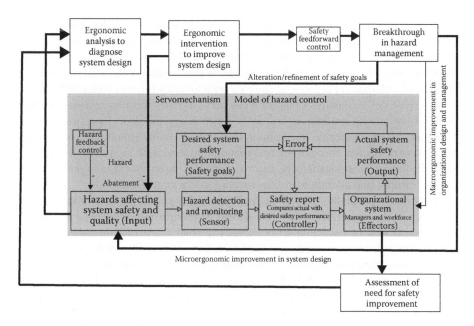

FIGURE 6.2 Closed-loop control model of hazard management with ergonomic analysis and intervention specified as keys to breakthrough in system performance in a feedforward control manner.

movement to new, higher levels of performance." The model in Figure 6.2 conceives of breakthrough in hazard management as a closed-loop process involving feedforward in addition to feedback control (Figure 1.7). The assumption of the model is that the impetus for breakthrough in organizational performance is some feedback indicator suggesting that the current level of performance is no longer adequate. This prompts the system to initiate breakthrough by feedforward projecting its behavior into the future to achieve a new, improved performance level. As noted above (Section 6.3), it is likely that often if not always, the root cause of inadequate system performance at any given level is some sort of design flaw, either in microergonomic design of the work process or environment, or in macroergonomic design of the organizational system, or both. In the safety field, these design flaws are termed hazards.

This is where ergonomics comes into play, given that the prominent focus of ergonomics is on design variability. As denoted by the system control elements outside the shaded region in Figure 6.2, ergonomic analysis can be used to detect system hazards. Ergonomic intervention then can be used to abate hazards and improve system design in order to facilitate the breakthrough process. Application of ergonomics can thus serve as a key breakthrough strategy by means of which the system elevates its safety performance from one control level to the next. In this manner, the organizational design incorporates ergonomic systems provisions for alteration of the safety goals themselves to deal with altered circumstances (new regulations, new technology, etc.).

Thus, the basic premise in Figure 6.2 is that an occupational ergonomics program meets this requisite system design need for initiating and managing breakthrough. This idea is aligned with the observation of Miller (1982), who notes that the philosophy of ergonomics is that most unsafe acts are design-induced and design-preventable. Methods of ergonomic analysis are admirably tailored for hazard assessment—results of this analysis therefore can serve as early warning sentinels for the need for initiating a breakthrough process. Once design problems have been identified, methods of ergonomic intervention then are admirably suited for contributing to problem resolution, through microergonomic improvement in system design features, including hazard abatement, macroergonomic improvement in organizational design features, and alteration or refinement of safety targets and goals.

A second implication of the model in Figure 6.2 is that variability in safety performance of the organizational system should be prominently influenced by design characteristics of the particular set of hazards at which the hazard management process is directed. In other words, the model predicts context specificity in organizational hazard management performance analogous to that observed with individual error performance (Section 6.3). This in turn suggests that there should be no magic bullet or completely generalized scheme for macroergonomic design of hazard management systems. Rather, the design of each system must be customized and specialized in relation to the particular mix of hazards being addressed. Moreover, as or when the design of the hazard environment is modified (through abatement of existing hazards, introduction of new ones, or both), it may be necessary in turn to modify the macroergonomic design of the hazard management system itself to deal most effectively with the new hazard mix.

6.7.4 KEY PRINCIPLES OF HAZARD MANAGEMENT

The remainder of this section provides a synopsis of key principles of hazard management whose adoption might support breakthrough in hazard management success specified in Figure 6.2.

A. *Organizational best practices for safety management.* Cohen (1977) and M.J. Smith et al. (1978) have investigated factors and characteristics that differentiate highly successful from less successful occupational safety programs. Both studies found that relative to high accident rate companies and sites, low accident rate companies and sites are distinguished by the following macroergonomic approaches to safety (the order does not represent an importance ranking): (1) greater management concern and involvement in safety matters, reflected in regular inclusion of safety issues in plant meeting agendas, and routine, personal inspections of work areas by top plant officials, (2) greater management skill in managing both material and human resources, (3) full-time safety director who reports to top management, (4) greater emphasis on hazard control, with tidier work areas, more orderly plant operations, better ventilation and lighting, and lower noise levels, (5) more open, informal communication between workers and management, with frequent contacts between workers and supervisors on safety and other matters, (6) more stable workforce, with older, more experienced workers and less absenteeism and turnover, (7) well-defined employee selection, placement, and job advancement procedures, (8) more flexibility in discipline for safety violations, such as safety counseling in addition to suspensions and dismissals, (9) greater availability of recreational facilities for worker use during off-job hours, (10) greater effort to involve worker families in safety campaigns, (11) frequent use of safety incentives and promotions, (12) frequent use of accident investigations and formal accident reporting systems, and (13) formal safety training for employees and supervisors.

B. *Worker self-regulation of hazard management.* A fundamental problem with the characterization of accident and injury risk in terms of behavioral and physical-environmental hazards is that hazard management strategies directed at abating these hazards do not reliably and predictably result in accident and injury prevention. For example, Jones in 1973 documented the limitations of dependence on engineering controls and related enforcement measures in bringing down occupational injury rates to acceptable levels. Gill and Martin (1976) echo this finding with the claim that the engineering approach to safety management is not sufficient to prevent all accidents and that innovative hazard control strategies such as performance standards and worker participation should be considered to achieve further improvement in accident prevention. M.J. Smith et al. (1971) and Gottlieb and Coleman (1977) report that inspections carried out by the Occupational Safety and Health Administration (OSHA) are relatively ineffective in preventing accidents because most hazards cannot be identified by traditional workplace inspections. Results from these and other studies suggest that, in general, only 5 to 25 percent of accidents can be avoided by rigorous compliance with conventional safety standards (Ellis, 1975; K.U. Smith, 1979). OSHA itself has noted that, "OSHA's own statistics appear to indicate that 70–80 percent of all deaths and injuries each year are not attributable to a violation of any OSHA specification standard" (Occupational Safety and Health Reporter, 1976, p. 684).

These considerations suggest that along with physical and behavioral hazards, there is an entire additional domain of hazards related to macro-ergonomic design defects in the management of safety. The analyses of Cohen (1977), Ellis (1975), and M.J. Smith et al. (1978) suggest that in terms of their adverse impact on safety, this class of hazards may be more significant than either behavioral or physical-environmental hazards. This perspective has prompted the development and testing of a range of macro-ergonomic strategies that emphasize the role of workers in self-regulating their interaction with workplace hazards and with the occupational safety system. Various aspects of this topic have been reviewed by Cohen et al. (1979), Coleman and Sauter (1978), K.U. Smith (1979, 1988, 1990), M.J. Smith (1994), M.J. Smith and Beringer (1987), and T.J. Smith et al. (1983).

The concept of worker self-regulation of hazard management is based on the premise that compared with managers and safety professionals, workers often know as much or more about their jobs, about job-related hazards that they encounter on a daily basis, and about how to reduce or eliminate those hazards. Four hazard management strategies based on this premise, three of which explicitly involve the participatory approach, are summarized below: (1) the worker hazard survey, (2) measures that encourage worker self-protection against workplace hazards, (3) worker involvement in safety program decision-making, and (4) ergonomic intervention to improve hazard management.

C. *Worker hazard surveys.* The purpose of the worker hazard survey, developed by K.U. Smith and colleagues (K.U. Smith, 1979, 1988, 1990; M.J. Smith, 1994; M.J. Smith & Beringer, 1987; T.J. Smith et al., 1983), is to discover what workers know about job hazards and to establish what knowledge, skills, techniques, and actions workers employ in detecting and controlling hazards. Survey information is acquired through short interviews conducted by unbiased interviewers. The worker interview should focus on (1) identification of job hazards perceived by the worker, itemized in order of severity, (2) worker rating of each hazard (high-, moderate-, or low-risk for causing accident or injury), (3) methods used by worker to detect and control identified hazards, (4) worker recommendations for reducing or eliminating hazards, (5) worker ideas and recommendations for improving performance skills required to control identified hazards, and (6) description of near, minor, or major accidents associated with identified hazards.

Generally, worker hazard survey results reveal that employees are one of the best sources of information about day-to-day hazards of their jobs. Major conclusions supported by this work can be summarized as follows (Cleveland, 1976; Coleman & Sauter, 1978; Coleman & Smith, 1976; Gottlieb, 1976; Kaplan & Coleman, 1976; Kaplan et al., 1976; Richardson, 1973; M.J. Smith, 1994):

- Workers generally identify hazards that most directly affect their personal safety. Unsafe conditions and operating procedures often are specified, but it is not uncommon for possibly hazardous actions or behaviors to be candidly cited.
- Worker-identified hazards largely comprise those of immediate self-concern. Whereas government standards are specification-based, tending to focus on physical-environmental hazards, workers tend to identify a broader range of hazards, often related to job performance, to dynamic operational circumstances, and/or to intermittent or transitory work situations or conditions for which no physical standards or standard operating procedures exist.
- Based on worker hazard survey results from six Wisconsin metal processing plants (Cleveland, 1976; Coleman & Sauter, 1978), employee-identified hazards bear a significantly closer resemblance to actual causes of accidents than do those designated by government specification standards or state inspections.
- Experience with the worker hazard survey method suggests that for every 100 hazards reported by employees, one is very serious and

should be addressed immediately, 24 require prompt attention to avert a potential accident, 50 represent less serious hazards that may require some minor action to improve working conditions, and 25 relate not to safety but to employee personal complaints (M.J. Smith & Beringer, 1987).

- Workers typically can clearly delineate their actions in avoiding, reducing, or controlling hazards. Workers tend to identify health hazards in terms of their acute toxic effects rather than their long-term potential for causing illness or disease. There is substantial awareness of engineering defects in ventilation or personal protection that may cause increased risk of toxic exposure.

- Employers cooperating in worker hazard surveys generally conclude that (1) employees can provide more and better information about job hazards than other available sources, (2) workers often identify hazards that management is totally unaware of, (3) survey results are of value in establishing priorities for safety and hazard management procedures, (4) the surveys do not tend to disrupt work, and (5) employee awareness of hazards typically increases following a survey.

Worker hazard survey results support the conclusion that workers represent a relatively untapped resource for hazard control. A few others in the safety community have reached this same conclusion. Swain (1974) refers to workers as subject-matter experts on their work situations whose input to management can serve as a key contributor to safety performance. He notes that worker hazard data represent a rare but effective source of information about near accidents and potentially unsafe situations. Hammer (1972, 1976) likewise presents evidence showing that workers have high awareness of job-related hazards and that asking them directly represents the best way of obtaining this information. Both authors note marked discrepancies between worker-provided hazard data versus those provided by more conventional specialist reports and post-accident investigations.

D. *Participation by workers in hazard management decision-making.* The thesis that workers should be actively involved in the process of making decisions about matters governing their own safety in the workplace is an implicit feature of the control systems model of hazard management in Figure 6.2. Evidence cited in point C demonstrates that workers have capabilities for both recognizing sensory feedback from hazards and developing behavioral proficiency in hazard feedback control. Therefore, from a cybernetic perspective (Juran, 1964; T.J. Smith & Smith, 1987a), it is unlikely that safety will be effectively served by a process that confines safety-related decisions exclusively to management, with no opportunity for worker input. As Viaene (1980, cited in Bryce, 1981, p. 1) puts it, "one of the most important trends that has emerged from the recent reforms is that of the transfer of power and responsibility on an ever-growing scale to those who are, fundamentally, the people who have to face these hazards."

Bryce (1981) and the U.S. Department of Labor (1989) note that the joint labor-management occupational safety and health committee represents one of the most effective avenues for worker influence over safety decision-making. Generally speaking, these committees (to varying degrees in different countries) are aimed at advancing the right of the worker to participate in occupational safety and health programs and in setting safety guidelines and standards, have access to job-related safety and health information and relevant training, and refuse to participate in unsafe work and to avoid penalty if this right is exercised (LaBar, 1990). The U.S. Department of Labor survey (1989) notes that there are thousands of such committees in the United States (most established after passage of the 1970 OSHA Act), and describes six case studies that document the relative success of these committees in achieving labor-management cooperation in dealing with safety problems.

E. *Ergonomic intervention to improve hazard management.* One of the premises of the model in Figure 6.2 is that ergonomic analysis and intervention can support breakthrough improvement in system safety by allowing the system to project the trajectory of its safety performance into the future in a feedforward manner through use of ergonomic analysis to identify and evaluate workplace and system hazards, followed by ergonomic intervention to abate targeted hazards. Behavioral cybernetic theory assumes that variability in safety performance should be influenced by the design of the performance environment with which the performing system interacts, just as is observed with individual and organizational performance generally (Section 6.4, point D). Ergonomic intervention strategies aimed at improving work design deficits therefore should beneficially impact variability in safety performance in a manner that complements and extends the impact of traditional safety programs. Various lines of evidence support this viewpoint.

Specifically, Cohen et al. (1997, pp. 114–118) review 24 ergonomic intervention studies reported since 1979, all of which were directed at reducing the incidence of musculoskeletal injuries and discomfort through implementation of microergonomic or macroergonomic redesign strategies. Tangible improvement in various measures of safety performance was cited in every study.

Similarly, Westgaard and Winkel (1997) review findings from 92 ergonomic intervention studies published between 1969 and 1995, all directed at improving musculoskeletal health. The studies are categorized into interventions to reduce musculoskeletal injuries using one of three strategies: (1) redesign of workstations, work methods, tasks, jobs, or work places to reduce exposure to mechanical hazards, (2) redesign of production systems, often with a dual aim of increasing productivity, and frequently involving participatory ergonomics, and (3) training/treatment methods, encompassing physiotherapy, health education, exercise, relaxation training, work technique training, or combinations thereof. Seventy-four of the 92 studies reviewed report reduction of musculoskeletal injuries and/or discomfort attributable to the intervention.

Subsequently, Karsh et al. (2001) reviewed results from 101 different studies published through 1998 to examine the efficacy of workplace ergonomic interventions to control musculoskeletal disorders. Eighty-four percent of all of the studies found positive results, although in a majority of these cases, results were mixed.

The most comprehensive investigation into the efficacy of different types of interventions—including ergonomics—in preventing occupational accidents is that of Guastello (1993, 2013). The earlier report reviews findings from studies of 48 different intervention programs, each of which employed one of eight different intervention strategies for accident prevention: personnel selection, technological interventions, behavior modification (safety training, incentives, goal setting), poster campaigns, comprehensive ergonomics programs, quality circles, safety audits, and exercise/stress management. Results from this analysis are shown in Figure 6.3. Ergonomic intervention was the most effective in accident reduction (51.6 percent average reduction in accidents). The remaining strategies were less effective.

The 2013 chapter of Guastello updates the earlier report with a more extensive analysis. A total of 158 studies of occupational accident prevention were evaluated, encompassing the first five of the eight intervention strategies cited above, plus safety committee, medical management, near-miss accident reporting, other management intervention, and governmental intervention strategies. Based on this updated analysis, technological interventions, behavior modification, comprehensive ergonomic interventions, and other management interventions had the highest success in occupational accident prevention, with average accident reductions ranging from 53 and 55 percent in each case (study numbers for each of these four strategies were five or higher). All other strategies featured notably lower accident prevention success.

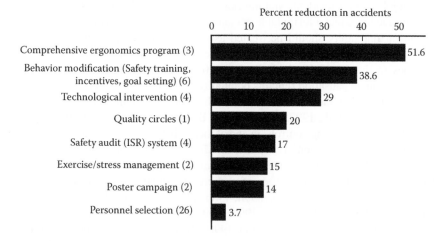

FIGURE 6.3 Effectiveness of different occupational accident prevention strategies (listed on the ordinate) in reducing the incidence (percent values on abscissa) of occupational accidents. Numbers next to each strategy on the ordinate refer to the number of studies evaluated to derive the average accident reduction percentage. (Graphical transformation of results tabulated by Guastello, S.J. (1993). Do we really know how well our occupational accident prevention programs work? *Safety Science*, 16, 445–463.)

Notably, two studies of ergonomic intervention targeting musculoskeletal injuries resulted in 77.5 percent reduction in accidents; this result complements the analysis of Karsh et al. (2001) cited above.

6.8 CONCLUSIONS

Few manifestations of human performance variability attract more attention on the part of the general public than the occurrence of error, particularly of major proportions. In the domain of human work, the ostensible objectives of occupational safety programs are to promote the establishment of relatively predictable and stable work environments, and thereby to achieve greater control over variability in error performance (Hollnagel, 2002) and in the incidence of occupational accidents and injuries. However, strategies adopted to achieve these objectives have featured mixed success.

Thus, some four decades ago Ellis (1975) surveyed the efficacy of five alternative approaches to occupational accident and injury prevention. Conclusions reached were that (1) sound evidence supporting the safety effectiveness of inspections is lacking, (2) complete abatement of all hazards identified by inspection would prevent only about one-fourth of all occupational injuries, (3) evidence regarding the safety benefits of training is mixed, (4) safety benefits of accident statistics feedback is uncertain, (5) accident and injury prevention outcomes of safety and hazard management programs are consistently positive despite a diversity of approaches, and (6) no solid evidence exists that workers' compensation yields safety benefits.

Many other observations and lines of evidence support the pattern of inconsistency and contradiction inherent in the findings of Ellis (1975). For example, there have been dramatic improvements in workplace safety over the course of the century, yet job-related accidents and injuries, particularly of the musculoskeletal variety, continue to extract a formidable personal and financial toll. There is widespread agreement that safety should be treated as an integral management function. Yet the philosophy of safety engineering, pioneered by Heinrich almost 70 years ago continues to dominate the field (Pope, 1990). The unsafe acts/unsafe conditions model of accident causation remains popular (Section 6.6). Engineering controls are emphasized as a key factor in the success of both ergonomic intervention and safety programs (Cohen et al., 1997; M.J. Smith & Beringer, 1987), yet various observers conclude that engineering controls, in and of themselves, do not dramatically improve safety performance (Cohen, 1977; Gill & Martin, 1976; Jones, 1973). Behavioral modification through safety incentive programs (promotions, slogans, and/or performance feedback) as a key strategy for improving safety performance, encounters praise (Bruening, 1989, 1990a; Kendall, 1986, 1987; Krause et al., 1990; McAfee & Winn, 1989; Moretz, 1988; Reason, 1990; Topf & Preston, 1991), conditional support (LaBar, 1989), or skepticism (Bruening, 1990b; Kohn, 1993). An objective survey of the efficacy of various accident prevention strategies shows a mixed record for different behavioral modification approaches (Guastello, 1993, 2013). Safety training is lauded as a key to effective safety performance (Cohen, 1977; Reid, 1987; M.J. Smith & Beringer, 1987; M.J. Smith et al., 1978), despite evidence showing that training does not compare favorably with other macroergonomic strategies for safety improvement (Ray et al., 1997).

From a behavioral cybernetic perspective, these contradictions and inconsistencies are strongly suggestive that safety performance, like individual, social, and organizational performance, is context-specific (Section 6.3), specialized in relation to design features of particular work environments. That is, variability in safety performance of different operations should be prominently influenced by distinctive design contexts of their physical, operational, and organizational environments. Interventions directed at altering the design of the work environment and/or the work system for purposes of hazard abatement thus should differentially affect safety performance of different operations in different ways. This is exactly what is observed.

However, unlike the analysis under controlled experimental conditions of cognitive performance variability presented in Chapter 3, the human error observations outlined in this chapter are not underpinned by definitive scientific evidence regarding the role of context specificity in human error performance variability. As such, the evidence cited must be considered less scientifically definitive and subject to a broader range of possible interpretations than that compiled under more controlled circumstances.

There is no intent in this chapter to dismiss the concept of human error. As Dekker (2007) points out:

> To the naive realist, the argument that errors exist is not only natural and necessary—it is also quite impeccable. The idea that errors do not exist, in contrast, is unnatural. It is absurd. Those within the established paradigm will challenge the legitimacy of questions raised about the existence of errors and the legitimacy of those who raise the questions: 'Indeed, there are some psychologists who would deny the existence of errors altogether. We will not pursue that doubtful line of argument here' (Reason & Hobbs, 2003, p. 39).

Instead, the intent is to suggest, based on arguments advanced in previous sections, that the emphasis on "human" in the term human error, stigmatizing the worker or operator as a major or exclusive contributor to error causation, has reached its expiration date. The alternative viewpoint advanced here is that design factors are implicated in error causation in many if not most cases, organizational and management designs bear a major responsibility for error causation in the preponderance of cases, and system design factors contributing to error causation represent hazards that typically can be effectively managed and controlled through proven hazard management methods and procedures. To put it succinctly, human error and design error are two sides of the same coin.

In summary, major conclusions from the foregoing analysis are (1) behavior is organized as a closed-loop, feedforward control process, allowing the individual to predict the sensory consequences of behavioral activity, (2) error control is inherent to behavioral control—if design factors in the performance environment compromise operator ability to predict sensory consequences of behavior, then the likelihood of error increases, (3) variability in behavioral control is tightly referenced to what is being controlled; namely, sensory feedback from design factors in the performance environment, (4) behavioral performance variability, including modes of variability associated with error outcomes, consequently is highly context-specific, prominently influenced by design features of the task, workplace, and/or system (two sides of the same coin), (5) context specificity in error performance implies that improving design features of the performance environment represents a potent strategy for reducing error, and (6) the potential for greater variability in behavioral control, and therefore for error control, is demonstrably higher in social and organizational contexts relative to individual tasks—patterns and levels of error thus are prominently influenced by how performance is socially and organizationally designed.

Implications for understanding human error based on these conclusions are

- Generalized domino (Heinrich, 1941) and Swiss cheese (Reason, 1990) metaphors should be replaced with a more rigorous control systems understanding of human error.
- Given that behavioral and error control are inherently variable, safety programs that specify zero error targets may be deemed practically unattainable and scientifically untenable. A quality management approach that emphasizes continuous improvement, coupled with statistical tracking of hazard and error reduction strategies, is far more realistic.
- Human reliability models assume that the likelihood of human error can be predicted in different contexts (Kirwan, 1992a, b, 2005; Miller & Swain, 1987). However, evidence regarding context specificity in error performance (Section 6.3) calls this assumption into question, in that patterns of behavioral variability observed in one design context do not necessarily predict those observed in other design contexts.
- Managing hazards through ergonomic intervention rather than behavioral modification programs typically offers a more promising strategy for reducing human error.
- As Maurino (2000) notes, human error must be confronted and understood in its social and organizational contexts.

7 Variability in Affective Performance

Thomas J. Smith

7.1 INTRODUCTION

This chapter deals with an area of behavioral expression and performance variability that has a far more ancient pedigree than modes addressed in previous chapters. According to Webster, (McKechnie, 1983), "affect" is defined (in the psychological sense) as feelings or emotions. This chapter focuses on personality and emotions. The former term is defined (in the psychological sense) as the sum total of the physical, mental, emotional, and social characteristics of an individual, resulting in the organized pattern of behavioral characteristics of the individual. The term "emotion" is defined as an affective state of consciousness (as distinguished from cognitive and volitional states), usually accompanied by physiological changes as well as overt manifestations of behavioral expression. Well into the twentieth century, a text on psychology by one of the leading scientists in the field (Woodworth, 1938), devotes attention to personality and emotions, but no mention is made of cognition.

Other modes of affective behavior are considered to be distinct from, yet at the same time to embody, emotional expression. *Moods* may share behavioral-physiological manifestations with emotions, but are assumed to have longer durations and lower intensities (Ekman, 1992). *Feelings* are understood as subjective, internalized representations of emotions. *Attitudes* are defined as manners, dispositions, feelings, and/or positions toward a person or thing.

There is a vast literature on affective behavior. A search of books on Amazon (as of August 2013) yields over 36,600 hits for the term "personality," over 33,500 hits for the term "emotions," almost 10,500 hits for the term "mood," and over 42,800 hits for the term "attitude."

This chapter will limit itself to a consideration of three basic questions pertaining to behavioral expression linked to emotions and personality: (1) what is the nature of performance variability, (2) what are the sources of performance variability, and (3) can performance variability related to the behavioral expression of personality be understood from a behavioral control systems perspective? The analysis addressing the second question will include an inquiry into whether context specificity contributes to performance variability observed.

These issues are examined in the remaining sections of this chapter, and provide a historical perspective (Section 7.2), personality and performance variability (Section 7.3), performance variability in emotional behavior (Section 7.4), and conclusions (Section 7.5).

7.2 HISTORICAL PERSPECTIVE

The terms "personality" and "emotion" each have a Latin etymology, suggesting origins dating back one to two millennia. Historical interest in modes of behavioral expression implied by these terms has equally ancient origins.

Personality. The origin of the idea that differences in behavioral expression are associated with what came to be called personality is attributed to Hippocrates (ca. 460 BC–ca. 370 BC) (Goldberg, 2006), with his theory of the four humors—black bile, yellow bile, phlegm, and blood. Hippocratic medicine maintained that these humoral substances are in balance when a person is healthy. Burch (2013) points out that the practice of bloodletting to treat illness was introduced because, among these four humors, blood levels were the most amenable to adjustment to restore humoral balance, and thus to restore health. This practice persisted for some two thousand years. For example, Burch suggests that the death of George Washington in 1799 likely was hastened, if not caused, by excessive bloodletting to "treat" his severe throat infection.

About half a millennium after Hippocrates, Galen (131–201 AD) (Bendick, 2002) proposed the first explicit categories of personality traits by associating what he called different *temperaments* with the four Hippocratic humors. Specifically, a sanguinary temperament was linked to blood humor, choleric temperament to yellow bile humor, melancholic temperament to black bile humor, and phlegmatic temperament to phlegm humor.

The first reasonably comprehensive effort to categorize human personalities into different particular traits was that of Galton (1883), who perused *Roget's Thesaurus* in an attempt to count the number of different words used to express the more distinctive aspects of human character (termed the psycholexical approach). He determined that there were one thousand such character traits with somewhat separate, but also somewhat overlapping, meanings (Allport & Odbert, 1936, p. 22).

Emotions. Theories in ancient Greece about the existence and roles of different types of emotions, as was also the case for personalities, likely were inspired by the humoral theory of Hippocrates (Goldberg, 2006) (above). Aristotle (384–322 BC) believed (Irwin, 1999) that emotions (which he termed passions) corresponded to appetites, and represented an essential component to virtue. Arguably the next major advance in this area was Darwin's evolutionary theory of emotions (1872), advancing the view that emotions have adaptive, and therefore survival, value.

An emotion is the only mode of behavior that has been recognized as both a fundamental national right and as a national aspirational goal. The pursuit of happiness is specified as one of the unalienable rights in the U.S. Declaration of Independence. It is likely that the lofty ideals of the French enlightenment provided the inspiration for Jefferson to specify this right (Kors, 1998, Lecture 21). In the Kingdom of Bhutan, Gross National Happiness is tracked by the government as a national aspirational goal.

7.3 PERSONALITY AND PERFORMANCE VARIABILITY

What is the nature of variability in human personality? No area of psychology has appealed more to the investigative impulse for categorization into different varieties than affective behavior. In the case of both personality and emotions, trait-based analyses abound, with their origins in Hippocrates, Aristotle, and Galen as noted above. A trait-based classification scheme for personality categories that has attracted much attention in recent decades is the so-called big five or five factor model of personality traits: openness to experience, conscientiousness, extraversion, agreeableness, and neuroticism. Digman (1990) suggests that this scheme represents what he calls a "grand unified theory of personality." However, the model has attracted criticism for its lack of scientific validity (Block, 2010) for a series of reasons that include (1) it is not based on underlying theory, (2) its exclusive reliance on factor analysis for classifying personality traits, (3) a lack of scientific consensus as to what the factors mean, and (4) the existence of other personality traits not subsumed by the model.

Another popular personality trait-based classification scheme is the Myers-Briggs Type Indicator (MBTI) test (Myers & Myers, 1980). The MBTI is based on the original theory of Carl Jung that there are four basic modes of behavior-environment interaction (sensation, intuition, feeling, and thinking) by which the world is experienced. This idea has evident parallels with the original idea of Locke that all knowledge is derived from experience (Chapter 3, Section 3.2). Myers and Briggs expanded on Jung's theory (in the 1940s) by postulating four dichotomous pairs of psychological types; namely, extraversion (E)-introversion (I), sensing (S)-intuition (N), thinking (T)-feeling (F), and judging (J)-perception (P). Self-report answers to the 93 forced-choice questions on the MBTI test are intended to reveal which member of each dichotomous pair is favored more prominently by the respondent. Test results are scored by identifying the preferred member of each pair to yield one of sixteen (2^4) possible type combinations (examples: ESTJ or INFP) that are assumed to be related to the dominant personality type of the respondent.

McCrae and Costa (1989) have correlated results from personality categorization using the five factor model (above) with those from the MBTI test. The analysis showed that four MBTI indices do measure aspects of four of the five major big five dimensions of normal personality. The authors conclude that the five factor model provides an alternative basis for interpreting MBTI findings within a broader conceptual framework.

As with the five factor model, the MBTI test has attracted criticism for lack of validity. Observations supporting this criticism include: test scores for the different MBTI dichotomous pairs (above) show a normal rather than the expected bimodal distribution, suggesting a lack of clear distinction between the traits specified in each pair (McCrae & Costa, 1989), and the construct of the test is such that respondents may not only have the latitude, but actually may be motivated, to fake their responses (Furnham, 1990). Validity concerns are nicely encapsulated in the following conclusion by Krznaric (2013):

> The interesting—and somewhat alarming—fact about the MBTI is that, despite its popularity, it has been subject to sustained criticism by professional psychologists for over three decades. One problem is that it displays what statisticians call low "test-retest reliability." So if you retake the test after only a five-week gap, there's around a 50% chance that you will fall into a different personality category compared to the first time you took the test.

Yet a third popular trait-based personality classification scheme is the Minnesota Multiphasic Personality Inventory (MMPI). Unlike the five factor model and the MBTI test discussed above, the MMPI features more of a focus on psychopathological personality assessment. The origins of the MMPI date back to the late 1930s, but the test was substantially revised and restandardized in the late 1980s and termed the MMPI-2 (Rogers et al., 2006). The original MMPI, devised by Hathaway and McKinley (1940), featured 504 true-false test items grouped into 25 personality scale categories. The MMPI-2 features 564 true-false test items grouped into nine scale categories.

Two key criticisms of the MMPI have been raised, namely: (1) questions of comparability between the original version and MMPI-2 in terms of reconciling results obtained with the two test versions, and (2) consistently higher psychopathologic personality scores for blacks relative to whites on the test.

Another personality trait-based classification scheme worthy of mention because it does not rely on any of the foregoing instruments plus being aimed specifically at the HF/E community is that of Schuster and Guilford (1964). The purpose of this instrument was to profile personality differences across 24 different personality trait categories based on responses to 395 forced-choice questions in order to discriminate between problem drivers and better-than-average drivers. Results from administration of this instrument to some 2000 drivers support the conclusion that useful degrees of discriminatory prediction were demonstrated using the personality scales employed.

It should be emphasized that the investment by personality psychologists in delineating personality taxonomies has proved to be of more than academic and clinical interest. As Hough and Schneider (1996) point out, after a series of decades in which personality trait-based analysis for purposes of predicting job performance was relegated to the wilderness of industrial and organizational psychology, this approach has enjoyed a renaissance since the early 1990s. Based on their analysis, these authors (Hough & Schneider, 1996, pp. 75–76) reach the following conclusions regarding the link between personality and job performance:

> Both basic personality traits and compound personality traits predict job performance criteria. For example, conscientiousness, achievement, integrity, and service orientation predict performance across many occupations. Several other personality traits, however, correlate substantially with job performance when other factors are considered. Type of job criterion construct, criterion measurement method (objective versus subjective), validation strategy (for several variables), and type of situation affect personality-performance relationships.

The foregoing synopsis of alternative approaches to the analysis of human personality differences based on trait-based criteria supports at least three key conclusions. First, there is no close agreement across the different approaches as to the nature and extent of variability in the behavioral expression of personality insofar as personality traits are concerned. The analysis of Digman (1990) seemingly contradicts this conclusion with the claim that the five factor model represents a "grand unified theory" of personality, such that personality traits identified by other observers can be mapped onto the five personality factors specified in the model. The analysis of McCrae and Costa (1989), showing that MBTI traits can be correlated with those specified in the five factor model, appears to support this claim.

However, the five factor model, along with the MBTI and the MMPI constructs of personality differences, each has encountered recurrent criticism as to lack of validity, raising the question as to whether any of these approaches (as well as others not addressed in this section) provide meaningful insight into variability in the expression of personality.

Nevertheless, reservations on the part of some observers about trait-based approaches to personality assessment have not hindered other observers from adopting such schemes. These efforts encompass an immense range of trait assignments. At the minimalist end, there are the Galen's four temperaments and the five factors of the eponymously named model. A bit more ambitious are the four dichotomous trait pairs, or the eight distinct traits, of the MBTI. The MMPI-2 assumes nine personality trait categories, less than half of the categories assumed in the original MMPI. Schuster and Guilford (1964) assume 24 personality trait categories.

Thereafter, the number of different personality trait assignments mounts rapidly. Allport and Odbert (1936, pp. 22–23) report personality trait numbers assumed by six different analyses that range from 1000 to 3000. First prize goes to Allport and Odbert themselves, who list a total 4054 different personality traits. These numbers strike me as a scientific dedication on steroids to categorizing behavioral performance variability. Whether four or four thousand, the obvious question is, how shall we account for the behavioral manifestation of different personalities?

What are the sources of variability in human personality? As with the impulse for behavioral categorization, affective behavior also is the poster child for the tension between nativist versus empiricist perspectives (Goldhaber, 2012) on sources of behavioral performance variability. In the case of personality, this question prompted one of the more acrimonious debates in the modern era of behavioral science. Specifically, battle lines were drawn between those who believed that personality traits are immutable, grounded in innate biological and genetic mechanisms (the nativist perspective), and those who believed that personality traits largely if not exclusively arise as a consequence of individual interaction with what researchers in the field termed "situations" (i.e., context) (the empiricist perspective).

Well over a century ago, James (1890) set the stage for the nativist versus empiricist views on personality variability with two observations that appear to present conflicting perspectives on these views. His nativist perspective, seemingly favoring the idea that personality is largely immutable, is embodied in the following quote:

> Habit is thus the enormous fly-wheel of society… Already at the age of twenty-five you see the professional mannerism settling down on the young commercial traveler, on the young doctor, on the young minister, on the young counselor-at-law. You see the little lines of cleavage running through the character, the tricks of thought, the prejudices, the ways of the "shop," in a word from which the man can by-and-by no more escape than his coat-sleeve can suddenly fall into a new set of folds… in most of us, by the age of thirty, the character has set like plaster, and will never soften again.

His empiricist perspective, seemingly favoring the idea that personality is more malleable, is embodied in the following quote:

> Many a youth who is demure enough before his parents and teachers, swears and swaggers like a pirate among his "tough" young friends. We do not show ourselves to our children as to our club-companions, to our customers as to the laborers we employ, to our masters and employers as to our intimate friends.

It is fair to say that these somewhat conflicting views on the nature of personality persisted for many decades after James. What has changed more recently is recognition that, essentially, both views are correct.

I rely on two reports that articulate this more recent perspective. Kenrick and Funder (1988) introduce their report by noting the following:

> For the past 2 decades the person–situation debate has dominated personality psychology and had repercussions in clinical, social, and organizational psychology. This controversy puts on trial the central assumption that internal dispositions have an important influence on behavior. According to emerging views of scientific progress, controversy serves the function of narrowing the field of competing hypotheses. We examine 7 hypotheses that arose during the course of the person–situation debate, ranging from most to least pessimistic about the existence of consensual, discriminative personality traits. The evidence fails to support the hypotheses that personality traits are (a) in the eye of the beholder, (b) semantic illusions, (c) artifacts of base-rate accuracy, (d) artifacts of shared stereotypes, (e) due to discussion between observers (who ignore behavior in favor of verbal self-presentation or reputation), or (f) by-products of situational consistencies. Evidence also fails to support the hypothesis (g) that although traits are related to behavior, the relationship is too small to be important.

A more recent synopsis by Funder (2001) adds the following additional observations:

- Personality psychology is as active today as at any point in its history
- Old empirical issues such as the person-situation debate…continue to simmer
- [Nevertheless]…the person-situation debate, concerning whether consistencies in individuals' behavior are pervasive or broad enough to be meaningfully described in terms of personality traits…can at last be declared about 98 percent over
- …behavior of a sample of individuals observed in one situation correlates with their behavior in a second situation with a magnitude that routinely reaches $r = .40$ or greater
- Even in the darkest days of the person-situation debate, personality trait constructs found an appreciative audience and useful application in industrial and organizational settings
- The long-standing and controversy generating dichotomy between the effect of the situation versus the effect of the person on behavior, therefore, is and always was a false dichotomy
- Biology and evolutionary theory have also attained the status of new paradigms for personality

With regard to the last point, Funder (2001) points out that (1) anatomical sites within the brain have been located that are important for personality traits; for example, the frontal lobes for foresight and anticipation…and the amygdala for aggression and certain types of emotionality, (2) even more impressive have been the contributions of physiology that show how, for example, the hormone testosterone is important for sociability and positive affectivity as well as aggressiveness and sexuality…, and the neurotransmitter serotonin is important for affect regulation; (3) behavioral genetics has documented, without a shadow of a remaining doubt, that personality is to some degree genetically influenced…identical twins reared apart have similar traits; (4) the evolutionary perspective on personality is that humans with certain behavioral propensities were particularly likely to survive and leave descendants; this perspective has been questioned on several grounds—for example, evolutionary theorists seem quick to assume that specific behavioral patterns are directly determined by biological mechanisms, yet no such mechanism that would allow genetic or physiological determination of human behavior to such a precise degree has ever been specifically identified; and (5) the *tabula rasa* view of personality as a blank slate at birth that is written upon by experience, for many years a basic assumption of theories of all stripes, is wrong.

More recently, the nativist perspective on personality has been extended to the idea that competency and success in the performance of work have a biological basis. This idea is summarized by *The Economist* (2010a) in relation to administrative achievement, and by Shane (2010) in relation to leadership prowess.

In contrast to these arguments for a nativist perspective on personality variability, the analysis of Otto and Smits (2011) may be cited in favor of an empiricist perspective. These authors review the results from dozens of published population-based and clinical studies and conclude that the beneficial effects of regular exercise on mild to moderate forms of depression is similar to the effect of cognitive behavioral therapy. Beneficial effects of exercise on mood and in reducing anxiety also are noted. Choosing to engage in exercise, of course, represents a design factor.

The collective impression left by recent research on the psychology of personality, embodied in the reports cited above is that context specificity as well as innate neural, physiological, and behavioral genetic factors both contribute to the overall variability observed in the behavioral expression of personality. This conclusion is aligned with comparable conclusions reached in preceding chapters dealing with cognitive, children's learning, and human error performance. That personality is context-specific, at least in part, implies that a feedback relationship exists between the development and expression of different behavioral manifestations of personality and the design of the environment in which behavior takes place. The implications of this idea are explored below.

Can personality be understood from a behavioral control systems perspective? From a control systems perspective, the contribution of context specificity to performance variability is attributed to behavioral control of sensory feedback from design factors in the performance environment, resulting in specialization of behavior (Chapter 1, Section 1.3.2). K.U. Smith and Smith (1973, p. 373) apply this interpretation to the development of personality using a social cybernetic (see Chapter 8) analysis. These authors point out that, "a person to some extent regulates his own personality development by using his perceptual-motor mechanisms to track the social objects in his environment and thus control his own social learning." Three types of habits are assumed to result from self-governed social learning; namely (1) individualized modes of physiological, emotional, and overt behavioral responses, (2) open-loop tracking of others without engaging in mutual social interaction (audience behavior is an example), and (3) closed-loop social tracking of others involving mutual social interaction (teacher-student, team sport, sexual, and occupational interactions are examples). Collectively, these modes of self-regulated social behavior—all of them context-specific—likewise are assumed to operate in a feedback fashion to guide the development and establishment of distinctive manifestations of personality.

Bandura (2001) essentially recapitulates this control systems idea of the role of social learning in personality development with what he terms the social cognitive theory of personality. This author emphasizes the role of self-regulation in defining personality as a result of the interaction of the person and his or her environment, which allows self-control through self-reward and self-punishment—a possible basis of moral behavior. However, the earlier social cybernetic perspective of Smith and Smith (1973) that originated and articulated what essentially amounts to the same control systems interpretation of personality development, is not acknowledged by Bandura (2001).

7.4   EMOTION AND PERFORMANCE VARIABILITY

It is appropriate to open this appraisal of variability in human emotions by quoting Plutchik (2001), who introduces his review on the nature of emotions as follows:

> Almost everyone agrees that the study of emotion is one of the most confused (and still open) chapters in the history of psychology. By one estimate, more than 90 definitions of "emotion" were proposed over the course of the 20th century. If there is little consensus on the meaning of the term, it is no wonder that there is much disagreement among contemporary theoreticians concerning the best way to conceptualize emotion and interpret its role in life.
>
> In everyday human existence we conceive of an emotion—anger, despair, joy, grief—as a feeling, an inner state. The internal experience of emotion is highly personal and often confusing, particularly because several emotions may be experienced at the same time. Imagine, then, how difficult the objective study of emotion must be. Most of us often censor our own thoughts and feelings, and we have learned to be cautious about accepting other people's comments about their feelings. The empirical study of a psychological phenomenon so complex and so elaborately cloaked cannot help but present a special challenge.

Given this cautionary note, the analysis below, as with that for personality variability in the preceding section, will confine itself to a consideration of the nature and sources of variability in human emotions and to an inquiry as to whether emotional variability can be understood from a behavioral control systems perspective.

What is the nature of variability in human emotions? Unlike perspectives on personality variability (Section 7.3), there appears to be a broader scientific consensus regarding what expressions of behavior represent what we call emotions. As with the case for personality (above), James (1884) ushered in modern scientific inquiry into emotions with his psychological theory of emotion. The substance of this theory is addressed below as part of the analysis of sources of variability in human emotions. In developing his theory, James specified two different classes of emotions, namely those that do not have a distinct bodily expression versus those that do. In the former category he includes four emotions: pleasure, displeasure, interest, and excitement. In the latter category, whose members he terms "standard" emotions, he specifies seven emotions: anger, curiosity, fear, greed, lust, rapture, and surprise.

So-called basic or primary emotions recognized by other observers display some differences, but also a degree of overlap, with the earlier formulation of James (1884). A survey of specifications of the basic emotions by a selected sample of authors is provided below. In some cases, additional types of emotions, claimed to be related to the basic emotions, also are specified.

- Russell (1980) specifies eight basic emotions: arousal, contentment, depression, distress, excitement, miserable, pleased, and sleepiness. He also specifies 20 additional emotions that subjects in his study were asked to associate with the eight basic ones.
- Ekman (1992) specifies 10 basic emotions: anger, contempt, disgust, embarrassment, enjoyment, fear, guilt, sadness, shame, and surprise.
- Cacioppo and Gardner (1999) specify seven basic emotions: anger, disgust, fear, guilt, joy, sadness, and shame.
- Plutchik (2001) specifies eight basic emotions: admiration, amazement, ecstasy, grief, loathing, rage, terror, and vigilance. He also specifies 34 additional emotions that he associates, to varying degrees, with the eight basic ones.
- Lottridge et al. (2011) specify six basic emotions: anger, disgust, fear, happiness, sadness, and surprise.

Plutchik (2001) notes that, "Over the centuries, from Descartes to the present, philosophers and psychologists have proposed anywhere from 3 to 11 emotions as primary or basic. All the lists include *anger, fear* and *sadness*; most include *joy, love* and *surprise*." How do specifications of basic emotions by the authors cited above align with this claim? All authors cited above except Russell (1980) specify *anger, fear,* and *sadness* (Plutchik himself (2001) equates *rage* with *anger, terror* with *fear,* and *grief* with *sadness*). Only Cacioppo and Gardner (1999) and Plutchik (2001) specify *joy* (Plutchik himself equates *ecstasy* with *joy*). Only Plutchik (2001) specifies *love* (which he equates with both *ecstasy* and *admiration*). Ekman (1992), Plutchik (2001), and Lottridge et al. (2011) specify *surprise* (Plutchik equates *amazement* with *surprise*).

This synopsis suggests that there is some, but not universal, agreement regarding what constitutes a basic emotion. Arguably, the observation of Plutchik (2001) that, "If there is little consensus on the meaning of the term, it is no wonder that there is much disagreement among contemporary theoreticians concerning the best way to conceptualize emotion," represents a reasonably accurate characterization of the current state of affairs. Nevertheless, when it comes to human emotions, it also is evident that the impulse for taxonomic categorization on the part of observers in the field is more constrained than is the case for human personalities (Section 7.3).

What are the sources of variability in human emotions? As with personality (Section 7.3), inquiry into the development and expression of emotion has featured both nativist and empiricist perspectives. Unlike the case for personality however, debate regarding sources of variability in emotional behavior has not been nearly as acrimonious, arguably because both perspectives have enjoyed support by prominent individuals in the scientific community for many decades. In his seminal 1872 treatise on human emotions, Darwin advocated what amounts to the nativist view with the argument that emotions evolved via natural selection and therefore feature innate and universal characteristics across different cultures.

Yet just over a decade later, William James (1884)—whose influence on the emergence of modern psychological science matches that of Darwin on the emergence of modern evolutionary science—advocated a theory of emotion that argued for the empiricist perspective. This view is encapsulated in the following statements from James' 1884 report:

> ...the only emotions I propose expressly to consider here are those that have a distinct bodily expression... My thesis...is that *the bodily changes follow directly the* PERCEPTION *of the exciting fact, and that our feeling of the same changes as they occur* IS *the emotion.*

In linking the development and expression of emotions to perception, the theory of James essentially extends the Lockean doctrine (1632–1704) (1974) about cognitive behavior (thinking arises entirely from experience) to emotional behavior.

One of the most dramatic lines of evidence supporting the original contention of Darwin (1872) regarding the universal nature of emotions comes from the study of facial expressions (Lottridge et al., 2011). This research shows that basic facial expressions, including happiness, sadness, fear, disgust, surprise, and anger, are consistently identified, regardless of the culture, ethnicity, and age of the participant or the subject being portrayed.

Theory and empirical evidence that emotional expression and physiology are closely linked (Cacciopo & Gardner, 1999; Darwin, 1872; Lottridge et al., 2011) has prompted investigation into the neural substrates of emotion. This topic is reviewed by LeDoux (1995). The emotion of *fear* receives the most attention in the analysis, that points out that neurophysiological study of fear dates back to the original work of Pavlov. LeDoux's neural model of emotion identifies the amygdala as a key brain center with an influence on emotion. However, this influence is distributed, given the reciprocal functional interactions of the amygdala with the sensory and perirhinal cortices, the hippocampus, and the nucleus basalis.

The idea that emotional behavior has a neural basis also has prompted recent study of cognitive-emotional interactions (Cacciopo & Gardner, 1999; Ekman, 1992; LeDoux, 1995; Lottridge et al., 2011). One point of view growing out of this analysis is that emotional behavior and cognitive behavior are inextricably linked (Helander & Khalid, 2006), and in fact may be one and the same. However, propagation of the idea that there is such a thing as emotional intelligence, measured as emotional quotient (EQ) (as opposed to the IQ measure of cognitive intelligence), suggests that emotion and cognition represent distinctly different modes of behavior (Lottridge et al., 2011). The EQ is broadly conceived as social and emotional awareness and interpersonal skills and is described as a set of abilities grouped into four components: (a) accurately perceiving emotions in the self and others, (b) using emotions to communicate and to facilitate thought, (c) understanding emotions, and (d) effectively managing emotions in oneself and others.

The synopsis above summarizes evidence supporting the nativist conclusion that emotional behavior has demonstrably innate or universal characteristics grounded in physiological, neural, and possibly even genetic (Caccioppo & Gardner, 1999; Ekman, 1992; LeDoux, 1995) mechanisms. However, in their summary of sources of individual differences in emotion, Caccioppo and Gardner (1999) point out that social and cultural (i.e., context-specific) factors also contribute to observed variability in the expression of emotions. These authors conclude their summary by noting, "...emotion is a short label for a very broad category of experiential, behavioral, socio-developmental, and biological phenomena."

This brings us to a consideration of the empiricist perspective on the role of context specificity in the development and expression of emotional behavior, for which supporting evidence, as noted in the preceding paragraph, also is compelling. The original theory of James (1884), that emotions are defined by behavioral interaction with design factors in the perceptual environment, has been bolstered by research over the ensuing 13 decades showing that perception of environmental design conditions consistently influences variability in the expression of emotional behavior (Caccioppo & Gardner, 1999; Lottridge et al., 2011; Plutchik, 2001; Russell, 1980).

An intriguing illustration of this point that is of particular relevance to the HF/E community is the review by Helander and Khalid (2006) of the link between product design and pleasurable emotional responses. Key points raised in this review include:

- During the past 10 years there has been a rapid growth in research concerning affect and pleasure (the terms *affect* and *emotions* are used interchangeably).
- That pleasure represents a fundamental emotion was recognized a full century ago.
- Emotions have become increasingly important and have gained significant attention in product semantics and interactive design.
- In emotional design, pleasure and usability should go hand in hand, as well as aesthetics, attractiveness, and beauty.
- Good design depends on both perceived usability and hedonic attributes involving the interplay between pragmatic attributes of usability, hedonic attributes of stimulation, attributes of goodness (i.e., satisfaction), and finally beauty.

- A review of the psychological literature on emotions shows that positive emotions such as joy, interest, contentment, love that share a pleasant subjective feeling have been marginalized in research compared to negative emotions.
- Affective evaluations provide a new and different perspective in human factors engineering. The research on hedonic values and seductive interfaces is a welcome contrast to safety and productivity, which have dominated human factors and ergonomics.
- Emotions are often elicited by products, such as art, clothing, and consumer goods; therefore, designers must consider affect and emotion in design. Emotion may be the strongest differentiator in user experience.
- There are many important reasons to consider emotion in product design, such as to increase sales and keep customers happy. This is done by maximizing positive emotions by while minimizing negative emotions. Understanding and reducing users' anxiety and fears (negative emotions) can help to increase satisfaction with products. Poor usability will also induce negative responses such as frustration, annoyance, anger, and even confusion.

The foregoing analysis indicates that both innate mechanisms and context-specific factors contribute to the overall variability observed in the behavioral expression of emotion. That emotions are context-specific, at least in part, implies that a feedback relationship exists between the development and expression of different behavioral manifestations of emotion and the design of the environment in which behavior takes place. The implications of this idea are explored below.

Can emotions be understood from a behavioral control systems perspective? The observation that perception and emotional expression are functionally coupled lends itself to a behavioral control system interpretation (Chapter 1, Section 1.3.3), that embodies a context-specific influence on variability in emotional behavior. Both Lottridge et al. (2011) and Plutchik (2001) recognize this point with closed-loop control system models of design-emotion interaction comparable to that illustrated in Figure 1.8. Figure 7.1 is a reproduction of Plutchik's (2001) model of this interaction. It assumes that a stimulus event (from a design factor in the performance environment) evokes an emotional effect behaviorally specialized for feedback control of the sensory stimulation. The fear response to a dangerous situation (LeDoux, 1995) or the pleasure response to an attractive design (Helander & Khalid, 2006) both

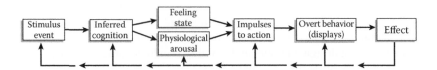

FIGURE 7.1 A control systems model illustrating a closed-loop feedback link between emotional expression (the effect) and one or more environmental design factors (stimulus event) evoking that expression. (Reproduced from Plutchik, R., *American Scientist, 89*(4), 344–350, 2001. With permission.)

illustrate this type of behavioral control systems coupling. In Figure 7.1, note that Plutchik (2001) infers a cognitive influence on the emotional response.

7.5   CONCLUSIONS

There is agreement among many students of affective behavior that affective performance and cognitive performance are interrelated. At the same time, there is broad consensus in this community, with Darwin (1872) providing a persuasive influence, that affective behavior represents a distinctive and fundamental class of behavioral expression that stands apart from other modes. As such, understanding the nature and sources of variability in affective performance represents an appropriate, indeed essential, focus of this book. The analysis in this chapter has explored this question, focusing on human personality and emotion as the archetypes of affective behavior. This analysis supports the following key conclusions:

1. Scientific interest in personality as a distinct type of behavior dates back some 2400 years. Popular approaches to assessing variability in behavioral expression of personality include the five factor model, the MBTI, and the MMPI. The number of personality traits specified by different observers range from a handful to over 1400.
2. Scientific interest in human emotions has a history almost as venerable as interest in personality. However, the impulse for categorizing different emotions has not been as dramatic as that for personality. Different observers specify 10 or fewer basic (primary) emotions, but various studies specify different types of emotions that total from about one to three dozen.
3. Both innate biological factors and design (context-specific) factors contribute to total variability observed in the behavioral expression of both personality and emotion.
4. Behavioral control systems analysis can be applied to account for the contribution of context specificity to variability observed in behavioral expression of both personality and emotion.
5. Both Darwin (1872), and more recently Plutchik (2001) with his psychoevolutionary theory, assume that human emotions have definite survival benefits, and therefore are subject to natural selection.

8 Social Cybernetics of Team Performance Variability

Robert Henning, Andrea Bizarro,
Megan Dove-Steinkamp, and Clark Calabrese

8.1 INTRODUCTION

The complex nature of social behaviors poses significant challenges to researchers and practitioners alike interested in understanding and managing sources of team performance variability. Teamwork is generally defined as a group of individuals working together to accomplish a common goal or purpose. Paradoxically, while most human factors researchers are comfortable attributing goal-directed behavior to teams, there is a general reluctance to delve into the scientific basis of these controlled behaviors. The inconsistent way that controlled behaviors are conceptualized at the team level versus individual team-member level also reveals the need for a more coherent systems approach. A working premise of this chapter is that a more consistent application of behavioral systems principles in modeling team behaviors can address these shortcomings and go a long way toward improving our understanding of sources of team performance variability.

Social cybernetics, an area of emphasis within behavioral cybernetics (T.J. Smith & Smith, 1987a), provides a comprehensive systems approach for identifying the key control processes and task design factors that contribute to team performance variability. As summarized below, empirical research studies have identified fundamental ways that individual team members engage in joint control activities during teamwork, and how these design factors impact team performance variability (this evidence shows that context specificity influences variability in team performance, recapitulating the theme of performance-design interaction (Chapter 1, Section 1.1.2) addressed in previous chapters). The social cybernetic model also provides a means to conceptualize the inherent trade-off between behavior that is controlled by individual team members or jointly controlled by the team. A key to understanding this trade-off is to understand the origins of behavioral control from a multilevel perspective. Systematic examination of these multilevel control relationships through the lens of social cybernetics can help guide human factors design efforts to limit team performance variability.

The primary author of this chapter has a longstanding familiarity with the social cybernetic model, having been involved in basic research and field applications in the 1970s under the direction of Professor K.U. Smith at the University of Wisconsin-Madison, and subsequently having applied the social cybernetic model in field studies of occupational health and safety and also in laboratory studies of the social psychophysiology of teamwork. The following themes and topics are covered in this chapter as a way to introduce the social cybernetic model and demonstrate its utility in the study of team performance variability: Section 8.2 is titled the Cybernetic Fundamentals of Feedback and Feedforward Control; Section 8.3 is the Longstanding Scientific Reservations about Cybernetic Psychology; Section 8.4 is the Social Cybernetics and Teamwork; Section 8.5 is titled the Social Cybernetic Studies of Social Interaction and Teamwork Through 1994; Section 8.6 is the Social Cybernetics Research since 1994; Section 8.7 is the Design Imperatives for Homeokinesis at the Team Level; and Section 8.8 is titled the Assessment of Homeokinesis at the Team Level.

8.2 CYBERNETIC FUNDAMENTALS OF FEEDBACK AND FEEDFORWARD CONTROL

Principles of behavioral cybernetics have already been introduced in Chapter 1 (Section 1.3.3), and so only those principles critical to understanding the social cybernetics of teamwork are covered here. To begin with, behavioral cybernetics is grounded in principles of control theory, including feedback control (compensatory control in response to error) and feedforward control (anticipatory or projective behaviors taken in advance to prevent error). Empirical evidence indicates that the control (or regulation) of social behaviors normally depends on a combination of both feedback and feedforward control (T.J. Smith & Smith, 1987a).

Feedback control (or error control) is relevant even in simple social behaviors. For example, when two people reach out to shake hands with each other, feedback control of hand position occurs in relation to the instantaneous distance between their hands. Poor quality feedback, such as not being able to see this instantaneous distance clearly, is referred to as noncompliant feedback because it can perturb or impair the ability to control the behavior involved. Feedback can be noncompliant due to spatial displacements (e.g., rotation, inversion) or temporal displacements (e.g., time delays), both of which have been found to significantly impair performance and increase performance variability (detailed in Chapter 5). Other transformations of feedback that can perturb behavior include discontinuity and lack of coherence with other sources of feedback.

Feedforward control (or projective control) of behavior is analogous to skilled steering behavior, where control actions are taken in advance of error actually occurring. Anticipating a slip or fall by reaching out to provide physical support to a person prior to them falling is an example of feedforward control in social behavior. Feedforward control is based, in part, on the participant's memories of past events and experiences. Even simple examples of feedforward control in social behaviors normally also involve some dynamic closed-loop feedback control of error. This enables timely (i.e., just-in-time) corrective adjustments to social behaviors that are largely guided by feedforward control.

8.3 LONGSTANDING SCIENTIFIC RESERVATIONS ABOUT CYBERNETIC PSYCHOLOGY

Since the scientific revolution, behavioral scientists have not been all that receptive to the idea that individual behavior can be both self-regulated and purposeful. According to Greenwood (2009), the originator of the cybernetic model for human-machine systems, Norbert Wiener, worked hard to convince skeptics that the concept of self-regulation had scientific standing. Wiener pointed out that it was possible to build a machine that self-regulated its behavior through use of feedback control, showing that self-regulated behavior in no way relies on immaterial or supernatural causes (Rosenblueth et al., 1943). The continued reluctance on the part of the behavioral science community to embrace the concept of self-regulated purposeful human behavior can be attributed, in part, to the great success of the physical scientist Sir Isaac Newton. His universal law of gravity made no reference to purposeful behavior, and this set the standard for what was considered a good scientific model (Greenwood, 2009). Even so, purposeful behavior was central to many of the behavioral models of early notable psychologists; William James promoted concepts of self-improvement and the importance of taking charge of your own life, John Dewey's functionalism demanded that motor-sensory control always be considered in the context of purposeful behavior, and Edward Tolman's neobehaviorism emphasized purposeful molar behaviors over low-level environmentally driven behaviors.

Nonetheless, simplified stimulus-response animal models that assumed environmental determinism still dominated much in the field of psychology and discouraged scientific consideration of the basis of cybernetic self-regulation. Even present-day cognitive models of behavior typically lack specific mechanisms to explain how behavior is organized and controlled, focusing instead on various ways that information can be processed. For example, Bandura's work on the social cognitive theory of self-regulation (1991) provides no adequate physiological mechanism to explain self-regulation. Furthermore, most social psychologists elected to use a reductionist approach when modeling group behavior by assuming that individuals react to social stimuli in the same ways that individuals react to other complex environmental stimuli. However, such explanations deny the possibility for emergent behaviors in social systems, a position that is contrary to modern systems science (see Chapter 2, Section 2.4, for a discussion of emergent behavior in self-organized systems).

One shortcoming in Wiener's original cybernetic model that may have contributed to behavioral researchers choosing reductionism over principles of self-regulation was the lack of a mechanism to explain how self-regulation in closed-loop systems could be sustained by living organisms. This shortcoming has been addressed in the more comprehensive behavioral cybernetics model (T.J. Smith & Smith, 1987a) where the primary mechanism that makes self-regulation possible is the reciprocal impact of motor (muscle) activity on both sensory stimulation and internal physiological states. Claiming that behavior can be used to regulate behavior may at first seem circular; however, the many reciprocal effects that motor activity has on current and subsequent behavior does provide a viable means for the self-regulation of ongoing behavior. For example, motor behavior that one uses to manipulate an object in the environment immediately produces self-generated sensory feedback through dynamic changes to the environment as well as from internal kinesthetic and other sensory systems linked to muscle and limb positioning. Also important to this control process, motor activity depends on metabolism, and this results in motor activity dictating the majority of energy use in the body during active behaviors. Many physiological systems also automatically respond to any change in metabolic activity, including cardiorespiratory functions, energy storage or outlay by the liver, and even neural-hormonal systems in which hormones are released to have multifold effects throughout the body (K.U. Smith, 1973).

The widespread reciprocal effects of motor behavior on internal physiological state enable what is referred to in behavioral cybernetics as bioenergetic control, which impacts both ongoing and subsequent behaviors, including central nervous system functioning that is already affected by self-generated sensory feedback. This highlights the importance of ongoing dynamic motor activity for the effective self-regulation of behavior—representing the concept of homeokinesis first introduced by K.U. Smith and Smith (1966). Homeokinesis can be contrasted with homeostasis, which is largely associated with passive and steady-state forms of physiological functioning during periods of rest and inactivity.

One final scientific concern that helps explain why behavioral researchers remain hesitant to adopt cybernetic models of behavior is that the origination of self-regulated behavior in organic systems is not readily apparent nor easily determined. For example, any compensatory control based on error feedback must first have a reference target or set point for error. There is no simple answer for where these set points come from in the behavioral cybernetics model. As explained above, self-regulation of behavior not only involves feedback control based on error, it also involves feedforward control that consists of control actions that occur prior to error occurring. Thus, projective bioenergetic control and projective control of self-generated sensory feedback are necessary to establish and sustain self-regulation of behavior.

In behavioral cybernetic theory, this suggests that the level of control can progressively increase over time in a nonlinear fashion once behavior is underway; for example, as reflected in the process of waking up and getting underway every morning. Those seeking a causal chain of events to explain these closed-loop behaviors may never be fully satisfied with this explanation, but conducting research on how purposeful team behavior is controlled and what environmental and task designs best support this control should not be avoided for this reason. Moreover, research scientists in other fields have not become scientifically immobilized over similar unanswered questions relating to causality; for example, the lack of knowledge of events prior to the Big Bang has not prevented physicists from conducting research on physical dynamics that are occurring after this event. As summarized below, there is ample scientific evidence that the human factors design of task environments affects social cybernetic control relationships as well as performance, and the social cybernetic model can be used to help understand and manage sources of team performance variability by systematically examining how behavior is controlled.

8.4 SOCIAL CYBERNETICS AND TEAMWORK

A basic form of social tracking occurs when one person actively attends to (or follows or tracks) the activity of another person. Even the simplest social tracking behaviors require some degree of skilled behavior on the part of the observer. In the case of visual social tracking for example, there is usually a need for both feedback and feedforward control of head and eye position in order to continuously track the activities of another person. Feedforward control in this case could involve anticipating which activity another person might initiate next and watching for the earliest signs of that activity. Feedback control in this case could involve compensatory control of eye movements when reacting to some unexpected activities of the other person. When two people are tracking each other simultaneously, the cybernetic aspects of the social behavior become much more complex and interesting. Any motor-sensory activity by one person serves as a new source of sensory feedback for the other person, which can then serve as the basis for further motor-sensory tracking behavior on the part of the first person, and so on, as depicted in Figure 8.1. In behavioral cybernetics, this linking up of behavior through joint motor-sensory control activity is variously referred to as joint motor-sensory control, integrated motor-sensory control, yoked motor-sensory control, or simply mutual control or mutual social tracking.

Depending on the nature of the social interaction, social tracking may demand an increased degree of effort and social skill. For example, when two individuals are in the process of completing an interdependent task, each individual is not only engaged in controlling aspects of his or her own behavior based on sources of self-generated feedback; each individual must also control some aspects of his or her behavior in relation to the new sources of sensory feedback that are generated while interacting with the other person. Multiple team members trying together to accomplish a shared goal or purpose further adds to the complexity of these motor-sensory control

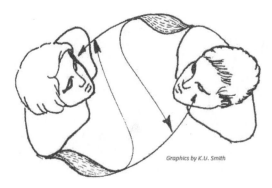

Graphics by K.U. Smith

FIGURE 8.1 Mutual social tracking.

relationships, impacting the bioenergetic control of individual team members and their ability to regulate their own behaviors effectively within the team context.

8.4.1 MODES OF SOCIAL TRACKING

Three distinct modes of social tracking involving closed-loop motor-sensory feedback control and/or feedforward control have been identified and verified through empirical testing (T.J. Smith & Smith, 1987a): (1) imitative social tracking, (2) parallel-linked social tracking, and (3) serial-linked (or series-linked) social tracking. In combination or in isolation, these three modes of social tracking can be used to characterize all forms of teamwork, providing a basis to consider how homeokinesis at the team level can be achieved and maintained via the feedback and feedforward control behaviors of team members.

8.4.1.1 Imitative Social Tracking

This mode of social tracking occurs when one team member imitates the behavior of another team member, as depicted in Figure 8.2. As shown, to be successful the person imitating his or her teammate's behaviors (the imitator) must actively work to minimize the error between his or her own behavior and that of the teammate. In training situations, a trainer would normally also track the behavior of the imitator in a closed-loop manner and adjust his or her behavior using feedback and feedforward control so that the imitator is able to keep pace and duplicate the target behavior without serious error. Furthermore, team leaders may often serve as a model or target of imitative social tracking. More effective team leaders would provide verbal and other sources of feedback to individual team members to reduce error and variability during imitative tracking. For example, human factors design can further support feedback and feedforward control during imitative tracking by assuring that adequate sources of compliant feedback are always available.

8.4.1.2 Parallel-Linked Social Tracking

The parallel mode of social tracking occurs when all team members can exert direct control over a system they are responsible for controlling and also receive sensory feedback about how the system is responding to their control inputs. This process is

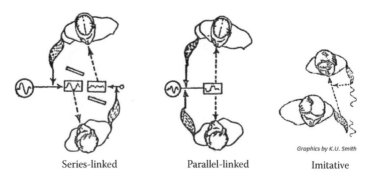

Series-linked   Parallel-linked   Imitative

Graphics by K.U. Smith

FIGURE 8.2 The three modes of social tracking.

also depicted in Figure 8.2, in which two team members are responsible for maintaining the status quo of a system. Disturbances to the system are depicted as a random waveform on the left side of the diagram, which each team member is able to help compensate for via control actions. An advantage of parallel social tracking is that control redundancy across team members can sometimes increase system reliability. However, a disadvantage is that a team member can have difficulty discerning whether a change in system status is due to his or her control inputs or the control inputs of other team members. This uncertainty in parallel-linked control may result in team members working at cross purposes, with control actions canceling out or adding together in undesirable ways. The resulting confusion can contribute to role ambiguity and teams failing to adequately respond to system disturbances. Human factors design can support this mode of social tracking and significantly reduce performance variability by making it possible for team members to track the control actions of other participants in addition to system responses.

8.4.1.3   Serial-Linked (or Series-Linked) Social Tracking

The serial-linked mode of social tracking occurs when the control actions and/or the feedback information necessary to achieve closed-loop control is somehow transferred through team members in a chainlike manner. In the case of series-linked tracking by a two-person team, only one team member can exert direct control over the system under their control, and only the second team member can receive sensory feedback about system responses to control inputs and determine what control actions are needed. As depicted in Figure 8.2, the random disturbance to the system depicted at the left must be compensated for by the control actions of the team. Many military command–control relationships as well as supervisor–subordinate teamwork relationships in the workplace are examples of series-linked social tracking. Advantages to series-linked tracking include role specificity and clear divisions of labor. Disadvantages include the risk of delayed actions and responses propagating across the chain of participating team members that can amount to substantial temporal displacements in feedback control, contributing to performance variability due to control instability. Apart from delayed actions or delayed feedback, errors can also propagate down the serial chain of participants and not be immediately detected.

8.4.2   Feedback Perturbation of Social Tracking

All modes of social tracking are susceptible to ergonomic design factors that impact performance variability by affecting the quality of feedback control relationships discussed earlier. Imitative social tracking performance can be perturbed and/or degraded by low-quality sensory information or by some form of noncompliant feedback described earlier because these compromise the imitator's ability to accurately and reliably match behaviors. For example, poor lighting would make it difficult for the imitator to clearly see the actions of the social target or to accurately detect any error differential between their own behavior and that of the social target. In the case of parallel social tracking, the lack of a communication pathway to provide information about the control actions being taken by other team members could result in team members duplicating control inputs and working at cross purposes.

Serial-linked tracking is dependent on a series of yoked control behaviors that can be seriously perturbed or degraded by time delays, such as a team member being slow to respond to a request for immediate action.

8.5 SOCIAL CYBERNETIC STUDIES OF SOCIAL INTERACTION AND TEAMWORK THROUGH 1994

Findings from earlier social tracking research are reviewed in detail elsewhere (T.J. Smith & Smith, 1987a, 1988b). Cybernetic research on sources of social performance variability related to teamwork was initially reviewed and summarized by T.J. Smith et al. (1994a), as detailed below.

8.5.1 DESIGN FACTORS AND VARIABILITY IN SOCIAL TRACKING

The preponderance of variability and specialization in social tracking performance is attributable not to a learning effect but rather to the effect of the human factors design of the tracking task and interface (Beare et al., 1985; Ting et al., 1972). Design factors of significance that may be specified include sensory feedback control parameters and conditions, mix of tracking modalities employed, temporal and spatial properties of sensory feedback, and the level and pattern of interpersonal, group, institutional, and/or human-system social relationships.

8.5.2 SENSORY FEEDBACK MODALITY AND SOCIAL TRACKING SKILL

Relative to visual-manual social tracking, accuracy and proficiency are greater for tactile, kinesthetic, and auditory social tracking (Cherry & Sayers, 1956; Rothe, 1973). Social tracking based on nonverbal sensory feedback often is more effective than verbal tracking in promoting social learning and communication (K.U. Smith & Smith, 1973). For all sensory modalities, the accuracy of interactive social tracking is comparable to that of individual tracking except under feedback delay conditions when the latter is superior to the former at all delay levels. Mutual social interaction entailing reciprocal exchange of sensory feedback (i.e., series- or parallel-linked tracking) is more effective than purely imitative tracking.

8.5.3 LEARNING OF SOCIAL TRACKING SKILLS

Even with provision of real-time feedback of tracking performance, social learning of specific social tracking tasks is highly variable, relatively limited and inconsistent, and unstable (Kao & Smith, 1971; Sauter & Smith, 1971).

8.5.4 PHYSIOLOGICAL FEEDBACK EFFECTS IN SOCIAL TRACKING

Movement compliance during interactive social tracking entails a concomitant synchronization of physiological functioning of the social partners (M.J. Smith, 1973; T.J. Smith & Smith, 1987a). Computerized social research has provided concrete evidence for such interactive physiological tracking for both cardiac (M.J. Smith, 1973) and ventilatory (Sauter, 1971) activity.

8.5.5 Effects of Sensory Feedback Perturbations on Social Tracking

Real-time temporal delays or spatial displacements in sensory feedback severely degrade the accuracy of social tracking performance (Probasco, 1969; Rothe, 1973; K.U. Smith, 1974; K.U. Smith & Smith, 1970; Yates, 1963), just as they do for individual tracking performance. For example, a feedback delay of 0.2 second decreases social tracking accuracy by 50 percent. As with individual behavior, the integrity of social behavior likewise relies on effective control by each partner of the temporal and spatial qualities of sensory feedback generated during the social tracking process.

8.5.6 Social Tracking in Group Interaction

Because of the introduction of additional sources of sensory feedback to control, the demands and complexity of social tracking in groups involving more than two people rapidly escalate (K.U. Smith, 1974; K.U. Smith & Smith, 1973).

8.5.7 Social Cybernetic Basis of Cognitive Behavior and Communication

Interactive social communication, with its pronounced motorsensory feedback control demands, is central to all modes of cognitive behavior and represents the principal determinant of effective learning of cognitive skills (K.U. Smith, 1974; T.J. Smith & Smith, 1987a, 1988b).

8.6 SOCIAL CYBERNETICS RESEARCH SINCE 1994

Since the above review, further research using the social cybernetics model has been conducted on the feedback control relationships during social interaction and teamwork.

8.6.1 Feedback Control Compliance During
Parallel-Linked Social Tracking

The influence of a wide range of ergonomic design factors on compliance has been investigated under laboratory conditions. Building on previous research conducted by Sauter and Smith (1971), two-person teams performed a cooperative computer-mediated task that involved both feedback and feedforward control in order to steer a simulated object through a complex path. Each team member used a joystick to control 50 percent of both the horizontal and vertical positioning of the simulated object during parallel-linked social tracking, requiring team members to coordinate their control inputs to perform the task. When control inputs to the object were delayed for only one team member, the team member without delayed control was able to compensate, showing that teams adapt to feedback control disparity by shifts in workload to the team members less burdened with feedback control perturbations (Li, 1998; T.J. Smith et al., 1998). These findings provide support for the joint feedback and feedforward control processes involved in social cybernetics. The less burdened team member was able to detect the error differential resulting from the partner's control delay and compensate in a proactive manner.

In another study, parallel-linked control of an inertial object occurred in a simulated microgravity environment. Each team member was able to exert horizontal and vertical forces on the object through use of separate joysticks (Glynn et al., 2001). In an effort to augment compliant social feedback between team members, force feedback was provided to each team member based on the direction and amplitude of their fellow team member's joystick inputs. Force feedback was found to improve team performance, demonstrating the value of providing additional sources of social feedback regarding control inputs beyond what is normally available regarding system response during the parallel-linked mode of social tracking.

In an investigation of team-controlled rest break activity, teams performed a joint computer-mediated task that required regular exchanges of information and text entry on separate computer workstations while also meeting a target level of rest break activity over the work period. Separate experimental conditions were used to examine parallel-linked and serial-linked control of short rest breaks. Parallel-linked control over discretionary rest break behavior was found to benefit productivity and wellbeing but also increased the demand for work coordination (Henning et al., 1997). Results from this study demonstrate the importance of team coordination for both team effectiveness and individual wellbeing.

8.6.2 Delayed Feedback in Serial-Linked and Mutual Social Tracking

To further investigate the impact of delayed multimodal feedback in social tracking, a series of studies were conducted using video disk technology as a means to systematically introduce audio delays and/or video delays in communications between two team members located in separate sound-isolated rooms. These experiments provided evidence for the critical nature of social control dynamics regardless of the mode of social tracking or modality through which feedback is provided. In one experiment, two-person teams performed the NASA Multi-Attribute Task Battery (MATB) that was adapted to require serial-linked social tracking. Audio delay of the voice commands was used to systematically perturb control of the task. Delays longer than 2 seconds were found to have nonlinear deleterious effects on team performance (Armstead & Henning, 2007).

In another study, two-person teams performed a simulated fire-fighting task that was adapted to require serial-linked social tracking. Audio delays were introduced in various schedules (e.g., fixed versus variable) to determine if practice on some schedules would promote better team learning or adaption. Results showed that practicing under longer feedback delays accelerated team training (Dove-Steinkamp & Henning, 2012), a finding that is consistent with related research showing that other types of feedback control perturbations also benefit team training (Gorman et al., 2010). Data from the simulated fire-fighting task were also analyzed to determine if social tracking was relevant to team outcomes other than performance; namely, stress. Results showed evidence of effects at the team-level analogous to individual-level effects in the job demand control model of stress (Karasek, 1979) as well as effects of team-level coordination, providing support that the emergent properties of teams can have a significant impact on the stress levels of individual team members (Calabrese & Henning, 2013).

Effects of delaying both audio and visual feedback on social interaction was investigated using video disk technology to experimentally delay (1 s) both audio (speech) and visual (upper body) feedback between two individuals located in separate rooms while they conversed with each other on assigned topics. In general, even short communication delays are known to have a perturbing effect on social interaction. These short imperceptible audio-visual delays were found to benefit perception of the emotional state of fellow participants in the first 10-min conversational period. However in the second 10-min conversational period, participants experienced increased frustration without incurring this same benefit (Powers et al., 2011). In a related study, participants conversed in either mixed-race or same-race dyads. Whereas mixed-race dyads reported greater anxiety and less interest in contact under audio-visual delay, same-race dyads reported less anxiety under delay conditions than in real-time (nondelayed) conversation (Pearson et al., 2008).

8.6.3 AUGMENTED TEAM COGNITION

Augmented cognition refers to a focused effort in the field of human factors and ergonomics to develop systems that significantly augment an operator's cognitive performance through use of physiological assessment of operator state in real time. Most research efforts in this area have been guided by linear information processing models, and attempt to address presumed bottlenecks to human information processing that would limit operator performance. A behavioral cybernetic model for augmented cognition was introduced by T.J. Smith and Henning (2005) as an alternative, in which the motor-sensory control of individual operators is augmented to benefit cognitive functioning. Additionally, measures of social-relational tracking were suggested for possible use in systems designed to augment team cognition; for example, using such measures to help teams establish the best pace for teamwork activities (Henning et al., 2005; T.J. Smith & Henning, 2006). The feasibility of this behavioral cybernetic approach to augmenting team performance had been demonstrated earlier when synchronicity in heart rate variability between two team members was used to create task-integrated feedback in real time that team members then used successfully to pace a projective tracking task involving parallel-linked control (Henning et al., 2000). In a later study, a communication delay between team members located in separate rooms was systematically introduced in an attempt to isolate physiological changes most closely associated with social-relational tracking for use in systems to augment team cognition (Henning et al., 2007).

8.6.4 SOCIAL CYBERNETICS IN PARTICIPATORY ERGONOMICS PROGRAMS

Principles of feedback control play a crucial role not only in small social systems, such as a team, but also in larger social systems, such as work organizations. As part of the Total Worker Health™ initiative by the National Institute for Occupational Safety and Health (NIOSH), the Center for the Promotion of Health in the New England Workplace and researchers from the University of Connecticut and University of Massachusetts Lowell have been involved in a multidisciplinary research-to-practice effort to introduce integrated workplace interventions to benefit worker health protection and promotion. Social cybernetics principles were applied by the first author to help guide the development of a programmatic approach based on participatory ergonomics. This program arranges for frontline employees to regularly participate in the planning and design efforts for these workplace interventions (Henning et al., 2009b). This participatory approach provides for organized communication from frontline employees to management regarding the need for interventions, as well as feedback from the organization to employees regarding interest and efforts to intervene. An Intervention Design and Analysis Scorecard (IDEAS) planning tool was developed to provide employee design teams with a structured approach to design and propose workplace interventions (Robertson et al., 2013; Henning & Reeves, 2013). Successful field-tests of the IDEAS tool and the overall programmatic approach were guided, in part, by the behavioral cybernetic model for organizational learning introduced by Haims and Carayon (1998). A survey tool to aid management in tracking the health of participatory ergonomics programs (Matthews et al., 2011) was adapted for program evaluation. Social tracking relationships in management systems have also been shown to impact the effectiveness of safety systems more generally (T.J. Smith & Larson, 1991; T.J. Smith, 2002).

8.7   DESIGN IMPERATIVES FOR HOMEOKINESIS AT THE TEAM LEVEL

It seems likely that homeokinesis at the team level may be as critical to the effective control of team behaviors as homeokinesis is at the individual level for the effective control of individual behaviors. Homeokinesis at the team level would occur when team members establish and maintain an adequate level of yoked social tracking behavior necessary for team members to function effectively as a social system. Such yoked behaviors would have significant reciprocal effects on the physiological functioning of all team members.

One reason homeokinesis at the team level may be critical to effective team function is that responding to task demands as a team depends on the effective use of the team's resources. The availability of team resources, such as the readiness of team members to engage in parallel social tracking in response to an immediate task demand, would require that individual team members are already tracking the task situation together via yoked motorsensory control. Establishing homeokinesis at the team level may also be necessary for desirable forms of mutual control to emerge as systems phenomena; for example, creative brainstorming as a team. In contrast, a lack of homeokinesis at the team level would be detrimental to team performance because the team would be unable to mobilize and coordinate its own resources effectively. Therefore, one human factors design imperative for effective teamwork is to specify the nature and extent of interdependent task activities necessary for achieving and maintaining homeokinesis at the team level.

A second human factors design imperative regarding homeokinesis for effective teamwork is to specify the types of activities team members should be involved in when not actively engaged in teamwork activities involving yoked motorsensory control. Such activities are necessary so that individual team members can regulate their own behaviors effectively during slow or downtime periods that lack interdependent teamwork tasks.

A third human factors design imperative regarding homeokinesis for effective teamwork is to systematically design team tasks and individual tasks to support smooth transitions between them from the standpoint of homeokinesis. This requires that team members not actively engaged in a mode of social tracking must be sufficiently engaged in alternate tasks that are selected and designed to make it possible for team members to temporarily function independently while also tracking teamwork activities. Otherwise, individual team members will have difficulty reengaging with interdependent tasks and social tracking activities when needed.

Together, these three human factors design imperatives offer a more comprehensive systems approach to the design of teamwork than is normally considered. The focus of human factors design efforts is expanded beyond the interdependent tasks that represent core team activities. Consideration of design factors to promote homeokinesis at both the team and individual levels and through the transition periods as well as in complex combinations is offered here as a coherent systems design approach for managing a key source of team performance variability.

The nature and extent of homeokinesis at the team level versus homeokinesis at the individual team member level impacts the availability of team resources in specific ways. In general, homeokinesis at either the team level or individual team member level can be considered as two primary means of creating resources during teamwork. While multiple resource models have been helpful in examining human capabilities and limits (e.g., Wickens, 2002), there has been no clear mechanism offered to account for how these resources are managed or replenished during teamwork. For example, crew resource management techniques for training high-performance teams are largely limited to prescribing standard ways that individuals should behave as team members, such as being more assertive when needed. There is also training on standard ways to handle problems through use of stepwise approaches, such as first recognizing a problem, then defining it, then identifying probable solutions, and then implementing a solution. The limited nature of these approaches may partly explain why a recent metaanalysis of 28 crew resource management training studies revealed a lack of widespread empirical support for the efficacy of crew resource management (Salas et al., 2006). It can also be noted that only a few of these training studies emphasized aspects of crew coordination that would reflect modes of social tracking.

Human factors approaches need to provide a level of design specificity necessary to promote homeokinesis at both the team and individual team member level. For example, one design recommendation for a specific team task might entail predictable periods of parallel-linked social tracking by the whole team as a means to regularly transition to homeokinesis at the team level. Team members could also regularly participate in consensus decision-making activities or engage in other joint activities with some team members. The division of labor during parallel-linked tracking has been shown to dictate the nature and extent of teamwork involved (Sauter & Smith, 1971); for example, control can be divided up equally so that all team members must work together to address the parallel task at hand. Additionally, serial-linked tracking activities can be employed to maintain close social tracking relationships among smaller numbers of team members over long periods of time. Imitative social tracking activities can involve job shadowing or interning for progressive training purposes. Individualized (non-team) tasks could include tracking team task activities by monitoring communications within the team or by preparing for future teamwork activities in ways that maintain yoked motorsensory control and readiness for upcoming teamwork demands.

Monitoring of individualized task activities by supervisors would help assure that these activities are engaging enough for homeokinesis, promote the needed readiness for engaging in future teamwork activities, and that switching between individualized task activities and teamwork activities is not unduly stressful. The popularity of dashboard systems to assist the management in monitoring social and individual activities to better guide or steer organizational behaviors suggests that many organizations already recognize the need for dynamic social tracking activities when managing sources of team performance variability. Therefore, the human factors design imperatives introduced above may be relatively easy to assimilate into existing management practices.

Consideration of the above human factors design imperatives and translation of social cybernetic principles can occur in the context of well-established human factors design techniques for system development (e.g., Chapanis, 1996). To begin with, a task analysis could provide a complete listing of all of the tasks that need to be accomplished. These tasks can then be allocated according to the human factors design imperatives for homeokinesis listed above. Similar to the challenges found in allocation of functions between human operators and automated machine systems (T.J. Smith & Smith, 1988b; T.J. Smith et al., 1995), tasks are allocated depending on the homeokinetic needs of the whole team, a subgroup of team members, or the individual team members. Functional flow analysis combined with timeline analysis can then offer a more systematic approach for allocating task activities to meet homeokinetic needs at any moment in time.

8.8   ASSESSMENT OF HOMEOKINESIS AT THE TEAM LEVEL

The design imperatives introduced above all center around establishing and maintaining teamwork in relation to homeokinesis. However, the level of homeokinesis at the team level may not be readily apparent to an outside observer or even to the team itself. Therefore, an objective means of assessing homeokinesis at the team level could greatly benefit the human factors design efforts in system development and would also benefit human factors research on the sources of performance variability in teamwork.

As explained earlier, there are numerous reciprocal effects of motor activity on physiological functioning. Team members' interdependent task behaviors during social tracking and teamwork therefore have predictable reciprocal effects on the physiology of each participating team member due to yoked motorsensory control,

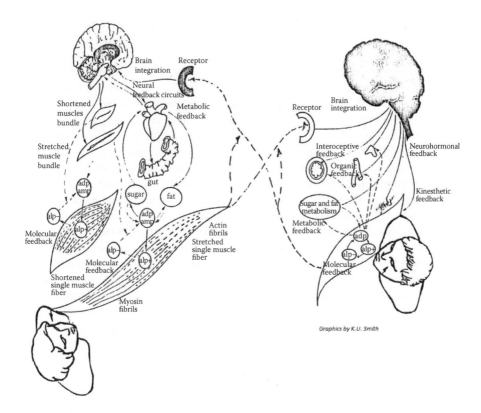

FIGURE 8.3 The reciprocal effects of yoked motorsensory control on the physiological functioning of two team members during social tracking.

as depicted in Figure 8.3. The term "physiological compliance" was originally introduced to represent the way that control over the participants' physiology becomes integrated during social tracking (T.J. Smith & Smith, 1987a). In as much as social physiological compliance reflects yoked motorsensory control during teamwork, it should be possible to detect this in the physiology of team members in order to monitor social-relational dynamics on a continuous basis. For example, sustained mutual control over social behavior during teamwork would be expected to be accompanied by an extended period of social physiological compliance among team members. Efforts have been made to test candidate physiological measures as predictors of subjective and objective team performance outcomes as well as participant ratings of their own teamwork effectiveness. For example, when team members used joysticks in a computer-mediated task to guide a simulated object through complex path, Henning et al. (2001) found that several social physiological compliance measures based on cross correlation and weighted coherence analyses of breathing pattern, electrodermal activity, and heart rate variation were each predictive of team tracking performance. An analysis of data from the same study also showed that social physiological compliance predicted team members' ratings of performance effectiveness (Gil & Henning, 2000).

Social physiological compliance would also appear to be a good candidate measure for assessing homeokinesis at the team level in task contexts. This could be helpful in determining which aspects of task design promote homeokinesis at the team level or place the team at risk for increased performance variability if homeokinesis at the team level is limited or absent. There is some evidence that social physiological compliance may be episodic during teamwork (Henning et al., 2009a), suggesting that maintaining high levels of homeokinesis at the team level for long periods may be difficult for teams and/or even counterproductive to teamwork. One can speculate that long-duration team tasks may require alternating periods in which either homeokinesis at the team level or individual team member level dominates in order to minimize performance variability or to prevent team burnout.

The nature and the extent of social physiological compliance and homeokinesis at the team level can also be expected to be impacted by the mode of social tracking team members are engaged in. Highly synchronous changes in the physiological states of team members can be expected during imitative social tracking because the motor behaviors of the participants are closely matched and synchronous, resulting in a very high level of team homeokinesis. In the case of parallel-linked social tracking, social physiological compliance and homeokinesis at the team level could vary depending on the degree to which team members engage in any of the following activities: close coordination of control inputs to the system, shared motor-sensory tracking of system responses to control inputs, and social tracking of the activities of fellow team members, including any communications between team members about the task at hand or about social-relational tracking itself. In the case of series-linked social tracking, social physiological compliance can be expected to vary depending on the nature and extent of serial communication behaviors that are necessary among team members for either control actions or for tracking system response to these control actions. Time delays in any of these communications would be cumulative in serial-linked social tracking, requiring that any substantial delays be factored in to prevent gaps in joint activity that might become problematic for maintaining homeokinesis at the team level.

In another study of social tracking, two-person teams performed the projective tracking task described above that involved steering a simulated inertial object through a complex path. Unexpected changes in the vertical and horizontal control dynamics of the input joysticks occurred (e.g., the sudden inversion of the vertical control inputs) that required the team to adapt quickly to these changing task demands. Social physiological compliance was found to be a significant predictor of how well a team would perform immediately following these unexpected events (Henning & Korbelak, 2005). These results provide further evidence that physiological compliance serves as an indicator of mutual control by demonstrating improved ability of team members to recover from disruptions once physiological compliance during teamwork is established.

Patterns of social physiological compliance were studied in a preexisting four-person team of graduate students who met weekly for research group meetings over two months. As the students planned their research project together, social physiological compliance was examined in a pairwise fashion between team members to match up with conversational exchanges. Subjective ratings of team performance at the end of each meeting were found to be negatively correlated with increased social physiological compliance (Henning et al., 2009a), replicating the findings of a study on married couples engaged in difficult counseling sessions (Levenson & Gottman, 1983).

Contrary to the positions and evidence presented in this chapter thus far, these results suggest that high levels of physiological compliance during social interaction may result in negative experiences. One possible explanation is that team member dissatisfaction or the observed negative affect during marital interaction do not necessarily imply the absence of effective mutual control during social interaction. Rather, team members may have slightly negative reactions to the loss of individual control that is inherent to periods of social physiological compliance. It is also possible that sustained high levels of homeokinesis at the team level are problematic and contribute to performance instability whenever individual team members are unable to take a break from teamwork or have insufficient time to fulfill tasks not involving teamwork.

Recent teamwork research provides empirical support for both of the interpretations offered above. One research study reported direct, positive relationships between social physiological compliance, measured through heart rate variability and respiratory sinus arrhythmia, and team performance on a combat video game (Elkins et al., 2009). More recently, Chanel et al. (2012) also found evidence that social physiological compliance measured as weighted coherence of breathing patterns was indicative of rich social interaction during more challenging task periods in a maze videogame. Additionally, social physiological compliance in two-person teams, based on ventilatory drive measures, was predictive of objective team performance in a shared decision-making marketing and production task (Bizarro, 2013). In this latter study, those teams able to maintain a high level of social physiological compliance made production and marketing decisions that resulted in higher simulated profits than teams that did not maintain a high level of social physiological compliance. These studies provide empirical evidence that social physiological compliance does have a positive impact on objective team performance.

In combination, the above-mentioned studies lend support to the conclusion that social physiological compliance can serve as an indicator of the mutual control over interdependent task activities as predicted by the social cybernetic model of teamwork. Physiological compliance measures have the potential to provide objective measures of team coordination that are not possible to obtain with subjective measures of team coordination and may also provide a means of studying homeokinesis at the team level.

8.9 CONCLUSIONS

Social cybernetic principles offer a coherent systems approach for conceptualizing how behavior is organized and controlled in teamwork. The three distinct modes of social tracking that make up all forms of social interaction can be used to systematically analyze the control dynamics involved in specific forms of teamwork. Empirical studies show that design factors that affect feedback control or feedforward control in social tracking can significantly impact social and team performance variability. Therefore, identifying the context-specific design factors in teamwork tasks that determine feedback compliance in the social tracking activities making up these tasks represents a scientific way to control for sources of team performance variability. Another key cybernetic consideration in the design of team tasks and task environments is maintaining homeokinesis at both the team and team member levels, resulting in a set of human factors design imperatives that can be met through use of established system development methods in the field of human factors. For example, allocating tasks to the team as a whole or to individual team members would be done more systematically so that transitioning between interdependent team tasks and individual tasks is not problematic. Social physiological compliance based on shared physiological changes, such as the synchronization of heart rate variability among team members, reflects yoked motorsensory control among team members and provides a potential means for assessing social-relational tracking and homeokinesis at the team level. Application of social cybernetics principles in program development efforts at the organizational level has been helpful in minimizing team performance variability as well as optimizing the wellbeing of individual team members.

9 Variability in Human Work Performance

Interaction with Complex Sociotechnical Systems

Thomas J. Smith

9.1 INTRODUCTION

This chapter deals with a mode of performance variability that subsumes modes addressed in all other chapters in this text, namely variability in human work performance. Human work behavior defines the human condition (T.J. Smith, 2001). In a feedback fashion, it has guided the course and pace of human evolution, as well as the emergence and elaboration of human civilization (Chapter 11). It forms the basis of the development and functioning of all patterns of human social, technological, organizational, institutional, economic, governmental, and nation-state designs. The central concern of the field of HF/E science is human work performance—Murrell first proposed the term *ergonomics* to refer to the natural laws (*nomos*) of work (*erg*) (Konz, 1995, p. 12). In the HF/E community, broad agreement has emerged that such laws refer to design factors, conditions, and strategies that accommodate the capabilities and limitations of the worker.

As another example, the term "economics" is defined as the science concerned with the production, distribution, and consumption of goods and services, or the material welfare of mankind; all of these concerns represent the consequences of human work. Around the globe, economic output typically is gauged using gross domestic product (GDP) as an index—implicit in this measure is the assumption that "product" refers to the outcome of human work. More broadly, the key to the competitiveness, and not infrequently the survival, for different specific instantiations of the various complex systems cited above is the productivity of the workforce.

Customarily, analysis of variability in the performance of different complex sociotechnical (ST) systems has not been considered in the context of behavioral variability in human work performance. An exception is the perspective of K.U. Smith (1965a), who introduces an entire book devoted to this idea by noting (with reference to economics) that (Chapter 1, p. 1-1):

> Traditionally, the sciences of behavior and economics have gone their own ways, with little or no intercommunication. The laws of economics have taken no account of the nature of human behavior, and the science of behavior, in its "pure" form at least, has remained detached from the hurly-burly of the marketplace. Yet economics is the

business of mankind—its main business—and one can understand neither the worka-
day world nor the human beings who live out their lives in it, without reconciling
the laws of economics with the principles of human behavior. Economic factors and
human work are fundamentally interdependent, and the study of one has but limited
validity unless it takes into account our knowledge of the other.

Almost 50 years later, Pentland (2013, p. 80) reiterated the same point with the
observation that, "...many economists and social scientists still think about social
systems using Enlightenment-era concepts such as markets and classes—simplified
models that reduce societal interactions to rules or algorithms while ignoring the
behavior of individual human beings."

Although these perspectives must be tempered by the recent emergence of behavioral economics as a distinct branch of economic analysis (Kahneman & Tversky,
1979), their validity remains intact in many important respects.

A series of reasons may be cited for addressing variability in human work performance and the reciprocal links to variability in complex systems performance in this
book. First, as noted above, regardless of the nature and extent of system complexity,
variability in the performance of complex ST systems ultimately reflects, and depends
on, variability in the work performance of human individuals and teams that comprise
such systems. In turn, in today's society, the availability of work (with rare exceptions) is
dependent on the effective functioning of such systems. The link between work performance variability and complex system performance thus represents a legitimate target
for this book. Finally, it can be argued that since the emergence of organized patterns of
human civilization some 10,000 years ago, a major influence on variability in individual
and group performance has been variability in the complex ST systems developed and
organized by these human agents. This is manifestly true of today's society.

The literature in this area is vast. An October 2013 search of books on Amazon
retrieved 2.339 and 2.383 million titles for the terms "business books" and "business," respectively, 1.908 million titles for the term "economics," and 686,405 titles
for the term "government." Jackson and Parry (2011, p. 2) note that a 2011 search of
Amazon books retrieved 57,666 titles for the term "leadership books," an October
2013 search for the same term retrieved 99,948 titles (effective leadership is widely
assumed to represent a major determinant of organizational, institutional, and governmental performance effectiveness).

A major domain of HF/E science that has come to be known as "macroergonomics" (Hendrick & Kleiner, 2002; Kleiner, 2008) is concerned with the design and
management of organizational and institutional systems. On the other hand, factors
that influence variability at the level of economic performance have largely been
ignored by the HF/E profession. This chapter attempts to remedy this oversight as
the third rationale for an integrated treatment of variability in both human work and
complex systems performance.

To document this point, I searched four journals published by the HFES for the
occurrence of the term "economics" in the title of reports in these journals. Based
on searches in September 2013 for October 1974 through September 2012 publications, titles of five *Proceedings of the HFES* reports featured the term, and all but
one dealt with the economics of ergonomics (the remainder dealt with the economics

of human gestures), and no reports published in *Human Factors* (September 1958 through August 2013 issues), in *Reviews of Human Factors and Ergonomics* (Volumes 1 through 7), and in *Ergonomics in Design* (January 1993 through July 2013 issues) featured the term in report titles. A charitable conclusion from these results is that the design and performance of economic systems represents a promising frontier for future HF/E inquiry.

The HF/E field has paid somewhat closer attention to the design and performance of governmental systems. The most comprehensive approach of this sort is embodied in the field of community ergonomics, defined as the collective set of system design factors that define the socioeconomics of a community, as well as to the study of these factors (J.H. Smith et al., 2003). Two other selected examples are: (1) Kleiner (2008) notes that macroergonomic principles are applicable to the analysis of governmental systems performance; and (2) Moore and Barnard (2012) note that realization of sustainable human development will require application of HF/E principles and practices at governmental levels.

I am not an economist, a political scientist, or a business executive. Instead, like my father cited above (K.U. Smith, 1965a), I am a student of human work. Consequently, and also in light of the large literature, the perspectives on work performance variability and complex systems performance offered in this chapter may strike the reader as somewhat selective. The chapter will recapitulate themes related to performance variability emphasized in earlier chapters, namely; the nature of variability, sources of variability, nativist versus empiricist perspectives on sources of variability, and application of control systems principles to the understanding of sources of variability. Sections below will apply these themes to an analysis of how and why work performance variability is linked to the performance of organizations and institutions (Section 9.4), and economies and nation-states (Section 9.5). These sections are preceded by a historical perspective (Section 9.1), and a background section devoted to a systems analysis of human work (Section 9.2).

9.2   HISTORICAL PERSPECTIVE

Throughout human evolution, variability in human work behavior has been inextricably linked to the social and societal contexts in which it occurs. For example, there is widespread agreement that throughout most of human evolution, all hominids and their early modern human successors engaged in a hunting and gathering mode of existence. As Lee (2005) puts it, "Hunting and gathering was humanity's first and most successful adaptation, occupying at least 90 percent of human history. Until 12,000 years ago, all humans lived this way." This term, of course, connotes a particular pattern of variability in human work behavior with a division of labor design (male hunters, female gatherers) anticipating by eons the analysis of Adam Smith (1776, 1977), who called attention to the economic advantages of this type of work design.

As another example, the major ages of man spanning over two million years— paleolithic, neolithic, bronze, iron—imply progressively more refined modes of human work behavior centered on mining mineral resources and their subsequent fabrication into tools and other artifacts. Consider also all the different terms that have emerged over the centuries to refer to different types of human work behavior—farmer, laborer, merchant, noble, peasant, priest, scholar, serf, warrior, and so forth—that anticipate the many hundreds of occupational categories that exist today. In the realm of religion, there is a clear Judeo-Christian respect for manual labor; thus, it is not God who saves humankind from the flood, but Noah, following God's instructions, who relies on his own labor to build the arc.

Although we owe our existence as a species to human work performance, it is fair to say that some forms of work, particularly of the manual kind, have not always been viewed with respect. Thus, during the medieval period, as Daileader (2001) points out, three basic societal classes were commonly recognized: (1) those who fought (the nobles), (2) those who prayed (the monks), and (3) those who worked (the peasants and townspeople). Of course, in reality everyone worked, but the third class of workers definitely was the lowest on the totem pole.

Some of the key historical milestones in the progressive elaboration of complex ST systems and their effects on patterns and variability in human work performance may be summarized as follows:

1. Commerce, defined as the collective operation and interaction of legal, economic, political, social, cultural, and technological systems, represents the bedrock of economic functioning. The origins of commerce are ancient. Watson (2005) dates the history of long-distance commerce to around 150,000 years ago. We may surmise that the progressive expansion and sophistication of commerce and trade since then has been accompanied by the emergence and elaboration of different modes and patterns of human work that support the different complex ST systems underlying commercial operations.

2. Of the 282 laws specified in the Code of Hammurabi (Harper, 1904), 36 (in my judgment) deal explicitly with the nature and/or consequences of work performance (identified on request). (Harper [1904] dates the code to about 2250 B.C., but Wikipedia cites a date of about 1772 B.C. and goes on to outline the uncertainty regarding the chronology of the ancient Babylonian period.) Among the myriad laws and regulations promulgated since then, many similarly deal with human work. For example, one of the major administrative bodies of the U.S. Government is the Department of Labor.

3. Arguably, so-called modern innovations in work design such as the assembly line, specialization of labor, and executive leadership have their origins with work of the ancient Egyptians, Greeks, and Romans, with their pyramid and trireme building projects and impressive engineering edifices, respectively. Landes (1999, Chapter 4) points out that the Middle Ages, the interlude between the fall of Rome and the explosion of new modes of work design and innovation with the Renaissance, saw the introduction of a remarkable variety of technological innovations to support human work. Examples that he cites include the water wheel, eyeglasses, the mechanical clock, printing, and gunpowder. Landes identifies new types of work, new patterns of work organization, and advances in work productivity made possible by each of these innovations that heretofore had not been possible.

4. The origins of the institution that today we call the "company" or the "firm" date back well over 1400 years. A tabulation by *The Economist* (2004) of the world's oldest family companies still in business today lists 15 firms

founded between 578 and 1530 AD: three in France, two in Germany, eight in Italy, and two (the two oldest) in Japan. The Wikipedia entry for "list of oldest companies" (http://en.wikipedia.org/wiki/List_of_oldest_companies) based on a 2008 survey notes 5586 companies older than 200 years—3146 in Japan, 837 in Germany, 222 in the Netherlands, and 196 in France. However, it was not until well into the twentieth century, with the seminal paper of Coase (1937), that the first systematic account was published of why firms actually exist. The essence of Coase's argument (see also Free Exchange, 2013e and Coase, 1993), contrary to the prevailing economic wisdom of the time that market prices adjust automatically based on supply and demand (dating back to Adam Smith's invisible hand), is that (1) transaction costs, linked to contract negotiation requirements, prevent market price mechanisms from working smoothly (i.e., it may be costly for buyer and seller to agree on a final price), (2) with a firm, prices typically are set, not by the market but rather in accord with central planning decisions of corporate management, (3) the central planning nature of firms thus makes it cheaper and easier to coordinate contract agreements, and (4) the foregoing points explain why firms exist. Although Coase's original analysis has been subject to subsequent revision (Demsetz, 1993), its essential validity remains intact.

5. One of the earliest efforts to develop a socioeconomic profile of an entire country was that of William the Conqueror. In 1086, 10 years after his conquest of much of England and parts of Wales, he commissioned *Domesday Book* (Martin & Williams, 2005), one of the main purposes of which was to determine how much each landholder possessed in land and livestock and what they were worth. One of the consequences of this effort was to more clearly delineate a social and economic caste system that persists, to some degree, in Great Britain today. An example is the emergence, after establishment of the parliamentary system, of two main political parties that historically relied primarily for their political support on aristocrats—the Conservatives (Tories)—or on unionized workers—Labour.

6. The late fifteenth century to the mid-sixteenth century featured two seminal milestones in the exploration of human work. The art of Leonardo da Vinci (1452–1519) included extremely detailed depictions of human work activity, some of the first highly accurate representations of the human body at work (K.U. Smith, 1965a, pp. 1–12). In 1556, Agricola published the seminal text on human work, with his detailed account of the scientific, technological, social, safety and health, managerial, and organizational features of work in underground metal mines in Northern Europe. This work anticipated a number of the micro- and macroergonomic HF/E concerns about work design and performance that emerged some four centuries later.

7. In a nine-decade span from 1776 to 1867, two works were published that had a foundational influence on modern understanding of economic thought, the history of labor, and human work performance; namely, those of Adam Smith (1776, 1977) and Karl Marx (1867, 1990). With his publication of *The Wealth of Nations* in 1776, Smith's influence on the subsequent progression of economics as a distinct field of inquiry down to today can hardly be

overstated. As Nasar (2011) points out: (1) Smith's publication corresponds precisely with the dawn of the industrial revolution (Ashton, 1948), which saw the emergence of the idea of "the economy" as a distinct entity whose activities could be monitored and measured, and (2) histories of economics thus tend to start with Adam Smith and his 1776 publication. Smith's contributions have been reviewed in recent publications by Rothschild (2001) and by Phillipson (2010), and in a delightful synopsis by Gopnik (2010).

Smith's seminal contributions to economics and work science may be summarized as follows: (1) Smith's views of human beings as inherently moral beings, and of sympathetic behavior arising therefrom as a form of *work* (rather than an internal reflexive affect) articulated in his 1759 book, *The Theory of Moral Sentiments* (Smith, 1759, 2010), led to his establishing a connection between the work of being a social being and the work that social beings do; (2) the connection delineated in point 1 prompted Smith to consider the market as a sympathetic community design system comprised of people working in uncanny harmony, each aware of the desires of others and responding to them, but also acting in their own self-interest; (3) in their efforts to enrich themselves in the marketplace through self-interested work, workers through their collective efforts insensibly promote an outcome of which they are not aware (Smith's invisible hand); namely, the wealth of the entire nation; (4) increased productivity of human work thus is the sole source of a nation's wealth, all owed to the labor of its people and the technology they possess; (5) in terms of their work productivity, agricultural, manufacturing, and service industries are equally important in contributing to a nation's wealth; (6) a nation's wealth is based on what its workers make (through growing, manufacturing, or serving) rather than on what it holds; and (7) division of labor is key to increasing the productivity of human work.

In support of the last point, Smith leads off his book with the following illuminating account that describes the contributions of dozens if not hundreds of workers necessary for the fabrication of a single product:

The woollen coat, for example, which covers the day-labourer, as coarse and rough as it may appear, is the produce of the joint labour of a multitude of workmen. The shepherd, the sorter of the wool, the woolcomber or carder, the dyer, the scribbler, the spinner, the weaver, the fuller, the dresser, with many others, must all join their different arts in order to complete even this homely production.... Let us consider only what a variety of labour is requisite in order to form that very simple machine, the shears with which the shepherd clips the wool. The miner, the builder of the furnace for smelting the ore, the feller of the timber, the burner of the charcoal to be made use of in the smelting-house, the brick-maker, the brick-layer, the workmen who attend the furnace, the mill-wright, the forger, the smith.... Without the assistance and co-operation of many thousands, the very meanest person in a civilized country could not be provided, even according to what we very falsely imagine, the easy and simple manner in which he is commonly accommodated.

The impact of Smith's idea about the division of labor on patterns of variability in work performance that emerged over the ensuing two-plus centuries is profound. One consequence was refinement of Smith's idea into the concept of specialization of labor, with a given worker limited to the performance of one (or at most a few) repetitive tasks. In the early twentieth century, the logical zenith of the specialization of labor idea was realized with factory implementation of assembly-line work by Ford and development of principles of scientific management by Taylor (1911). Taylor's assumption was that if task demands for a given worker were rigidly prescribed, then it should be possible to accurately predict, and thus mandate, how much production to expect from that worker over a given time period. As addressed in a subsequent section of this chapter, as a reaction to Taylor's approach (his task demand prescriptions were physically impossible to meet for most workers, coupled with the fact that he fudged his data, lied to his clients, and inflated the record of his success [Lepore, 2009]), efforts to encourage more human-centered approaches to work design emerged later in the century.

I have addressed Smith's treatise in some detail to underscore the point that, throughout, Smith fundamentally is concerned with the patterns and nature of human work and with the factors that contribute to variability in human work performance. This recapitulates the perspective offered in the introduction to this chapter that there are intimate links between the concerns of economics, the prosperity of nations and states, and the interests and outcomes of human work performance.

8. With his 1867 publication of *Das Kapital* (termed by Engels the bible of the working class), Karl Marx (1867, 1990) established himself as a second major influence on modern economic thought and work design. Considered one of the fathers of socialism, and the archenemy of capitalism, Marx inspired one of the most profound and far-reaching experiments in the modern era of work design; namely, communism. However, the worker's paradise that this movement promised turned out somewhat differently. As a consequence of both purges and collectivization, it is a matter of record that between them, Stalin and Mao killed millions of workers.

Some tenets of Marxism are discredited today. However, Magnus (2011) points out that other of his predictions regarding the shortcomings of capitalism have been realized. In particular:

- Companies' pursuit of profits and productivity would naturally lead them to need fewer and fewer workers, creating an industrial reserve army of the poor and unemployed. The process he describes is visible throughout the developed world, particularly in the United States. Companies' efforts to cut costs and avoid hiring have boosted corporate profits as a share of total economic output to the highest level in more than six decades, real wages are stagnant, and the postrecessionary unemployment rate has yet to recede to its prerecessionary lows.
- Income inequality, meanwhile, is by some measures close to its highest level since the 1920s.
- Marx also pointed out the paradox of overproduction and under consumption. The more people are relegated to poverty, the less they will be able to consume all the goods and services companies produce.

When one company cuts costs to boost earnings, it is smart, but when they all do, they undermine the income formation and effective demand on which they rely for revenues and profits.

More broadly, the special concern of Marx about how workers relate to that most fundamental resource of all, their own labor power, remains highly relevant today. This concern is aligned with my own first principle of occupational science; namely, that work is essential to health (T.J. Smith, 1980).

9.3   THE NATURE AND SIGNIFICANCE OF WORK

Portions of this section are adapted from T.J. Smith (2001).

9.3.1   INTRODUCTION: HUMAN CONTROL OF THE BEHAVIORAL ENVIRONMENT THROUGH WORK

Work is the most ubiquitous form of human activity on the planet. In the strict physical scientific sense of the term (application of energy to drive chemical reactions [internal work] or to move body effectors [external work]), human engagement in work is synonymous with life itself, beginning at the moment of conception and terminating only with death. In the behavioral sense, all conscious human activity represents work directed at some sort of goal or another. In the occupational sense, all humans routinely engage in purposeful task or job work activity throughout most of their lives. In the evolutionary sense, since emergence of the species work activity has served both as the engine of the human condition and as the means of human self-selection.

Encyclopedic treatments of the term "work" typically focus only on its physical scientific meaning, ignoring the much broader human perspective. Dictionary definitions (the term has relatively recent Old English origins dating back less than 1500 years) hint at its broader implications—Webster (McKechnie, 1983) provides over 30 definitions of "work" used as a transitive or intransitive verb and a dozen definitions of its use as a noun. However, it is with the historical perspective that the true human meaning of the term begins to emerge. As noted in Section 9.2, all of the so-called prehistorical and historical ages of human development and civilization—stone, copper, bronze, iron, classical, medieval, renaissance, industrial, information—refer explicitly or implicitly to distinctive modes and expression of human work activity.

Work serves also as a primary focus of human cultural, religious, intellectual, political, and scientific thought. The emergence of human civilization essentially rested on a series of transformations in how work was organized. The first words of the Bible—"and God created heaven and earth"—refer to work of the Almighty. Human work was an object of reverence for the ancient Hebrews, a Judeo-Christian tradition that persists to this day. The organization of work—in warfare, commerce, cultural and religious expression, and society—has represented a primary focus of the state from its earliest origins. Political thought and philosophy revolve around the best approach to such organization. Certain forms of government such as communism exalt work as the centerpiece of their design. The evolution of technology reflects the application of increasingly sophisticated artifacts to aid human work performance. All academic, intellectual, and scientific disciplines concerned with human performance—from art to athletics, from business to engineering—are concerned in one way or another with how human work is organized and expressed.

Given the extremely broad scope of the topic, it would be impractical in this section to consider all of the possible human implications of the term "work." Sections below therefore touch upon scientific theories of work, landmarks in the evolution of work, general conclusions about laws and customs of work, and the future of work.

9.3.2 THEORIES OF WORK

What is the purpose of human work? In light of the universal significance of work performance for defining, organizing, and guiding the human condition, it is remarkable that no generally accepted theory of human work has yet been promulgated. Scientific concepts of work essentially have not been developed much beyond conventional views of labor as a means of production and source of wealth (Adam Smith, 1776, 1977), as a limited aspect of technology (Marx, 1867, 1990), as an exploited competitor in class struggle (Marx & Engels, 1848, 2013), or as essentially a mechanistic component of work process subject to "scientific management" (Taylor, 1911). A variety of sources can be cited (Galbraith, 1958; Horowitz, 1998; McGregor, 1960; K.U. Smith, 1962b, 1965a) to support the conclusion that all of these viewpoints have been largely discredited as they apply to motivational behavior in the performance of work.

It is equally remarkable that despite its putative role as *the* major scientific discipline concerned with human behavior at work—as noted above, the term "ergonomics" refers to the natural laws of work—HF/E science has been largely silent in addressing this question. For example, no theoretic treatment of work is to be found in handbooks concerned with HF/E (Salvendy, 1982, 1987). Moreover, work receives short shrift as an object of formal academic focus by HF/E science. Among 74 North American graduate programs in HF/E catalogued by the U.S. Human Factors and Ergonomics Society in 2013 (accessible from http://www.hfes.org//Web/Students/grad_programs.html), only one (Department of Work Environment, Lowell University, Lowell, MA) has the term "work" in the title.

The perspective offered here is that purposeful work represents the principal behavioral activity by means of which humans self-control their interaction with their environment, self-define themselves as individuals, create and human factor new modes and dimensions of work technology and design, and have self-selectively guided their evolutionary trajectory as a species (K.U. Smith, 1965a). That is, with its goal of human factoring the ecosystem to achieve environmental control, human performance of work has a mutually adaptive influence both on the nature of the performer as well as on the nature of work itself.

The major assumptions of this control systems theory (K.U. Smith, 1965a; T.J. Smith & Smith, 1987a) of human work are summarized in Figure 9.1. The theory views work as a closed-loop process in which the worker employs various behavioral strategies to control sensory feedback from the physical, social, and organizational design attributes of the work environment. In so doing, the worker self-organizes and self-defines his or her individual biosocial identity as a distinct member of society, both psychologically and physiologically.

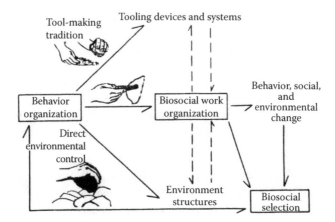

FIGURE 9.1 Control systems theory of human work. Throughout evolution, humans have endeavored to control their behavioral environment through the use of work to fabricate tools and other types of technology, and to organize and structure their society. Work activity of this sort tends to catalyze behavioral, social, technological, and environmental change, which in turn feeds back to influence the primary features of behavioral organization of the individual. (Reproduced from Smith, T.J., In Karwowski, W. ed., *International encyclopedia of ergonomics and human factors*. New York: Taylor & Francis, 2001.)

In addition to individual self-definition, another major feedback influence of work behavior assumed by the model in Figure 9.1 is on the nature of work design itself. That is, interaction with the work environment prompts consideration on the part of the individual worker and the organizational work system as to how work design features can be modified and human-factored to facilitate work performance. In this manner, work itself has a positive feedback effect on cognitive creation of new modes and dimensions of work design and work technology.

Finally, the model in Figure 9.1 assumes that the behavioral feedback effects of work represent the principal mechanism of human self-selection in evolution. This idea will be explored in more detail in Chapter 11.

A number of lines of evidence can be cited to support the control systems model of work depicted in Figure 9.1 (K.U. Smith, 1962b, 1965a). Various observations from behavioral science indicate that both the distinctive personalities as well as the cognitive and motor behavioral proficiencies of different individuals are influenced in a major way by their daily work activity throughout life. Indeed, the motivational behavior of work preoccupies human attention and time to a far greater extent than other sources of motivation such as eating and drinking, aggression, activity-rest cycles, sexual behavior, or relaxation. These context-specific (Chapter 1) behavioral feedback effects of work augment whatever contributions innate biological differences may make to defining individual differences, and we may speculate that such effects date back to the dawn of the species.

What is distinctive about human work? No one who has observed a bee or ant colony can doubt that nonhuman species are entirely capable of intense, highly organized, and socially coherent patterns of work. What then distinguishes the special nature and outcomes of human work? Suggested answers to this question include human capacity for verbal communication and cognition, factors that also are implicated in addressing the related question of "humanness." An undeniably unique feature of human work performance that both complements and underlies its other distinct attributes relates to the use of work to track and thereby control time (K.U. Smith, 1965a). Human capabilities in the control of time encompass perception and comprehension of the past, the future, time continuity, time persistence, historical time, relative time (early, late, then, before, etc.), time interruption, time marking and recording, temporal change, clock time, space-time interaction, time prediction, and space-time relativity.

Humans rely on these capabilities to control spatiotemporal feedback from prevailing conditions in the present and past, and to thereby feedforward control their actions into the future. No other species displays comparable sophistication in temporal awareness and control. However, to attain such control, humans have had to create work-derived artifacts for marking and tracking time because innate comprehension of time apparently differs little among humans and other animals. For example, languages of extant aboriginal cultures contain no words for temporal concepts such as early, late, then, now, past, future, history, and so forth. One of the key events in the evolution of modern humans therefore was the augmentation of space-structured behavior (common to humans and many other animals) with capabilities for time-structured behavior through fabrication of artifacts for telling time. Marshack (1964) offers evidence that such artifacts date back some 30,000 to 40,000 years, anticipating by thousands of years the massive Neolithic and Bronze age megaliths and temples erected in Egypt, on the Salisbury Plain, and elsewhere for tracking time.

Through fabrication of such artifacts, human control of work and of time have become inextricably integrated. From a control systems perspective therefore (Figure 9.1), one of the key self-selective influences of work behavior in human evolution has been the use of work to human factor technological and social means for tracking time, thereby allowing projection of work-derived artifacts, knowledge and skill into the future to benefit generations yet unborn.

9.3.3 GENERAL LAWS AND CUSTOMS OF WORK

Laws and customs of work—the definition of "ergonomics"—refer to work design features and characteristics that conform to the capabilities and limitations of the worker. As evidenced in diverse HF/E handbooks, textbooks, and publications, there are literally hundreds of laws and customs of work that have been promulgated in the form of guidelines, recommendations, and standards. Customarily, these take the form of specifications pertaining to the design of any and all components of the work system: tools, machines, equipment, work stations, software or hardware interfaces, material handling, work layout, production flow, work environment, jobs, or organizational systems. The underlying premise of such specifications is that modifications of work design to conform more closely to the behavioral abilities and skills of the worker will enhance work performance through improvements in productivity, quality, safety, health, and/or efficiency of the workforce and the work system.

The analysis below addresses a series of general conclusions that can be drawn about laws and customs pertaining to the interaction of work design and work performance.

1. *Work performance is highly context-specific.* Extending a theme introduced in Chapter 1, this conclusion means that the behavior of the worker is specialized to a substantial degree in relation to the design context of the workplace and the work environment. Evidence in support of the idea of context specificity is summarized in Chapters 3–6. Essentially all HF/E guidelines and standards pertaining to musculoskeletal performance at work reference work capacity and injury risk to design factors—weight and height of lift, task repetition and force requirements, and so forth. In the realm of work physiology, guidelines regarding both physical work capacity and fatigue susceptibility are referenced to task design factors such as work load and task duration. Recommendations pertaining to scheduling effects on work performance reference design factors such as when the shift is scheduled, length of shift, shift start and stop times, and so forth. One corollary of this conclusion therefore is that "laws" of work cannot and should not be generalized across all work design contexts—instead they must be customized for different work environments and types of work. Another corollary is that insight into sources of variability in work performance for a particular type of work should be based on observations of that performance in the work design context under consideration.

2. *Work motivation also is context-specific.* This means that there is no universal law that explains why people want to work. Although evidence on this point is less comprehensive, suggestive support is provided by observations that among a range of factors implicated in work motivation (related to opportunities that work provides for gainful return, for social interaction, for purposeful activity, for pride in accomplishment, for self-fulfillment, etc.), no single factors or set of factors uniformly applies to all workers and work sectors. For slavery—a class of work quite prevalent for thousands of years until recent times—fear and intimidation were key motivators. For early humans, plus the millions of workers living in poverty today, simple survival through acquisition of food and avoidance of harm likely represents the most important motivational influence for work. The only generalization that appears to apply is that human motivations for work and for self-control of the environment essentially are identical (Section 9.3.2).

3. *Effects of new work design on work performance are not necessarily predictable.* This conclusion follows from point 1. If work performance is context-specific, new work designs (of tools, machines, interfaces, jobs, etc.) will evoke new patterns of variability in work performance that cannot necessarily be predicted from patterns associated with existing designs. To deal with this issue, there is growing acceptance of iterative design testing in which some measure of usability (user acceptance, productivity, quality, safety risk, etc.) of a prototype design is assessed and the design modified based on the results (Gould, 1990). This process is repeated until an acceptable level of performance with the new design is achieved.

4. *The operational validity of work performance models rests on their recognition of context specificity.* Some models of work performance, such as
   ecological or various lifting and fatigue models, are based on the premise
   of performance-design interaction, and therefore build context specificity
   in performance directly into the framework of the model. Others, such as
   information processing, mental workload, human reliability or time-motion
   assessment and prediction models, disregard such interaction in their formulation. Context specificity in work performance is a matter of empirical
   record. The operational validity and consequent predictive power of models
   that ignore this record may therefore be questioned relative to models that
   are context-sensitive.

5. *Organize work as closed-loop system.* Cybernetic or closed-loop control of
   behavior represents an inherent biological feature of work performance at
   the level of the individual (T.J. Smith & Smith, 1987a). The same cannot
   be said of organizational behavior. Instead, closed-loop linkages between
   work effects (outcomes) and work investment (inputs) have to be built into
   the design—the macroergonomics—of work organizations (T.J. Smith,
   1999). There is ample observational and economic evidence to show that
   if yield and investment of a work system (the ratio of which defines work
   productivity) are not closely coupled, the system ultimately is destined to
   flounder if not fail. The need to close the loop between the effort and effects
   of work undoubtedly was recognized by early hominid work groups, was
   manifest in ancient Sumerian and Egyptian work systems, and represents
   an organizational design imperative for successful complex work systems
   of today (Juran, 1964; Kramer, 1963). Closed-loop control of work may be
   implemented for a broad range of organizational design alternatives from
   authoritarian to unsupervised. However, macroergonomic observations on
   organizational design and management (ODAM) point to strategies commonly associated with success in achieving this goal. One is clearly defined
   decision-making authority—every control system needs a controller. A second is the participatory approach—workers represent a good resource for
   tracking both the effort and effects of work. A third is to customize input
   and output measures of work system performance to the particular type
   of system under consideration. A fourth is realistic accommodation of the
   inherent variability of a control system, particularly at the organizational
   level. Although this represents a key feature of modern quality management
   (QM) systems, the same cannot be said of other management domains.

6. *Human error and design error often are synonymous.* The invocation of
   human error as the underlying cause of work-related accidents and injuries
   is not uncommon in the HF/E and safety fields, a proclivity originating with
   Heinrich over 80 years ago (Heinrich, 1931). From the perspective of point
   1, however, as discussed in Chapter 6, we might expect that hazards in the
   workplace, manifest as work design flaws, should be a major contributor to
   errors in work performance. Indeed, a variety of field observations support
   the general conclusion that design factors, often management-related, are
   implicated in about half of all work-related accidents and injuries (T.J. Smith,

2002). Before invoking pure human error therefore, the possible contribution of flaws in physical and/or organizational work design to errors in work performance should be carefully scrutinized.

7. *Ergonomic standards addressing upper limb musculoskeletal injury risk should accommodate individual differences and be referenced to work design.* This conclusion is based on three considerations. First, motor behavioral theory tells us that there are essentially an infinite number of alternative trajectories that can be employed for guiding movements of the hands and arms from one point in space to another (Bernstein, 1967). This is because the total degrees of freedom in the combined muscle-joint systems of these effectors (7) exceeds that of the spatial environment in which movement occurs (6). Second, motor behavioral research tells us that the motor control system is inherently imperfect. This motor control variability ensures that no particular arm-hand movement is ever repeated with precisely the same spatiotemporal pattern by a given individual (Stergiou et al., 2006). Collectively, these two phenomena underlie the observation that upper limb musculoskeletal performance is highly individualized and context-specific (point 1), such that in the same work design context, different individuals may perform differently and the same individual may perform differently at different times. The idea of general ergonomic standards for upper limb musculoskeletal performance that are presumed to apply for all workers across all work design contexts therefore is scientifically untenable. Instead, such standards should be customized for different classes of workers (categorized by age, gender, anthropometry, etc.) and for different types of work.

8. *To modify work performance, modify work design.* As a cybernetic, self-regulated process, work behavior cannot be directly controlled by an outside force or agent. However, variability in work performance can be influenced by modifications to the design of the work environment in which the performance occurs (point 1). Therefore, modifications to work design (encompassing the design of training programs) offer the most realistic, and often the most cost-effective, avenue for modifying work performance. This conclusion represents the *raison d'etre* for HF/E science.

9.3.4 THE FUTURE OF WORK

From a biological perspective, the basic behavioral and physiological attributes and capabilities that humans bring to the performance of work likely have remained fairly consistent over previous millennia and are unlikely to change much in the millennia to come. Some historical support for this view is available. For example, the ergonomic guideline introduced by the ancient Greeks over 2000 years ago that packs carried by Hoplite soldiers be limited to 70 pounds is retained by armies of today.

What has changed is the amount of knowledge and information that has accumulated about the organization and conduct of work and the number of technological and communication innovations developed to facilitate the work process (Chapter 11, Figures 11.2 through 11.4). However, given that variability in work performance is context-specific, the nature of the impact of new designs of technology and work systems on work performance is not entirely predictable (Section 9.3.3, points 1 and 3). Hancock (1997) suggests that given the numbing effect of much computer-based work on job satisfaction, more avenues for job enjoyment and opportunities for leisure must be introduced to sustain worker acceptance of this type of work. Drucker (1999) addresses the possible impact of the newest major technological innovation in work design—the Internet—on the future of work and expresses uncertainty as to what the outcomes will be. In support of such indecisiveness, he points out that the industrial revolution, which also involved major transformations in work design, had many unintended and unanticipated consequences (Section 9.3.3, point 3) for the performance of work.

Nevertheless, at least four predictions for the future trajectory of work can be offered with a reasonable degree of assurance, all of which already are somewhat in evidence. The first is that the persistence of new work designs will rest on the degree to which they augment work productivity, albeit with some lag built in. For example, many observers attribute recent productivity gains to the wholesale computerization of work, an effect that has taken some three decades to emerge. Work designs that do not benefit work productivity eventually will be discarded.

The second and third predictions are that work will become more disseminated on a regional basis but more integrated on a global basis. Advances in distributed communication such as the Internet will facilitate both of these transformations. The need for maintaining production facilities in discrete, physically defined locations for purposes such as manufacturing, mining, or agriculture should not change much in the foreseeable future, although the degree of remote control of such work likely will steadily increase. However, for many other types of work across the entire commercial sphere, from education to finance to retail, concepts of the "job" and of "going to work" almost certainly will undergo fundamental modifications.

Finally, the decline in population growth and progressive aging of the workforce in most western societies coupled with an unabated demand for workers suggests that respect for work capabilities of older workers will undergo a resurgence, and that the concept of retirement as an expected and accepted concomitant of aging may be fundamentally revised.

9.4 WORK PERFORMANCE VARIABILITY AND INTERACTION WITH COMPLEX SOCIOTECHNICAL SYSTEMS

The previous section sets the stage for the remainder of this chapter, which will be concerned with how and why variability in work performance is influenced by interaction with complex ST systems. In particular, this linkage is addressed in Sections 9.4.1 through 9.4.6 in relation to firms and in Section 9.5 in relation to economies and nation-states. As noted in Section 9.1, the sections below will recapitulate themes related to performance variability emphasized in earlier chapters; namely, (1) the nature of work performance variability, (2) sources of such variability, and (3) application of control systems principles to the understanding of sources of such variability, as influenced by interaction with the performance of complex ST systems.

9.4.1   WORK PERFORMANCE VARIABILITY AND THE DESIGN AND MANAGEMENT OF ORGANIZATIONAL SYSTEMS

In terms of the influence of complex organizational system performance on work performance variability, there are two key considerations, namely, the degree to which system performance (1) supports jobs of the workforce (which dependson the productivity of workers performing those jobs), and (2) delivers service to system customers (who after all, also are workers). At least three key reasons may be offered for focusing on this topic in this chapter. First, work performance variability equates in a major way to the quantity and quality of jobs, and the major source of jobs in the developed countries of the world is the private or public organization; as pointed out in Section 9.2, private companies have a long history of doing so. In the United States alone, there are about six million such companies employing some 120 million people and controlling roughly $30 trillion in assets (Schumpeter, 2013a).

Second, over many decades, a great deal of effort has gone into evaluating what factors explain high-performing organizations and their ability to support jobs and serve customers. Finally, over the past three decades, the design and management of organizational systems has become a major disciplinary concern of HF/E science, a concern termed macroergonomics (Hendrick, 2007). The remainder of this section thus considers the interaction of organizational and work performance variability. The next five sections address the success of ergonomics in benefiting the former in service of the latter.

9.4.1.1   How Are Organizational and Work Performance Variability Interrelated?

Formal inquiry into this questions dates back at least to Adam Smith (1776, 1977), and over the centuries the financial, social, and safety and health welfare of the worker have been closely linked to the performance of organizational systems responsible for work system oversight. Worker welfare across all of these parameters—the essence of work performance variability—has waxed and waned over the centuries in relation to the business cycle, technological innovation, political transitions at the nation-state, regional or local levels, market competition, natural disasters, and so forth, as well as the ability of organizational systems to cope with and control these externalities.

I will limit myself here to one example; namely, the impact of the so-called 2007–2009 Great Recession on the performance of workers in the United States. It is widely agreed that this event was precipitated by banks in the United States and elsewhere shouldering excess risk, primarily in the subprime housing market. When too many house owners fell behind on their payments, coupled with the bankruptcy of Lehman Brothers, the U.S. financial system collapsed like a house of cards, and the contagion spread around the world. However, it was U.S. workers and consumers, not the bankers, who subsequently suffered.

Figure 9.2 illustrates this point with plots of the total labor force, total employment, and private sector employment in the United States from 2003 to 2013, encompassing the period of the Great Recession, based on data from the U.S. Bureau of Labor Statistics (http://www.dlt.ri.gov/lmi/ laus/us/usadj.htm). During the downturn

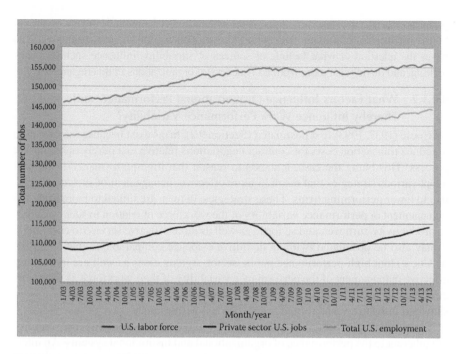

FIGURE 9.2   Effects of the 2007 'Great Recession' on U.S. jobs. (Based on U.S. Bureau of Labor Statistics data. Available at http://www.dlt.ri.gov/lmi/laus/ us/usadj.htm.)

period, although the total labor force didn't really drop (top plot in Figure 9.2), total employment decreased from a high of 146.6 million in November 2007 to a low of 138.0 million in December 2009, a loss of about 8.6 million jobs, and the unemployment rate increased from a low of 4.4 percent in May 2007 to a high of 10 percent in October 2009, an increase of 2.3-fold.

The Great Recession also caused total U.S. public sector employment to drop by over 580,000 jobs, the largest decrease in any sector (www.hamiltonproject.org/ papers/a_record_decline_in_government_jobs_implications_for_todays_ economy_ an/). This added workers directly to the unemployment rolls, translating to fewer paychecks and lower consumption, thereby magnified the effects of the recession.

An expanded profile by The Economist (2014, p. 54) of 12 major economic slumps in the U.S. between that of 1792 (Bank of U.S. crisis), and the 2007–2009 recession, focuses on stock market losses, with obvious adverse implications for work performance (as enumerated above) during each slump. Across these 12 slumps, peak market losses averaged 39.5 percent, with the Wall Street crash of 1929-33 (84 percent) and the Great Recession (56.5 percent) featuring the two largest peak losses.

For the typical worker transitioning from a job to the unemployment line during economic slumps, the frequent consequences are loss of financial security, often loss of a home, lower self-esteem, greater social isolation, increased stress, in many cases divorce, loss of technical skills, and perhaps most importantly, the loss of a sense of personal fulfillment and accomplishment, as well as social support and camaraderie, that many jobs provide. In other words, economic slumps embody the dark side of variability in work performance, as a consequence of decisions by hundreds of thousands of U.S. organizations to reduce (or eliminate) their workforces in the face of financial tsunamis. This raises the question of what factors, or sources of variability, influence such decision-making, and distinguish more from less-successful organizations in this regard.

9.4.1.2 What Factors Influence Organizational Performance, and Thereby Influence Work Performance Variability?

As Coese (1937, 1993) first postulated (Section 9.2), firms exist because they are able to reduce transaction costs for contract negotiations relative to a purely open market process. This being the case, it stands to reason that high-performance firms will adopt various strategies that result in more effective management of such costs relative to lower performing firms. These strategies, in turn, are linked to more effective management of performance variability of the workforce in relation to such variables as productivity, turnover, and so forth, as well as more effective service to customers. It may also be postulated (without definitive evidence comparable to that compiled by Coese) that high-performance public sector organizations as well are marked by effective management of the performance variability of their workforce as well as of customer service. This section provides a selected analysis of factors that have been shown to influence variability in organizational performance.

The popularity of the term "business cycle" suggests that variability and volatility are inherent to the performance of organizational and institutional systems. An interesting recent perspective on this situation (Schumpeter, 2013b) suggests that companies may not be shrinking but they are becoming more fragile, one reason being the advent of the Internet. Unlike older companies who have been able to rely and prosper on economies of scale (such as General Electric [GE] and General Motors [GM]), this strategy is unavailable to Internet giants who must constantly struggle to innovate to see off competitors. This analysis also points out that company leadership also is less secure—the composition of the top 1 percent is constantly changing.

A synopsis by Norris (2013) of changes in the top companies in the Standard & Poor's (S&P) 500 stock index underscores the stark nature of this fragility. This author notes that there were five technology companies among the top 25 companies (in terms of stock capitalization) in the S&P 500 in the year 2000 that lost more than 90 percent of their stock value over the ensuing 13 years. Two other of these companies in year 2013 are worth less than a third of their year 2000 value.

In Chapter 10, Prof. Fisher documents a dramatic and ominous consequence of organizational and institutional design fragility; namely, the predisposition under such circumstances to fracture-critical behavior. Examples cited include the unanticipated debt crises in the banking and home mortgage systems that precipitated the Great Recession of 2008, flooding of New Orleans after Hurricane Katrina and of the subway system in New York after Hurricane Sandy, and explosive failure of the nuclear power plant in Fukushima, Japan, after the earthquake and tsunami.

At the outset, based on analyses in previous chapters, we perhaps should not be surprised that both nativist and empiricist perspectives (Goldhaber, 2012) have been offered to explain business performance. In the nativist camp, based on twin studies, Shane (2010) concludes that genetic (biological) factors exert a demonstrable influence over which jobs people choose, how satisfied they are with those jobs, how frequently they change jobs, how important work is to them, how poorly they perform on-the-job, and the salary they earn (results suggest that around 40 percent of the variation between people's incomes is attributable to genetics). *The Economist* (2010a) also has reviewed this nativist evidence. Yet Shane (2010) points out that environmental design (i.e., context-specific) factors also influence the outcomes cited above, and that environmental and biological factors interact in subtle ways. In the remainder of this chapter, the emphasis will be on the influence of the former.

One source of information about factors that characterize strong organizational performance is the "top workplace" surveys that are conducted periodically in different U.S. cities and states. The 2013 Minnesota survey for example (Oslund, 2013) found, based on responses from 64,342 employees employed in 300 Minnesota companies, that the following six factors are deemed most important by employees in rating their workplaces (listed in order of decreasing importance): connection, direction, execution, my work, my manager, and my pay and benefits. Kennedy (2013) reports that a virtual portfolio comprised of 42 public companies from around the country recognized as top workplaces outperformed the S&P 500 index by 48 percent over the period 2008 through 2012.

Another source of information about factors that characterize top performing organizations is surveys that attempt to identify such factors relative to characteristics of less successful companies. I will use two examples, separated by some thirty years, to illustrate this approach. The prototype is *In Search of Excellence* by Peters and Waterman (2004), originally published in 1982. These authors surveyed 43 of America's best-run companies and identified the following eight principles that characterized the success of these organizations:

- A bias for action: a preference for doing something, rather than deferring decision-making
- Staying close to the customer: learning customer preferences and catering to them
- Autonomy and entrepreneurship: encouraging independent initiative by small groups within the organization
- Productivity through people: promote awareness that all employees contribute to company success
- Hands-on management: executives must stay in touch with essential business activities of the company
- Narrow focus: the company should stick to the business it knows best
- Few administrative layers
- Foster employee dedication to central values and mission of company

Some thirty years later, CedarCrestone (2013) engaged in a similar survey exercise and identified the following five generalized practices of top performing organizations, most focusing on human resource management:

- They standardize processes and use sophisticated change management practices.
- They avoid extensive customization of their human resource management system.

- They feature higher levels of adoption of employee and manager self-service and shared services.
- They have sophisticated business intelligence solutions in place and put these tools in the hands of managers.
- They have more human resource technologies in use and spend less on human resource technology per employee.

It is evident that there is little overlap between these two analyses of what distinguishes a top performing organization. One possible interpretation is that criteria have changed over thirty years. More probably, as Schumpeter (2009) points out, it is unlikely that these corporate hagiographies prove anything. The following observations provide support for this conclusion: (1) five years after *In Search of Excellence* first appeared, a third of its ballyhooed companies were in trouble, (2) Andrew Henderson of the University of Texas recently subjected "excellence studies" of corporations to rigorous statistical analysis and concluded that luck is just as plausible an explanation of their success as excellence, and (3) conventional wisdom holds that customer-relationship management is all about learning about and from your clients (see Peters and Waterman's principle number 2 above), but Henry Ford pointed out many years ago that if he had listened to his customers he would have built a better horse and buggy. Schumpeter (2009) sums up by noting that if management could indeed be reduced to a few simple principles, then we would have no need for management thinkers. But the very fact that the art of management defies easy solutions leaves managers in a perpetual state of angst, meaning that there will always be demand for more irritating management gurus.

My own take on the matter is that top performance of a given organization is in the eye of the beholder; as with individual performance (Chapter 3), organizational performance is highly context-specific, dependent on such factors as the culture, the market systems or operational frameworks, the sociotechnical characteristics, and so forth, in which a given organization operates and functions; and it therefore is challenging, and possibly highly unlikely (as Schumpeter [2009] suggests), to be able to identify universal design principles applicable to a broad range of different top performing organizations.

Nevertheless, an attempt to identify a selected sample of design factors that have been shown by different observers to influence variability in organizational performance, with consequent effects on work performance variability, is offered in Table 9.1. Paragraphs below expand on factors listed in the table.

9.4.1.2.1 Building Comfort

To my knowledge, some of the most definitive evidence demonstrating a link between organizational performance and work performance variability derives from occupancy quality studies of office buildings and the office workers working in these buildings. Specifically, these studies document a statistically significant relationship between office worker perceptions of the comfort of the building they work in, and their perceptions of how productive they are on-the-job. Key results are illustrated in Figures 9.3 and 9.4.

The most persuasive evidence regarding this relationship, based on a survey by Leaman and Bordass (2006) of workers employed in 151 different office buildings

TABLE 9.1

Sources of Variability in Organizational Performance: A Selective Survey

Factors Implicated in Effective Organizational Performance	Comments	Reference
Building comfort	Office worker perceptions of building comfort are correlated in a statistically significant manner with their perceptions of how productive they are at work.	Leaman and Bordass (1999) T.J. Smith and Orfield (2007)
Buyer's market	Increased unemployment resulting from the recent financial downturn has been accompanied by a drive for efficiency on the part of U.S. companies, working conditions reminiscent of a century ago, low worker engagement, and a collapse of trust in business.	Schumpeter (2013a) St. Anthony (2013) Semuels (2013)
Government research and tax policy	Through both state-sponsored research (that spills over into the private sector) and tax policy (that encourages private sector research), the state has played, and continues to play, a major role in promoting innovation and contributing to the success of technology-based businesses.	Mazzucato (2013) Schumpeter (2013g)
Intellectual capital	Intellectual capital (IC), comprising organizational assets such as talented and committed workers, cultural values, and long-term relationships among the firm and its stakeholders, plays a strategic role in gaining and sustaining competitive advantages, making management of IC a key issue in the management agenda.	Martín-de-Castro et al. (2011) Subramaniam and Youndt (2005)
Leadership	Good judgment is one of the most important attributes of effective leadership. Good corporate management (i.e., good leadership) boosts corporate performance (measured in terms of productivity, profitability, growth, and survival).	Schumpeter (2013d) Schumpeter (2014)
Limits to growth	A 2008 update of the original 1972 Club of Rome *Limits to Growth* analysis suggests that, in line with projections of the earlier analysis, industrial output per capita may collapse by the end of the twenty-first century.	Turner (2008)

(continued)

TABLE 9.1 (Continued)
Sources of Variability in Organizational Performance: A Selective Survey

Factors Implicated in Effective Organizational Performance	Comments	Reference
Pay for performance	Nearly 40 percent of top-paid CEOs in the United States headed companies that either failed or required a federal bailout during the recent financial crisis, or the CEOs themselves had to pay massive fraud-related fines or settlements, or were fired. Pay for performance of regular workers also may be problematic.	Anderson (2013) Free Exchange (2013c) Stafford (2013)
Rules for success	Ideas for business success, distilled from selected sources.	Raynor and Ahmed (2013) Schumpeter (2012) Schumpeter (2013f)
Technology	The impact of computerization and information technology on economic productivity has been decidedly mixed. There are a number of examples where negative effects on workers outweigh positive ones.	Brynjolfsson, E. (1993) Free Exchange (2013a) Schumpeter (2013c) Siegele, L. (2010) Strickler, J. (2013).

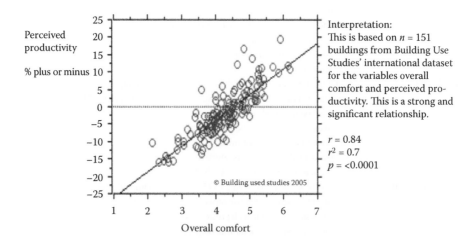

FIGURE 9.3 Office worker perceptions of their on-the-job productivity relative to their perceptions of building comfort based on a survey by Leaman and Bordass (2006) of workers employed in 151 different office buildings.

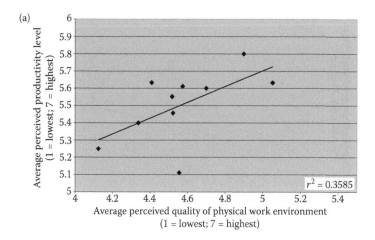

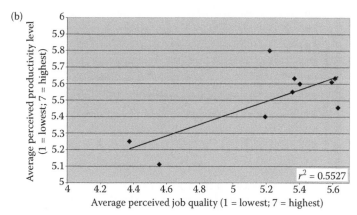

FIGURE 9.4 (a) Relation between average perceived productivity levels and average rankings for perceived overall physical environment quality, for the 10 sample cohorts. (b) Relation between average perceived productivity levels and average rankings for perceived overall job quality for the 10 sample cohorts.

located in a number of different countries, is shown in Figure 9.3. Each point in the figure represents results from one building; office workers in each building were asked to judge both the comfort of the building they were working in and their level of on-the-job productivity. Results were averaged across all employees in a given building to yield the plot in Figure 9.3. There is a statistically significant positive correlation ($r^2 = 0.7$; $p < 0.0001$) between worker perceptions of building comfort and their perceptions of how productive they are on-the-job.

Results in Figure 9.4a reinforce those in Figure 9.3 with findings from a survey by Smith and Orfield (2007) of workers employed in a much smaller sample of office buildings ($N = 10$); each point in the figure represents results averaged across all employees working in a given building. Figure 9.4a shows a statistically significant relationship ($r^2 = 0.36$) between employee perceptions of the overall quality of their physical environment, and their perceptions of on-the-job productivity.

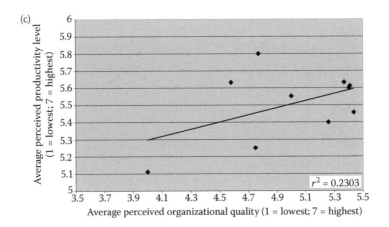

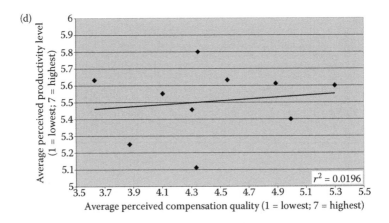

FIGURE 9.4   (Continued) (c) Relation between average perceived productivity levels and average rankings for perceived overall organization quality for the 10 sample cohorts. (d) Relation between average perceived productivity levels and average rankings for perceived overall compensation quality for the 10 sample cohorts.

Results in Figures 9.4b through 9.4f extend those in Figure 9.4a in showing how other occupancy quality factors influence office worker perceptions of their productivity. Specifically, correlations between worker perceptions of overall job quality (Figure 9.4b) and of overall organizational quality (Figure 9.4c) and their perceptions of on-the-job productivity, also are statistically significant ($r^2 = 0.55$ and $r^2 = 0.23$, respectively). On the other hand, no statistically significant relationships are observed between worker perceptions of their on-the-job productivity and their perceptions of overall compensation quality (Figure 9.4d), overall work station quality (Figure 9.4e), or overall employment quality (Figure 9.4f).

Collectively, results in Figures 9.3 and 9.4 indicate that office workers judge their on-the-job productivity levels to be closely aligned with the quality of the physical environment of the building they work in (particularly building comfort), and with the quality of their job and of their organization. In contrast, compensation,

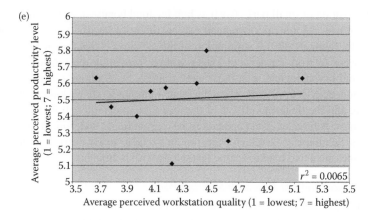

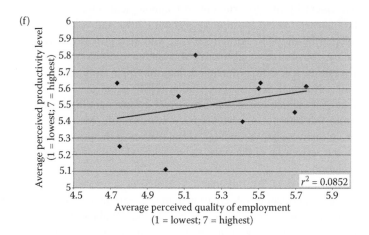

FIGURE 9.4   (Continued) (e) Relation between average perceived productivity levels and average rankings for perceived overall work station quality for the 10 sample cohorts. (f) Relation between average perceived productivity levels and average rankings for perceived overall employment quality for the 10 sample cohorts. (Reproduced from Smith, T.J., & Orfield, S.J., In *Proceedings of the Human Factors and Ergonomics Society 51st Annual Meeting.* Santa Monica, CA: Human Factors and Ergonomics Society, 2007.)

workstation, and employment quality are judged to have much less of an influence on worker perceptions of productivity. The inescapable conclusion is that organizations aspiring to high levels of performance should play close attention to the nature and degree of occupancy quality of buildings housing their employees.

9.4.1.2.2   Buyer's Market

Semuels (2013) points out that the relentless drive for efficiency on the part of U.S. companies consequent to the recent financial downturn plus a ready availability of workers due to high unemployment (Figure 9.2) has been accompanied by a harshness in the workplace reminiscent of assembly-line conditions a century ago: boring work, elimination of small perks (such as holiday food bonuses), extra hours without pay, lack of rest, rigorous performance quotas, threat of layoffs, video monitoring, and so forth. Semuels' (2013) summation of the situation is that businesses are asking employees to work harder and longer without providing the kinds of rewards, financial and psychological, that once were routine. St. Anthony (2013) notes that one consequence is low worker engagement—employee disenchantment is getting worse, and the lingering effects of the recession make it tough for managers to change things. A broader effect appears to be a general and widespread loss of trust in business; Schumpeter (2013a) reports that in a recent survey of trust in the professions, businessmen and bankers came last, along with politicians.

9.4.1.2.3 Government Research and Tax Policy

Mazzucato (2013) observes that technological innovation in the United States manifest in some of the world's top companies such as Apple and Google owes a great deal to investment by the government in technological research. Examples cited include the Internet, Global Positioning System (GPS), and voice activation, originally developed by the U.S. Armed Forces, who also provided early funding for Silicon Valley; touchscreen and the HTML language developed by academics in public universities; a $500K government loan to Apple in its early days before it went public; a National Science Foundation (NSF) grant to Google that enabled development of its search engine; and recurrent government big bets on new technologies from the Internet to revolutionary new aircraft. Mazzucato (2013) concludes that, in many cases, technological innovation in the private sector critically depends on the active role of the state, the unacknowledged enabler of today's electronics revolution.

Considering that as of 2013, Apple has created or supported 598,500 U.S. jobs (http://www.apple.com/about/job-creation/) and Google had well over 46,000 permanent employees (http://investor.google.com/financial/tables.html), the impact of the entrepreneurial state on both organizational and work performance variability can hardly be overstated.

Schumpeter (2013g) inserts a cautionary note by observing that there are numerous examples of entrepreneurial states throwing good money after bad by investing in misguided or failed technologies or companies. Mazzucato (2013) emphasizes the key role of competition and innovation networking in distinguishing successful from unsuccessful state entrepreneurial efforts. In any event, the key role of government in influencing the nature and extent of both organizational and work performance variability seems well established.

9.4.1.2.4 Intellectual Capital

IC is defined as a firm's intangible assets such as talented and committed workers, cultural values, patents, and long-term relationships among the firm and its stakeholders—customers, allies, suppliers, and society in general. Martin-de-Castro et al. (2011) maintain that from a strategic point of view, IC is becoming a crucial factor for a firm's long-term profit and performance in the knowledge-based economy. Subramaniam and Youndt (2005) argue that the development and application of knowledge remain the basis of economic growth and welfare, and that an organization's capability to innovate, as well as its performance, are closely related to its IC resources and to its ability to utilize its knowledge resources in an effective manner.

9.4.1.2.5   Leadership

As noted earlier in this chapter, an October 2013 search of Amazon books retrieved almost 100,000 titles for the term "leadership books." The premise of many of these books is that effective leadership represents a major determinant of organizational, institutional, and governmental performance effectiveness. Leadership is a common social design feature of many subhuman species, and undoubtedly has represented a pervasive feature of the human condition throughout human evolution. The nature of variability in both organizational and work performance and the welfare of workers the world over are fundamentally tied to the effectiveness of organizational leadership.

Schumpeter (2013d) offers a perspective on factors that contribute to, or detract from, effective leadership. This author first cites four recent books and an article on leadership to make the following points: (1) a popular point of view is that leaders should focus on their strengths, and (2) yet others point out that forceful, decisive leadership may be associated with a stubborn, close-minded, and possibly even bullying style, coupled with an arrogant, condescending approach to subordinates. Schumpeter (2013d) draws two key conclusions: (1) with reference to a major theme in this book, leadership skills are context-dependent—good leaders must adjust their leadership approach and style to the particular organizational and sociotechnical context in which they operate, and (2) judgment on the part of the leader is what matters most and is hard to measure—it takes judgment to know when to moderate your approach or when to pull out all the stops.

In a subsequent summary, Schumpeter (2014) cites recent evidence showing that good corporate management (i.e., good leadership) boosts corporate performance, measured in terms of productivity, profitability, growth, and survival.

9.4.1.2.6   Limits to Growth

In 1972, in one of the early applications of a computational approach to modeling system dynamics, the Club of Rome issued its infamous report, "The Limits to Growth" (Meadows et al., 1972). The analysis, by a team of analysts from the Massachusetts Institute of Technology (MIT), presented some gloomy scenarios for global sustainability based on a system dynamics computer model to simulate the interactions of five global economic subsystems; namely, population, food production, industrial production, pollution, and consumption of nonrenewable natural resources. Projections of the 1972 analysis are based on observed data covering the years 1900–1970.

Turner (2008) updates the 1972 analysis by comparing recently collated historical data for the years 1970–2000 with scenarios presented in the Limits to Growth (LtG) model. Turner's projections for the twenty-first century using historical data for the years 1970–2000 compare favorably with key features of a LtG business-as-usual scenario called the standard run scenario, which results in collapse of the global system by the end of the twenty-first century.

Results of the Turner (2008) twenty-first century projections for industrial output per capita are illustrated in Figure 9.5. The different plots in the figure depict 1900–1970 data that the LtG model relied on (in the shaded area to the left of the figure); effects of comprehensive use of technology (open triangles, top plot; stabilized world projections, assuming stabilizing behavior and policies (open squares); standard run scenario, extending the LtG projections (open diamonds); and historical data for the years 1970–2000 (solid circles), on which Turner based his updated analysis.

Arguably, industrial output per capita represents an economic variable, and projections in Figure 9.5 might better be considered in Section 9.5. However, industrial output ultimately relies on the collective efforts of multiple organizations and firms and the workers employed by these organizations. These results thus are discussed here as a direct manifestation of organizational variability and as an indirect manifestation of work performance variability. Based on results shown in Figure 9.5, Turner (2008) concludes that projections based on 1970–2000 historical data

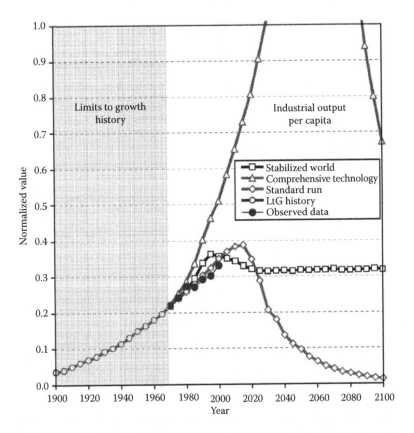

FIGURE 9.5 Comparison of observed data (solid circles ●) for industrial output per capita with the LtG model output for each scenario: "standard run" with open diamonds (◊), "comprehensive technology" with open triangles (△), and "stabilized world" with open squares (□). The calibrated model output over 1900–1970 is shown with open circles (O). (Reproduced from Turner, G.M., *Global Environmental Change, 18*, 397–411, 2008.)

compare most closely with the LtG standard run projection, and the data do not compare well with other scenarios involving comprehensive use of technology or stabilizing behavior and policies. If Turner's conclusions are even partially accurate (and some of the readers of this chapter likely will find out), all other possible influences on work performance variability fade to insignificance relative to the implications of the standard run projection shown in Figure 9.5.

9.4.1.2.7   Pay for Performance

Many companies claim to peg the compensation of their senior managers to company performance, a pay for performance scheme. The premise of this scheme is that, with the promise of higher compensation, CEOs will be motivated to boost the performance of their companies. Anderson (2013) and Stafford (2013) both report deep skepticism about this premise. The former author concludes that for more than two decades, corporations have gotten away with the pay for performance sham at the expense of workers, shareholders, and taxpayers. In support of this conclusion, Anderson (2013) looked at the record during the recent financial crisis of CEOs who had made the top 25 highest-paid lists in at least one of the past 20 years. Nearly 40 percent of these CEOs headed companies that either failed or required a federal bailout during the recent crisis, or the CEOs themselves had to pay massive fraud-related fines or settlements or were fired. Anderson observes that the pay for performance myth has been unraveling for years, and that it is time to put the notion to rest. Note that she identifies workers as one group victimized by the scheme.

Pay for performance of regular employees also may be problematic. First, there is ample evidence (such as that illustrated in Figure 9.4d) that, for many employees, pay may not be the strongest motivator for performing in an effective and productive manner. Second, particularly for office workers, it can be difficult to impossible to measure performance (Free Exchange, 2013c). Finally, this analysis also summarizes suggestive evidence that diligent efforts to monitor employee performance may be counterproductive, particularly for those engaged in creative or innovative work. The seemingly counterintuitive conclusion of the Free Exchange (2013c) analysis is that sometimes the wisest thing to do is to just let workers get on with the job.

9.4.1.2.8   Rules for Business Success

Prescriptions for business success by Peters and Waterman (2004) and CedarCrestone (2013) were summarized earlier to suggest that there likely are no generalized rules that apply to all companies and that a likely interpretation is that business performance is context-specific. Additional rules for business success are distilled here from selected sources to further underscore the point that there is little agreement among different observers as to what differentiates a successful company. Thus, Schumpeter (2012) cites various lines of evidence that successful companies cope with complexity by making great efforts to keep their business models as simple as possible, and this same author (Schumpeter, 2013f) also points out that company managers and employees would be better off if they reduced distractions and disruptive demands by doing less (such as rationing e-mail and convening fewer meetings) and thinking more.

Raynor and Ahmed (2013) extend this idea with their assertion that it is how companies think that distinguishes more from less successful companies. These authors reviewed the records of 25,000 companies over 45 years (1966–2010) and focused on 344 companies that produced exceptional results over this period. Based on this analysis, they distilled two (rather unsurprising) rules for success; namely, companies are more likely to succeed over the long run if they emphasize (1) better before cheaper—compete on quality or performance rather than on price, and (2) revenue before cost—drive up revenue (i.e., higher prices) rather than driving down cost.

9.4.1.2.9 Technology

As noted earlier in this chapter, feedback between technological innovation and human performance represents a fundamental concomitant of the human condition. By creating new technologies, humans have controlled the course of their own evolution (Chapter 11). Technological innovation is key to human productivity, and thereby contributes to economic prosperity and the wealth of nations. Given today's computerized ST systems and globally integrated computerized technology coupled with the advent of mobile (and increasingly wearable) smart systems, it is worthwhile asking whether the historical benefits of technology have been sustained in the modern digital era.

The answer to this question is decidedly mixed. The most systematic analysis of the link between computerization of society and economic productivity has revealed what is termed the "productivity paradox" (Brynjolfsson, 1993). This author first notes that productivity is the fundamental economic measure of a technology's contribution. He contrasts a claim from the mid-sixties that IT will usher in the biggest technological revolution in history with a flat assertion some 25 years later that computers do not boost productivity. His analysis then reviews various lines of evidence suggesting that delivered computing power in the U.S. economy increased by two orders of magnitude from 1970 to 1993, yet productivity in the service sector remained essentially flat during this period. Brynjolfsson (1993) points to four probable reasons why computers have not measurably improved productivity: (1) measurement error, (2) time lags between current IT costs and payoffs from IT investment, (3) use of IT for redistributive purposes by firms, making it privately beneficial without adding to total output, and (4) mismanagement—the lack of explicit measures of the value of IT is associated with misallocation and overconsumption of IT resources (this supports the arguments of Martín-de-Castro et al. [2011] and Subramaniam & Youndt [2005] that management of intellectual capital should represent a key issue in the management agenda [Table 9.1]).

Fast forward some two decades to the age of the Internet and smart mobile technology, and the situation has not changed much. Free Exchange (2013a) cites various studies that attempt to quantify the gains that the Internet has brought to consumers, with a resultant rough estimate of a net increase in economic growth of 0.39 percent of GDP since 2002. However, this analysis also points out that wasted time spent on the Internet detracts from productivity.

Buttonwood (2013b) reinforces this point with a summation of evidence pointing to the conclusion that the Internet has not yet produced hoped-for productivity gains. This analysis cites a productivity puzzle on both sides of the Atlantic (thus channeling Brynjolfsson, 1993), wherein employment following the financial crisis has recovered more strongly than GDP. Buttonwood (2013b) also points out that the Internet has made freelancing easier but it also has disrupted work-life balance for full-time workers who now may be expected to stay in touch with the office 24/7.

Implications of smart technology (i.e., smart phones, ubiquitous sensors) for the future of economic productivity also are uncertain and highly speculative at this point. Siegele (2010, p. 18) cites claims that smart technology will give a huge boost to productivity but will also lead to greater inequality. Given the failure of IT, dating back now well over four decades, to boost productivity, one is inclined to treat the former claim with some skepticism.

There are other reports of the dark side of today's IT beyond the negative impact of the Internet on work-life balance noted by Buttonwood (2013b). Thus, Strickler (2013) observes that the unprecedented technology-driven evolution of the workplace can leave older workers feeling obsolete. More broadly, and perhaps more ominously, Schumpeter (2013c) references recent projections suggesting that jobs currently held by knowledge workers will increasingly be displaced by smart machines.

9.4.1.3 Summary

One notable feature of the foregoing analysis is the lack of predictability with regard to the influence of the factors listed in Table 9.1 on organizational performance variability. Many organizations failed to predict the onset and the severity of the recent financial crisis, causing dramatic contraction of some and failure of others. Government promotion of business has resulted, in an unpredictable manner, in success of some projects, but failure of others. There appears to be some support for the accuracy of the original Club of Rome projections regarding limits to growth, but there is considerable uncertainty as to the trajectory of these projections in the decades to come. Paying for performance at the senior management level in many cases seems to be completely disconnected from the performance actually observed. So-called rules for business success vary dramatically across different observers. From an economic perspective, the vast organizational investment in technology over the past few decades has not resulted in anticipated work productivity benefits. Collectively, these observations recall conclusions of Karwowski (2012) and Silver (2012) that predicting the performance of complex ST systems is inherently difficult.

The nine selected factors listed in Table 9.1 and discussed above—each of which have been shown to influence variability in organizational performance and consequently in work performance—all are design factors. This suggests that variability in both organizational and work performance is highly context-specific despite claims (referenced earlier in this section) that there is a biological (nativist) contribution to variability in the skills and abilities of managers. Moreover, the idea that organizational design and organizational performance are intimately coupled attracted the attention of the HF/E community, resulting in a broadening of traditional microergonomic HF/E concerns (i.e., workstations, jobs, technology, etc.) to an analysis of the design and management of organizational systems, termed macroergonomics. The next section addresses this topic that today represents a major HF/E disciplinary area.

9.4.2 Macroergonomic Perspectives on Organizational Performance Variability

Macroergonomics emerged as a distinct area of emphasis within HF/E science in the early 1980s following a study commissioned by the U.S. Human Factors and Ergonomics Society calling for attention by the field to the design and management of organizational systems (Kleiner, 2008). Since its inception, macroergonomic professionals have been concerned with formal design characteristics of different types of organizations and how different organizational designs influence the functioning of both workers and managers (Carayon & Smith, 2000). The inspiration for this approach is the ST systems approach to work systems design (Emery & Trist, 1965; Hendrick, 1984) that emphasizes the need to jointly consider both technology and personnel when designing organizational systems.

A basic impetus identified early on for developing macroergonomics was the need for going beyond the 10- to 20-percent performance gains typically realized in ergonomics to the 60- to 80-percent improvements realized when attending appropriately to ODAM factors (Kleiner, 2008). This author refers to case studies in academia, industry, and government that demonstrate 60- to 90-percent performance impact and positive qualitative changes such as culture change. Hendrick (2007) introduces a series of results substantiating this claim. For example, with one project a new college based at the University of Denver was designed to accommodate 30 study centers in the United States and Germany. A macroergonomic approach achieved the following results compared with an earlier organizational effort at the University of Southern California that did not involve such an approach: a 27 percent savings in operating expenses, a 23 percent reduction in campus staffing requirements, a 20 percent reduction in off-campus study center administrative time (30 centers), and a 67 percent reduction in average processing time for student registrations, grades, and so forth, from off-campus locations.

The same report summarizes outcomes from other macroergonomic projects—at a petroleum company (see Imada [2002] for details), at Red Wing Shoes, and at L.L. Bean—that resulted in improvements in different dependent measures of organizational performance ranging from 51 to 97 percent. Similarly, Carayon et al. (2013) documents the variety of benefits that the macroergonomic approach can bring to the performance of health care systems.

From its beginnings, macroergonomics has adopted a formal approach for defining the parameters that characterize different dimensions of organizational design and a set of methods for evaluating these dimensions. This analytical approach delineates both the *nature* of organizational design variability (from a macroergonomic perspective), as well as how both organizational and work performance vary in relation to different organizational design dimensions (i.e., *sources* of variability). Table 9.2 summarizes this macroergonomic analytical framework based on summaries provided by Hendrick (1984, 1991, 2007) and Kleiner (2008).

The left column in Table 9.2 specifies six different organizational design parameters identified by Hendrick (2007) as "Dimensions of Work Design." With this heading, this author explicitly links organizational design with work design, recapitulating the theme adopted throughout this chapter. The right column summarizes the

TABLE 9.2

Macroergonomic Analytical Framework for Delineating Different Dimensions of Organizational Variability, and Their Influence on Organizational and Work Performance

Dimensions of Work Design	Impact on Organizational and Work Performance Variability
Complexity – Vertical Differentiation	Refers to the number of hierarchical levels separating the chief executive position from the jobs directly involved with the system's output. These levels typically increase as size of organization increases. Factors that affect span of control are degree of professionalism in employee jobs, degree of formalization (below), type of technology, psychosocial variables, and environmental characteristics.
Complexity – Horizontal Differentiation	Refers to the degree of departmentalization and specialization within a work system. Increases complexity because a greater number of methods of control, as well as more sophisticated and expensive methods, are required. Division of labor (Adam Smith, 1776, 1977) creates groups of specialists or departmentalization. The most common ways of designing departments into work systems are on the basis of (1) function, (2) simple numbers, (3) product or services, (4) client or client class served, (5) geography, and (6) process. Most large corporations use all six.
Complexity – Spatial Dispersion	Refers to the degree an organization's activities are performed in multiple locations. Three common measures are (1) the number of geographic locations, (2) the average distance of the separated locations from the organization's headquarters, and (3) the proportion of employees in the separated units in relation to the number in the headquarters. Complexity increases as these measures increase.
Complexity – Integration	Refers to the number of mechanisms designed into a work system for ensuring communication, coordination, and control among the differentiated elements. As the differentiation of a work system increases, the need for integrating mechanisms also increases. Integrating mechanisms that can be designed into a work system are formal rules and procedures, committees, task teams, liaison positions, system integration offices, and computerized information and decision support systems. Too few integrating mechanisms result in inadequate coordination and control among the differentiated elements. Too many integrating mechanisms stifle efficient and effective work system functioning and usually increase costs.

(continued)

TABLE 9.2 (Continued)

Macroergonomic Analytical Framework for Delineating Different Dimensions of Organizational Variability, and Their Influence on Organizational and Work Performance

Dimensions of Work Design	Impact on Organizational and Work Performance Variability
Formalization	Defined as the degree to which jobs within the work system are standardized. Highly formalized designs allow for little employee discretion over what is to be done, when it is to be done, or how it is to be accomplished. In highly formalized designs, there are explicit job descriptions, extensive rules, and clearly defined procedures covering work processes. With low formalization, employee behavior is relatively unprogrammed, the work system allows for considerably greater use of intellectual abilities of the employees, greater reliance is placed on the employee's professionalism, and jobs tend to be more intrinsically motivating. In general, the simpler and more repetitive the jobs are, the higher the level of formalization should be.
Centralization	Refers to the degree to which formal decision-making is concentrated in a relatively few individuals, groups, or levels, usually high in the organization. When the work system is highly centralized, lower-level supervisors and employees have only minimal input into decisions affecting their jobs. In highly decentralized work systems, decisions are delegated downward to the lowest level having the necessary expertise. Under conditions of low formalization and high professionalism, tactical decision-making (dealing with day-to-day operations) should be decentralized, whereas strategic decision-making (concerning long-range planning) may remain highly centralized.

Source: Based on Hendrick, H.W. (1984). In *Proceedings of the Human Factors Society 28th Annual Meeting* (899–902). Santa Monica, CA: Human Factors and Ergonomics Society; Hendrick, H.W. (1991). *Ergonomics, 34,* 743–756; Hendrick, H.W. (2007). In P.R. DeLucia (Ed.), *Reviews of human factors and ergonomics* (Volume 3), pp. 44–78; Kleiner, B.M. (2008). *Human Factors, 50*(3), 461–467.

definition of each of these dimensions, and the consequent nature of organizational design associated with each dimension. Of the six dimensions of organizational variability specified in the table, the first four are concerned with different design strategies adopted by organizations to deal with complexity. Design strategies adopted to deal with organizational formalization and centralization are described in the bottom two entries of Table 9.2.

A more recent perspective on how to manage organizational complexity, one that complements the analysis of Hendrick (2007) in Table 9.2, is provided by Schumpeter (2013h). This latter author points out that

- Given that modern telecommunication technology makes possible real-time access to information, as well as transmission of management decisions, business may not be more complex than it once was (due to communication lags of weeks or months). However, managing complexity remains near the apex of management concerns.
- Among contemporary businesses, two macroergonomic design strategies have been adopted for managing organizational complexity. One is to replace the command-and-control management (linear) approach with a new approach based on the idea of management built around nonlinear, self-organizing networks. This idea of managing organizational complexity through self-organization parallels observations earlier in this book that (1) self-organized behavior may represent a general property of complex technological and organizational systems (Karwowski, 2012; Chapter 1, Section 1.4), and (2) biological systems are thought to self-organize in order to find the most stable solution for producing a given movement in the face of degrees-of-freedom complexity (Chapter 2, Section 2.2).
- The second design strategy adopted for dealing with organizational complexity is to strive for organizational simplicity. Schumpeter (2013h) cites a well-known management consulting firm that has achieved success by advising companies to "design complexity out."
- The record shows that the biggest threat to business almost always comes from too much complexity. Examples are cited in support of this conclusion.

In summary, the implications of the analysis of Schumpeter (2013h) plus the information in the right column of Table 9.2 are that there are direct, observable links between how an organization is designed and the effect of this design on the performance of both its management and its workforce. In other words, the macroergonomic approach to the analysis of ODAM encompasses consideration of both the nature and the sources of organizational and work performance variability.

Hendrick (2007) goes on to discuss 13 different methods that have been developed to evaluate the macroergonomic design characteristic of a given organization in line with the framework outlined in Table 9.2. Foremost among these, for purposes of this chapter, is what is termed participatory ergonomics, a method that advocates direct involvement by employees in decision-making related to the application of ergonomics in a given organization. This method will be discussed further in the next section.

9.4.2.1 Summary

Tables 9.1 and 9.2 both summarize a series of factors that have been shown to influence variability in organizational and work performance. A comparison of these two tables, however, reveals little overlap between factors specified in Table 9.1 and those specified in Table 9.2. In other words, the selected environmental and economic design factors specified in Table 9.1 do not address the possible influence of organizational design factors, whereas the macroergonomic factors specified in Table 9.2 are minimally concerned with environmental and economic effects. This comparison provides further support for the idea introduced earlier that variability in both organizational and work performance is highly context-specific, influenced by a broad array of design factors that essentially are impossible to exhaustively delineate, and that furthermore continue to transform and evolve as new modes of sociotechnical, socioeconomic, and management design innovations regularly emerge.

Macroergonomics represents one in a series of strategies in the ergonomic toolkit that have achieved demonstrable gains in organizational quality and productivity as well as worker safety and health. The next section explores these effects in more detail by documenting the synergism between ergonomic design interventions and their influence on organizational and work performance variability.

9.4.3 SYNERGISM OF ERGONOMICS WITH SAFETY, QUALITY, AND PRODUCTIVITY OF ORGANIZATIONAL AND WORK PERFORMANCE

Sections above identify a series of environmental and economic design factors (Section 9.4.1) and macroergonomic interventions (Section 9.4.2) shown to influence variability in organizational and work performance. The present section extends this analysis with a consideration of how ergonomic design interacts with work design to synergistically influence the productivity, quality, and safety and health of work systems (portions of this section adapted from T.J. Smith, 1999). Control systems models are introduced to emphasize the synergistic, closed-loop properties of these design interrelationships. Evidence from field observations is summarized that supports these conceptual models. The section concludes with a discussion of how synergism between ergonomics and quality can support breakthrough in QM.

9.4.3.1 Organizational Cybernetics of Complex Sociotechnical Systems

The premise of the control systems perspective introduced here is that closed-loop models of social and group interactions (Chapter 8) can be extended and applied to the behavior of organizations, institutions, and societies operating together with advanced technology as complex ST systems. The basic thesis of this approach is that the level of productivity, quality, and safety and health performance of a complex ST system is predicated on the degree to which its organizational design incorporates elements of a closed-loop behavioral control system. When all of these elements are present, we may anticipate high-quality organizational performance.

This concept is depicted in Figure 9.6, which presents a control systems model of safety and QM of a complex ST system. Safety and quality feedback (analogous to sensory feedback with individual behavior) is provided by system ergonomics and system production (output) design factors. Safety and quality feedback control (analogous to sensory feedback control with individual behavior) is mediated by the mutual influence of the system managers and workforce on safety and QM.

The model assumes system ergonomics to be absolutely essential to effective organizational self-regulation of system safety and quality. If the work performance environment is poorly designed, feedback from design factors affecting safety and quality in this environment cannot be closely controlled. As with the behavior of an individual, all of the key determinants of organizational behavioral effectiveness are compromised under such conditions of impaired system feedback control;

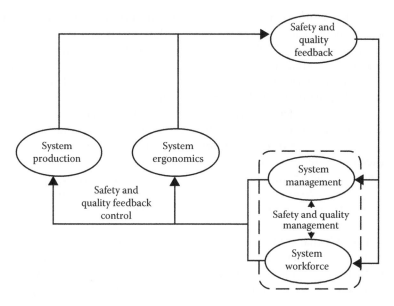

FIGURE 9.6   Control systems model of safety and quality management of a complex socio-technical system. (Reproduced from Smith, T.J., *International Journal of Occupational Safety and Ergonomics*, 5(2), 247–278, 1999.)

namely, quality, safety, health, efficiency, productivity, and competitiveness. The mutual influence of ergonomics, safety, and quality on one another, therefore, represents a manifestation of organizational cybernetics—it arises as an inevitable closed-loop consequence of effective self-control by an organization of its own quality performance.

A basic assumption of the model in Figure 9.6 is that synergism between ergonomics, safety, and quality observed in operational contexts relies on effective social tracking (Chapter 8) between its organizational, technological, and individual employee elements. That is, effective exchange and control of sensory feedback among individual employees or users, technological design features, and organizational and institutional design features of an ST system provides the operational foundation for overall system quality performance.

A key requirement for effective social tracking among ST system elements is employee or user involvement in system decision-making (Brown, 2002; Hendrick, 2007; Imada, 2002). From a social tracking perspective, participatory ergonomic (Haims & Carayon, 1998) and SQM programs are effective because they enable workers to control sensory feedback from job-related decisions or working conditions that affect them and to in turn generate sensory feedback for the control and benefit of other system participants. Conversely, lack of influence by system employees or users over decisions governing system design and operation essentially excludes them from social tracking interaction with the system. Under such conditions, when social tracking linkages between ST system elements are incomplete or ineffective, synergism between ergonomics, productivity, safety, and quality consequently is compromised and becomes difficult or impossible to achieve.

The social tracking significance of the participatory approach for integrating ergonomics, safety, quality, and productivity is illustrated in Figure 9.7 using manufacturing as an example (T.J. Smith et al., 1995). Figure 9.7 compares and contrasts social tracking opportunities available to a manufacturing system employee (center of figure) under technocentric (i.e., Tayloristic; Taylor, 1911) versus human-centered (i.e., sociotechnical) strategies for system organizational design and management. Four major areas of difference between the two ODAM systems are delineated in the figure. Starting at the top, under the technocentric ODAM strategy (left side of figure), the employee receives sensory information (one-way arrow) from decisions affecting performance but is able to exert little if any control over this process. Conversely, under the human-centered ODAM strategy (right side of figure), the employee both receives and controls decision feedback (dual arrows), thereby enabling direct employee behavioral influence over the process.

Under the technocentric approach, production line tasks typically are ordered serially (assembly line fashion), with the employee assigned to only one or a limited number of tasks paced by the technology (one-way arrow) and not by the employee. Conversely, under the human-centered approach, the employee may be assigned responsibility for a collection of different tasks whose pace and quality is under

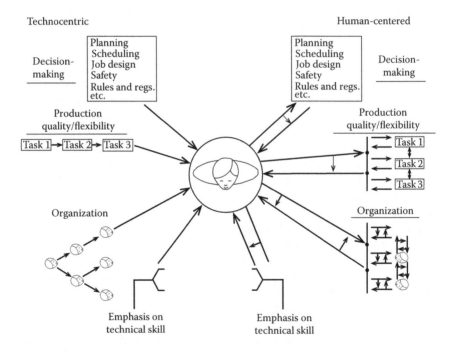

FIGURE 9.7 Social tracking patterns of technocentric versus human-centered strategies for organizational design and management of a sociotechnical system. Unlike the technocentric approach, the human-centered approach provides an array of opportunities for the worker to socially track system operations and control sensory feedback from system design factors. (Reproduced from Smith, T.J., *International Journal of Occupational Safety and Ergonomics*, 5(2), 247–278, 1999.)

employee control (dual arrows). This arrangement in turn promotes job enrichment and flexibility and also provides greater opportunity for the employee to directly influence product quality.

Under the technocentric approach, organizational design tends to be top-down and hierarchical in that the employee only interacts with one manager. Typically, this interaction takes the form of directives and orders governing employee job performance and behavior issued by management that cannot be controlled or greatly influenced by the employee (one-way arrow; Sheridan, 1992b, p. 339). Conversely, the human-centered approach typically is characterized by a flatter organizational structure built around self-managed teams. Team work facilitates mutual social tracking among team members (dual arrows) that in turn promotes communication, cooperation, integration, and efficiency in team performance through development of tight social yokes that bind the team together.

Finally, under the technocentric approach, the employee interacts with system technology primarily in a passive fashion—typically technical skills are considered neither a resource nor a target for upgrading or training. Conversely, both skill training and utilization of employees as technical resource specialists are emphasized under the human-centered approach. This type of projective tracking of future skill needs on the part of the organization promotes professionalism and fosters career development.

In summary, as suggested by Figure 9.7, relative to the technocentric approach the human-centered approach provides substantially enriched social tracking opportunities for the employee to both receive and control sensory feedback from coworkers and also from both organizational and technological design factors. Worker involvement in decision-making, worker control over the production process, and job enrichment enhance the degree of worker control over sensory feedback from the organization and the job and thereby enhance the overall level of worker self-control. Use of workers as resource specialists and emphasis on skill development encourage provision of more sensory feedback from the worker to the organization and thereby benefits organizational decision-making governing both organizational and technological change and the consequent integration of ergonomics, safety management, QM, and productivity of the system.

9.4.3.2 Supporting Evidence from Field Observations

What evidence can be cited to support a control systems interpretation of synergism between ergonomics, safety, quality, and productivity? Possible limitations to the interpretation of evidence that may be forthcoming should be recognized. First, such evidence will be forged in the crucible of operational environments under relatively uncontrolled conditions. Functioning of a complex ST system does not occur in the scientific isolation of a laboratory. Thus, documentation of a particular case of synergism between ergonomics, safety, quality, and productivity will not necessarily provide proof of underlying mechanism. Consequently, any evidence regarding the origins of such synergism necessarily will be inferential in nature and in most cases is likely to be based on retrospective rather than prospective analysis.

With these limitations in mind, three classes of evidence are considered here. The first concerns evidence for the existence of design or context specificity in the performance of complex ST systems. Availability of such evidence implies that their organizational structures allow for interaction between their design features and their performance, which in turn creates the potential for synergism between system ergonomics, system safety, system quality, and system productivity.

The second line of evidence concerns the degree to which well-designed ergonomics, safety, production management, and QM systems incorporate all of the requisite elements of a control system (Chapter 1). The assumption is that system design built around control systems principles establishes a basis for system self-regulation and thereby establishes a basis for integrating system design with system performance manifested as synergism between ergonomics, safety, quality, and productivity. Evidence in this category thus establishes a logical link between good design, control systems design, and quality and productive performance.

The third class of evidence concerns documentation of reciprocal influences between the performance of ergonomics, safety, and quality programs that may be observed in operational contexts. The existence of such reciprocal effects connotes between-program synergism, suggesting effective social tracking linkages between the programs.

Context specificity in social and organizational performance. Limited evidence suggests that, as with individual performance, much of the variability in social and organizational system performance appears to be attributable to system design factors. For example, findings from laboratory studies of social tracking between the participants of two-person teams indicate that a preponderance of variability in social tracking performance is attributable to the human factors design characteristics of the tracking task (K.U. Smith, 1974; T.J. Smith et al., 1994a). In their situational leadership model (whose validity is supported by some observational evidence), Hersey and Blanchard (1977) maintain that managerial performance improves when managers customize their leadership style for different subordinates (i.e., their social tracking behavior) in relation to both the differential capabilities of different subordinates (based on personal factors), as well as the differential demands of tasks being performed by these subordinates (based on design factors). Thus, the premise of this model is that variability in managerial performance should be referenced, at least in part, to task design conditions that prevail for different modes of subordinate-task interaction.

Finally, observations of two seminal contributors to the quality movement are worthy of note. Deming (1982) asserts that about 90 percent of the time, the success or failure of a QM program in industry depends on how the program is designed. For example, Deming believes that adherence to a closed-loop plan-do-check-act cycle of QM is the macroergonomic design linchpin to continuous improvement in quality. Juran (1954, 1964) advocates a series of macroergonomic universals in organizational design, which he believes must be adopted and implemented to achieve success in both control and breakthrough improvement in QM. In two later books he (1) introduces the concept of a spiral of progress in quality (Juran, 1992; Juran & Gryna, 1980) that bears some resemblance to the cycle favored by Deming, and whose operational effectiveness presumably depends on QM adherence to these design universals; and (2) equates success in quality outcomes and progress directly with the design of the QM process (Juran, 1992).

Control systems properties of well-designed ergonomics, safety, and quality programs. Using the analyses of Deming (1982) and Juran (1954, 1964) as starting points, we may ask what there is about the design of a particular program or system that distinguishes high-quality from low-quality performance. Based on field observations of the distinctive properties of well-designed systems, the answer offered here is that the success of ergonomics, safety, quality, and production programs can be equated directly with the degree to which program designs incorporate elements of behavioral control systems (Chapter 1, Section 1.3.3).

For example, as outlined in Table 9.3, the designated quality system requirements of the ISO 9001 QM standard (International Organization for Standardization [ISO], 1994), which has achieved international acceptance and credibility and whose performance benefits for thousands of organizations are a matter of record (Peach, 1994; Struebing, 1996), encompass all of the requisite elements of a closed-loop behavioral control system.

As shown in Table 9.4, the same can be said for successful safety programs. Cohen (1977) describes results from a survey of 42 pairs of companies in the U.S. state of Wisconsin matched in terms of industrial and geographic sector and workforce size but distinguished by low versus high work injury rates. The analysis identified 11 program factors that appeared to have the most bearing on safety program success of

TABLE 9.3
Behavioral Control System Elements of the ISO 9001 Standard

	ISO 9001 Specification	Behavioral Control System Element
4.1	Management Responsibility	Control Goals and Objectives
4.20	Statistical Techniques	Sensory Receptors
4.8	Product ID and Traceability	Sensory Feedback
4.10	Inspection and Testing	
4.19	Servicing	
4.2	Quality System	Learning and Memory
	• Quality Manual	
	• Quality Procedures	
	• Quality Planning	
4.5	Document and Data Control	
4.16	Quality Records	
4.18	Training	
4.3	Contract Review	Effectors
4.4	Design Control	
4.6	Purchasing	
4.9	Process Control	
4.11	Calibration	
4.15	Handling, Storage and Packaging	
4.13	Non-Conforming Product	Sensory Feedback Control
4.14	Corrective and Preventive Action	
4.15	Internal Quality Audits	

TABLE 9.4

Behavioral Control System Elements of Safety Programs in Companies with Low Work Injury Rates

Safety Program Factor	Behavioral Control System Element
Management commitment	Control goals and objectives
Safety motivation strategies	
Regular inspections and communication	Sensory receptors and sensory feedback
Accident and near-miss investigations	
Safety training	Learning and memory
Recordkeeping	
Safety committee	Effectors
Compliance with safety rules	
Emphasis on worker experience	
Low worker turnover and absenteeism	
Hazard control strategies	Sensory feedback control

Source: Cohen, A. (1977). *Journal of Safety Research, 9*(4), 168–178.

the low work injury rate companies. Table 9.4 suggests that these factors encompass all of the requisite elements of a closed-loop behavioral control system.

Finally, Table 9.5 indicates that well-designed ergonomics programs also encompass all of the requisite elements of a closed-loop behavioral control system. The ergonomics program elements specified in Table 9.5 are those recommended by NIOSH (Cohen et al., 1997) based on a survey and analysis of a large number of successful programs in the United States and elsewhere.

TABLE 9.5

Behavioral Control System Elements of Well-Design Ergonomics Programs

Ergonomics Program Element	Behavioral Control System Element
Management commitment	Control goals and objectives
Workplace hazard analysis	Sensory receptors and sensory feedback
Training in ergonomics awareness, job analysis, control measures, and problem solving	Learning and memory
Worker involvement	Effectors
Hazard controls	Sensory feedback control
• Engineering controls (i.e., design improvements)	
• Administrative controls	
• Personal protective equipment	
• Work practices	
Medical management	

Synergism of ergonomics with safety, quality, and productivity. A comparison of Tables 9.3, 9.4, and 9.5 reveals that in their mutual adherence to common criteria for a closed-loop behavioral control system, successful ergonomics, safety, and quality programs share a number of parallel design features. This analysis suggests that if the programmatic designs of different operational programs in an organizational system are patterned on a behavioral control systems model, we may anticipate effective social tracking and performance synergism between them. This section considers evidence from field observations for functional synergism of ergonomics with safety, quality, and productivity.

For each type of relationship, possible evidence for mutual synergism is considered. This approach differs from the paradigm commonly adopted by HF/E science, which assumes that performance is the derivative beneficiary of design improvements. However, if a given organizational system is linked with design as a mutually coupled social tracking system (Figure 9.7), there is no one-way, linear, cause-and-effect relationship between design and performance. Rather, inherent to the design of any closed-loop system, actions of performance are both the cause and the effect of those of design.

Operational synergism between ergonomics and safety. There is unequivocal evidence to support the conclusion that an emphasis on ergonomics benefits safety and accident prevention. For example, in an evaluation of results from 91 field studies in which the effectiveness of 10 different accident prevention strategies were considered, Guastello (1993) finds that comprehensive ergonomics programs were more effective than any other strategy in preventing industrial accidents. Cohen et al. (1997) cite 46 studies dating from 1971 to 1996 that document the effectiveness of ergonomic design improvements in reducing the risk of work-related musculoskeletal disorders (MSDs). In an analysis of the efficacy of workplace ergonomic interventions to control MSDs based on an investigation of 101 studies reported in the literature, Karsh et al. (2001) reports that 84 percent of all of the studies found some positive results, although the majority had mixed results.

Operational synergism between ergonomics and quality. There also is evidence to support the conclusion that an emphasis on ergonomics benefits quality performance. Riley and Bishu (1997) list three examples from industry of links between poor work design and poor-quality performance. Hendrick (1997) cites a number of examples from industry in which better ergonomic design resulted in improved quality performance. Based on findings from several field studies, Eklund (1997) concludes that about one-third of quality defects are related to or directly caused by workplace design deficiencies.

Considering the converse effect of quality on ergonomics, Eklund (1997) points out that a number of QM principles and practices may be antithetical to accepted principles of system ergonomics. As examples he lists just-in-time delivery, statistical process control, standardization and reduction in variability in work operations, and reward systems. Similarly, T.J. Smith et al. (2004) point out that the advocacy by the so-called green movement for sustainable design features indoor environmental quality specifications that are deficient across 11 different criteria relative to internationally accepted occupancy quality guidelines, based on ergonomic analysis of noise and acoustic, lighting and daylighting, thermal comfort, workspace privacy, and whole-body vibration in buildings environmental conditions.

These examples of apparent discordance between QM and sustainable design quality guidelines and design quality specifications developed through ergonomic analysis suggest that synergism between ergonomics and quality may be, at least in part, one-way rather than mutual, and that QM and quality control systems have yet to reap the full benefits of HF/E science.

Operational synergism between ergonomics and productivity. As with ergonomics and safety, and ergonomics and quality, there also is evidence for synergism between ergonomics and productivity. Section 9.4.2 cites examples of the positive impact of macroergonomic interventions on work system performance and productivity, with impressive improvements in selected outcome measures ranging from 20 to 97 percent observed in different field studies (Hendrick, 2007; Imada, 2002; Kleiner, 2008). Hendrick (1996) reviews cost benefits accruing from the introduction of various types of ergonomic design improvements in 10 different industries based on outcomes reported in a number of different field studies. In every case, substantial cost savings with concomitant implications for productivity gains were observed. Table 9.6 illustrates this point with a summary of findings from a subset of these applications.

One of the more dramatic examples of productivity benefits linked to the application of ergonomic design improvements has been documented by the 3M Corporation corporate ergonomics program. Each year this program invites submissions from 3M plants around the world describing ergonomic projects aimed at improving productivity and/or reducing safety and health risks associated with hazardous tasks or jobs at these plants (N. Larson, personal communication, 2013). From these submissions, a smaller number of projects are recognized with the Applied Ergonomics Innovation Award (AEIA), which is based on five ratings criteria.

TABLE 9.6

Organizational Performance and Productivity Benefits Attributable to Ergonomic Applications

Industry	Ergonomic Application Area	Cost Savings
Forestry	Leg protectors	$250,000/year
	Vehicle improvements	$ 65,000/year
Airline	C-141 cargo compartment redesign	$2 million savings in initial fleet costs
Steel mills	Manual materials handling –3 mill areas	10% productivity increase; payback periods of 15–18 months; 80% drop in maintenance costs
Product manufacturing firms	Forklift redesign	Stock price increase
	TV/VCR remote control redesign	Millions in sales
	CRT display	$2.94 million annually
	Rear brake light	$910 million in crash property damage

Source: Hendrick, H.W. (1996). In *Proceedings of the Human Factors and Ergonomics Society 40th Annual Meeting* (1–10). Santa Monica, CA: Human Factors and Ergonomics Society.

For the years 2010 and 2011, a total of 166 3M plant ergonomic projects were submitted for award consideration and 18 were recognized with the AEIA. Over the two years, productivity savings (staffing reductions plus operating efficiency gains) associated with the ergonomic design modifications described for the 166 submitted projects totaled $3.68 million, of which 42 percent was achieved by the 18 AEIA winners alone. Across all 166 projects, this works out to productivity savings of $22,167 per ergonomic project for two years, or $11,083 per year.

As for the possibility of a reciprocal influence of work system production systems on ergonomics, consider participatory ergonomics. The premise of this macroergonomic strategy is that workers should be involved in decision-making governing the application of ergonomics. As Hendrick (2007), Imada (2002), and Kleiner (2002) point out, one advantage of this approach is that production workers represent one of the best resources for identifying production system design flaws that might benefit from ergonomic applications. A similar potential exists with safety programs that involve workers in safety program decision-making. With this approach, there is ample evidence that frontline workers represent a valuable source of information about production system hazards amenable to mitigation by ergonomic applications (T.J. Smith, 2002).

9.4.3.3   Summary

A control systems perspective on the organizational design and management of ergonomic, safety, and quality programs suggests the possibility that there might also be closed-loop interactions or synergism between these organizational functions. Evidence provided in the foregoing analysis, extending also to production systems, indicates synergistic interactions of ergonomics with safety, quality, and production. The next section describes a case study supporting the conclusion that safety and quality also interact synergistically.

9.4.4   OPERATIONAL SYNERGISM BETWEEN SAFETY AND QUALITY

Figure 9.6 illustrates a control systems model that assumes a closed-loop relationship between safety management and quality management (Section 9.4.3.1). This section extends this model to empirical findings from a field study of one particular company that I believe provides rather convincing evidence for mutual synergism between safety and quality (Smith & Larson, 1991). The study documents the experience of a small manufacturing firm in the U.S. upper Midwest that has developed outstanding QM and safety management (SM) programs.

The firm in question specializes in the manufacture of industrial floor maintenance equipment. It has a successful safety program dating back well over two decades. In 1980, it applied Crosby's approach (Crosby, 1979; Hale et al., 1987) to install and develop a quality assurance program, which has achieved international prominence in the past decade (Youngblood, 1991). Separate line management is assigned to the two programs; namely, a quality, test, and reliability manager for the QM program, and a corporate facilities and risk manager for the safety program.

According to Hale et al. (1987), organizational design principles guiding management of the company's QM program are (a) management commitment, (b) employee involvement, (c) cooperative worker-manager relationships, (d) rewards for people, and (e) time, energy, and determination. According to the corporate facilities and risk manager (Smith & Larson, 1991), organizational design principles guiding management of the company's SM program are (a) management responsibility for safe working conditions and work practices, (b) individual employee responsibility for safe work performance, (c) no sacrifice of safety for production quality or quantity, and (d) full worker-manager cooperation There are obvious parallels between the organizational design principles for the two programs.

9.4.4.1 Method

The safety record of the firm over a 21-year period (1970–1990) was evaluated, with particular attention to possible changes in the lost-time injury incidence rate since 1980, when the firm's quality program was installed. The firm's quality performance record from 1980 to 1990 also was evaluated.

Questionnaires dealing with the firm's quality and safety programs were distributed to 14 senior, midlevel, and frontline managers and to 20 shop-floor workers. Respondents were asked 12 questions about possible reasons for each program's success and about perceived similarities and differences, as well as possible interactive effects between the two programs. The worker questionnaire included two additional questions dealing with the current and the desired level of worker involvement in decision-making in both programs.

The 14 manager respondents represented over 60 percent of manufacturing support managers. The 20 worker respondents represented less than 10 percent of shop-floor workers. The manager responses, therefore, are somewhat representative, but the low number of worker responses limits the generalizability of the results to the entire workforce.

9.4.4.2 Results

From 1980 through 1990, the quality and the safety performance of the firm showed concomitant improvement. Development of the firm's quality program during the decade of the 1980s was accompanied by over a twofold reduction in the 10-year average injury rate, from 3.8 lost-time injuries per 100 employees per year for the period 1970–1979 to 1.5 for the period 1980–1989.

Figures 9.8 and 9.9 illustrate relationships between the performances of the QM and SM programs from 1980 to 1990. In Figure 9.8, the yearly quality record expressed in percentage of defective parts at installation is plotted against lost-time injuries per 100 employees per year for each of the 11 years. In Figure 9.9, the quality record expressed as manufacturing rework hours is plotted against the safety record. In both figures, the latter years of the decade exhibit fewer lost-time injuries associated with fewer defective parts and rework hours. For both relationships, the correlation of improved quality with improved safety performance ($r^2 = .608$ in Figure 9.8; $r^2 = .784$ in Figure 9.9) is statistically significant ($p < .05$).

With the questionnaire responses, the three most common factors cited by shop-floor worker that contribute to the success of the firm's SM and QM programs are awareness (of quality or hazards), management commitment, and employee involvement. There are differing responses to questions dealing with subjective perceptions as to possible reciprocal effects of the firm's quality and safety programs on one

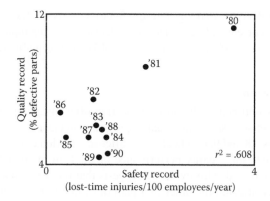

FIGURE 9.8 Company achievement in quality performance (percentage of defective parts) versus safety performance (lost-time injuries) for the years 1980–1990. (Reproduced from Smith, T.J., *International Journal of Occupational Safety and Ergonomics*, 5(2), 247–278, 1999.)

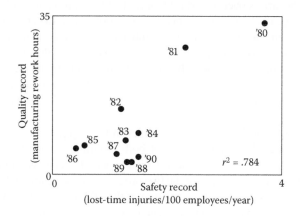

FIGURE 9.9 Company achievement in quality performance (manufacturing rework hours) versus safely performance (lost-time injuries) for the years 1980–1990. (Reproduced from Smith, T.J., *International Journal of Occupational Safety and Ergonomics*, 5(2), 247–278, 1999.)

another. Specifically, all of the manager respondents and 50 percent of the worker respondents believe that the safety program contributed to the success of the quality program. Conversely, only about one-third of both manager and worker respondents believe that the quality program contributed to the success of the safety program.

Using a five-level Likert rating scale, the questionnaire also assessed worker responses as to current and desired levels of decision-making (1 = no input; 5 = sole responsibility) with both the quality and safety programs in relation to seven quality program decision-making areas and eight safety program decision-making areas. For both programs, results are essentially identical in that workers perceive that they currently have some input (rating = 2.0) into the decision-making process, but they express the desire for a higher level of input between influencing and sharing responsibility for decision-making (rating = 3.3 to 3.5). By analysis of variance, differences between current and desired levels of decision-making input are statistically significant for both the safety ($F_{1,7} = 125.7$, $p < .001$) and the quality ($F_{1,6} = 106.9$, $p < .001$) programs.

9.4.4.3 Discussion

The observations outlined in the previous section offer intriguing insight into possible synergism between safety and quality programs in place at this firm. Objective evidence (Figures 9.8 and 9.9) indicates that after installation of the quality program in 1980, both safety and quality performance of the firm improved concomitantly as the decade progressed. This evidence suggests that the performance of each program reciprocally affected the other in a pattern of mutually beneficial synergism. Subjective responses of both managers and workers convey the view that the influence of the safety on the quality program is primarily responsible for this pattern.

One explanation for this finding is that individual responsibility for working safely and for participating in hazard management enunciated in the firm's safety policy naturally carries over to careful workmanship in producing defect-free products. The more general control systems interpretation is that the emphasis placed on employee involvement in both programs introduces intimate behavioral feedback links between the safety and the quality of work performance that leads inevitably to the pattern of program interaction observed. The linchpin of both quality and safety is worker performance, and management commitment to support and encourage self-responsibility in the effective execution of work can be expected to benefit results in both areas.

The HF/E basis for this assumption resides in the high degree of consistency in design principles guiding management of both programs, which was summarized earlier. Worker perceptions about success factors and about current and desired levels of input into decision making for the two programs likewise are highly consistent. In light of parallels in organizational design of the QM and SM programs instituted by the firm, parallels in performance variability of the two programs, therefore, are to be expected.

Given that the firm's quality and safety programs may have mutually benefitted one another (Figures 9.8 and 9.9) under conditions of separate line management for each program, what arguments can be raised for a more integrated management approach to both programs? One argument for this idea is the data presented in Figures 9.8 and 9.9 suggesting that accident prevention and defect prevention have intimate behavioral feedback links. Moreover, with their common emphasis on the participatory approach, both programs already have an appreciable degree of organizational design integration.

Most persuasive perhaps are comments from some worker respondents that the greater emphasis placed by company management on the quality program does not positively serve the safety program. Indeed, as noted above, only about one-third of both managers and workers feel that the quality program contributes to the success of the safety program. These points suggest that workforce perceptions as to how the programs dovetail with one another, as well as performance outcomes of the programs themselves, might benefit from a more integrated program management approach.

A control systems interpretation of the performance advantages of an integrated relative to a dual approach to managing safety and quality programs is given in Figures 9.10 and 9.11. With the dual program approach (Figure 9.10) employed by the company, the shop-floor worker must sense and control psychosocial feedback from three managers, organizational design sensory feedback from two programs, and ergonomic design sensory feedback from workplace design factors and conditions. It seems reasonable to suggest that the demands of these multiple sources of social and design sensory feedback on the behavioral control capabilities of the worker may be considerable and may compromise safety and quality performance at times.

Conversely, with an integrated approach to safety and quality management (Figure 9.11), social feedback from only two managers (production and HF/E managers, with the production manager also responsible for quality control) and sensory feedback from one set of organizational plus ergonomic design factors must be controlled by the shop-floor worker. It seems reasonable to suggest that this integrated approach simplifies behavioral control demands placed on the worker, which may in turn benefit both safety and quality performance in a mutually synergistic manner.

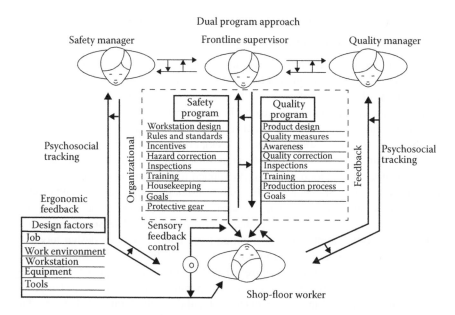

FIGURE 9.10 Social tracking and sensory feedback control demands on the shop-floor worker under a dual program approach to safety and quality management. (Reproduced from Smith, T.J., *International Journal of Occupational Safety and Ergonomics*, 5(2), 247–278, 1999.)

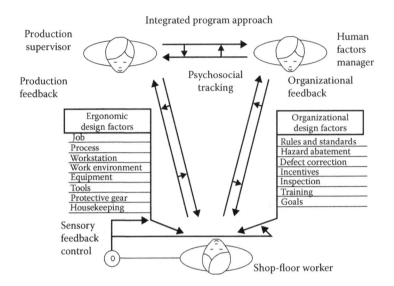

FIGURE 9.11 Social tracking and sensory feedback control demands on the shop-floor worker under an integrated program approach to safety and quality management. (Reproduced from Smith, T.J., *International Journal of Occupational Safety and Ergonomics*, 5(2), 247–278, 1999.)

9.4.5 A CONTROL SYSTEMS PERSPECTIVE ON ORGANIZATIONAL DESIGN AND MANAGEMENT: ROLE OF ERGONOMICS

Figure 9.6 introduced a control systems model of safety and quality management in which system decisions are guided by feedback from system performance attributes such as quality (Table 9.3), safety (Table 9.4), or ergonomic design (Table 9.5). The plan-do-check-act cycle of continuous quality improvement advocated by Deming (1982) also implies closed-loop linkages between control system elements of control goals and objectives (plan), sensory feedback (check), effectors (do), and sensory feedback control (act). Juran's spiral of progress in quality (Juran, 1992; Juran & Gryna, 1980) specifies similar closed-loop linkages. As applied to goods production, for example, the Juran spiral links market research, product development and design, and production planning (control goals and objectives), production (effectors), inspection and market outcomes (sensory feedback), and process control (control of sensory feedback).

The most explicit application of a closed-loop, control systems design model to organizational design and management is that of Juran. In 1954, he introduced a servomechanism model as a design universal that all managers employ (with varying degrees of success) for controlling their system operations, a concept that he later elaborated on in his seminal texts on managerial breakthrough (Juran, 1964, 1995). The Juran model assumes that effective management control of an organizational system is mediated by comparison of feedback from actual system performance with performance targets (standards or specifications). As with an engineering servomechanism, error between actual and desired performance is used to guide management decision-making directed at tightening system control and reducing performance error.

Given that Juran advanced his servomechanism model of management control over four decades ago, what new insights can be introduced to further refine his model? One answer is to call attention to explicit parallels between the cybernetic properties of individual behavior (Chapter 1, Section 1.3.3) and those of the behavior of complex sociotechnical systems (Figure 9.6 and Tables 9.3, 9.4, and 9.5). Support for such homology is provided by evidence indicating that the behaviors of both individual and organizational systems display context specificity in performance variability (Section 9.4.1) and social tracking properties (Chapter 8, Section 8.4), which can be used to account for social tracking performance synergism observed between organizational systems with compatible control systems designs (Sections 9.4.3 and 9.4.4).

The control systems model in Figure 9.6 suggests a second answer to the question posed above, in that its closed-loop design provides an explicit basis for interaction between ergonomics and variability in organizational performance. Deming (1982) does not appear to recognize this linkage. Juran and Gryna (1980, pp. 198–199) refer to the human factor as one of the variables affecting the reliability of quality performance but do not emphasize the central role that ergonomics can and should play in improving quality performance through improved design. Figure 9.6 suggests that managers and workers (effectors) guide their control of the organizational system based on feedback from quality and safety performance, which in turn is influenced by various attributes of system design (Juran's [1964, p. 187] term for these attributes is "variables affecting performance"). The application of ergonomics to improve system design (a major goal of HF/E science) therefore can be expected to benefit system reliability by reducing perturbing effects of inadequate design on organizational performance variability.

Yet another implication of the model in Figure 9.6 is that it points to a role for ergonomics in facilitating breakthrough in organizational quality performance. The key to maintaining and sustaining performance of a system at any given level is the operation of system control, defined by Juran (1995, p. 1) as "staying on course, adherence to standard, prevention of change." As he points out (1995), "under complete control, nothing would change—we would be in a static, quiescent world." As noted above, Juran (1995, pp. 199–206) assumes that managerial control is mediated as a closed-loop process in which feedback from actual performance is compared with desired performance in order to identify and minimize performance error. However, Juran (1995, p. 3) goes on to observe that "control can be a cruel hoax, a built-in procedure for avoiding progress—we can become so preoccupied with *meeting* targets that we fail to challenge *the target itself*—this brings us to a consideration of breakthrough."

Juran (1995, p. 3) defines managerial breakthrough as "change, a dynamic, decisive movement to new, higher levels of performance." He assumes that there is an unvarying sequence of events that occur in breakthrough from one level of performance control to a new, improved level; namely: (1) breakthrough in attitude, (2) Pareto analysis, (3) diagnosis, (4) cultural adaptation, (5) breakthrough in results,and (6) achieving control at a new performance level.

Unlike the model for system control defined by Juran as a servomechanism, his model for breakthrough in system performance is not presented as a cybernetic process. Nevertheless, it can be argued that breakthrough can be conceived as a closed-loop process involving feedforward rather than feedback (i.e., servomechanism) control. Feedforward control is ubiquitous among biological systems. It enables them to rely upon sensory feedback from present conditions to project their behavior into the future. Similarly, the impetus for breakthrough in organizational performance is some feedback or error indicator suggesting that the current level of performance (Juran's "the target itself") is no longer adequate, thereby prompting the system to initiate a breakthrough sequence to project its behavior into the future in a feedforward manner to achieve a new, improved performance level.

In many instances, the root cause of inadequate performance of a complex sociotechnical system at any given level is some sort of design flaw either in the microergonomic design of the work process or work environment, or in the macroergonomic design of the organizational system, or both. This is where ergonomics can lend a hand. Ergonomic analysis can be used to detect poor system design. Ergonomic intervention can be used to improve system design in order to facilitate the breakthrough process. The application of ergonomics can thus serve as a key breakthrough strategy by means of which the system elevates its behavioral performance from one control level to a more refined level.

This concept is illustrated in Figure 9.12, applied to the interaction of ergonomics and quality. The shaded region of the figure depicts Juran's servomechanism model of quality control (1995, p. 202) in which the organizational system of managers and workers acts on feedback from system design variables (sensed by inspection or quality audit) to effect a product or service shows actual level of quality is compared with a desired quality target or goal set by the system. Error between desired and actual quality (recorded in a quality report) is used in a quality feedback control manner to adjust system design variables with the aim of reducing quality errors. With a tightly control system, error is close to zero and the system (because of its apparent near optimal level of performance) consequently is highly resistant to change.

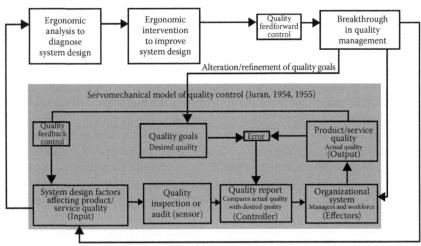

FIGURE 9.12 Model of breakthrough in quality management as a feedforward control process, with ergonomic analysis and intervention specified as key contributors to feedforward control. (Reproduced from Smith, T.J., *International Journal of Occupational Safety and Ergonomics*, 5(2), 247–278, 1999.)

However, as Juran points out (1995, Chapter 1), change is essential for system survival. System design, therefore, must include some provision for alteration or adjustment of the quality goals themselves, a process Juran terms "breakthrough." The basic premise in Figure 9.12 is that breakthrough is actually a process of feedforward control superimposed on the servomechanism or feedback control process. Feedforward control means that the system projects its performance into the future by relying on perceived inadequacies in system design as a predictive indicator that performance breakthrough will be required for the system to continue to prosper.

The further premise in Figure 9.12 is that ergonomics can greatly facilitate the breakthrough process in two major ways; namely, through ergonomic analysis and ergonomic intervention. Methods of ergonomic analysis are admirably tailored for detecting inadequacies or shortcomings in system design; results of this analysis, therefore, can serve as early warning sentinels for the need for initiating a breakthrough process. Once design problems have been identified, methods of ergonomic intervention then are admirably suited for contributing to problem resolution through microergonomic improvement in system design features, macroergonomic improvement in organizational design features, and alteration or refinement of quality targets and goals.

Finally, the model in Figure 9.12 assumes that ergonomics can and should serve as a key universal in the armamentarium of techniques that managers employ to guide the breakthrough process along with others specified by Juran (1995). In particular, ergonomic analysis can contribute in a major way to breakthrough in knowledge for purposes of diagnosis (Juran, 1995, Chapter 8) and ergonomic intervention can contribute in a major way to breakthrough in performance through action (Juran, 1995, Chapter 10). In this manner, managerial control of system ergonomics becomes tightly integrated with managerial control of system quality such that, on an operation level, the two control functions become functionally indistinguishable.

9.4.6 CONCLUSIONS

The foregoing analysis indicates that synergism between ergonomics, safety, quality, and production, observed with a variety of complex ST systems, can be understood in the context of control systems analysis. In particular, such synergism emerges as a social tracking consequence of closed-loop coupling (Figure 9.6) between behavioral performance of the system (safety and quality performance and management) and microergonomic and macroergonomic design features of the system.

From a behavioral control systems perspective, ergonomics may be considered as absolutely essential to effective organizational self-regulation. If the performance environment is poorly designed, sensory feedback from design factors in the environment cannot be closely controlled. As with the behavior of an individual, all of the key determinants of organizational behavioral effectiveness are compromised under conditions of impaired sensory feedback control; namely, quality, safety, health, efficiency, productivity, and competitiveness. The mutual influence of ergonomics, safety, and quality on one another, therefore, represents a manifestation of organizational behavioral control: it arises as an inevitable closed-loop consequence of effective management and employee self-regulation and control of sensory feedback from system ergonomics.

This interpretation rests on three basic assumptions: (1) the idea that behavioral performance is design- or context-specific, extensively documented in the case of individual performance (Chapter 3), can also be applied to the performance of complex ST systems; (2) the nature and extent of design specificity in the performance of a given complex ST system depends on the degree to which it self-regulates its own performance in that a self-regulatory system design that incorporates all of the essential elements of a behavioral control system (Chapter 1, Section 1.3.3) establishes, by its very nature, closed-loop linkages between system performance and system design; and (3) synergism between safety performance, quality performance, and ergonomic design of a system is defined and established by these linkages.

Some evidence from field observations (Section 9.4.3.2) can be cited to support these assumptions. However, at present, the evidence is inferential, indirect, and sparse. There is vast opportunity for further research to assess the applicability of a control systems model to complex ST systems and the validity of the aforementioned assumptions on which it rests. I believe that such research can be most productively applied to evaluating the heuristic value of the model by exploring such questions as, (1) To what degree is the safety, quality, and productivity performance of complex ST systems design- or context-specific? (2) Does design specificity in system performance grow out of its self-regulatory properties? (3) To what degree is achieving better safety, quality, and productivity performance of a system predicated on making improvements in system ergonomics? and (4) What ergonomic design factors have the greatest influence on variability in system safety, quality, and productivity performance?

Another key implication of the control systems model in Figure 9.12 proposing a feedforward role of ergonomics in mediating breakthrough management of organizational systems is that coupling feedback with feedforward control introduces inherently variable system dynamics (as opposed to the inherently stable and unvarying state of a pure error-minimization feedback control system). This is because both ergonomic analysis (to detect new design challenges) and ergonomic intervention (to address these challenges) impose feedforward control and consequent modification of the system state in a recurrent manner (Chapter 1, Section 1.4.2).

Both Karwowski (2012) and May (1976) suggest that this type of variability can best be understood from a dynamical systems perspective, and that variability in complex sociotechnical systems (such as organizations and institutions) is best understood as dynamic, complex, nonlinear, and deterministic. To review the perspective of these authors introduced in Chapter 1, Section 1.3, May (1976) cites variability patterns observed in population dynamics, prey-predator systems, ecological systems, economic systems, business cycles, learning, and social interaction as possible candidates that meet these criteria. In a similar vein, Karwowski (2012) cites eleven different complex sociotechnical system domains (with some overlap with May's candidates) for which the theory of complex adaptive systems can be applied; namely, (1) biology, including ecosystems, (2) business management, (3) social interactions, (4) economics, (5) disease prevention, (6) health care management, (7) human service systems, (8) information and software engineering, (9) manufacturing operations and design, (10) medicine, nursing, and medical practice, and (11) occupational biomechanics. The nature and sources of performance variability, for the particular cases of economic and nation-state systems, are addressed, is addressed in the next section.

9.5   WORK PERFORMANCE VARIABILITY AND PERFORMANCE VARIABILITY OF ECONOMIES AND NATION STATES

This section extends the analysis of performance variability of organizations and institutions provided in Section 9.4 to that of entire economies and nation-states. The rationale for considering performance variability at this level of sociotechnical complexity rests on two basic considerations. First, economic and nation-state systems are the ultimate arbiter of variability in the performance of human work. To put it simply, as systematically documented by Adam Smith (1776, 1977), the viability of a given economic and nation-state system depends on the productivity of its workforce, and in turn, the prosperity of a given workforce depends on the degree to which its productivity is supported by the economic and nation-state system in which it performs.

This concept of a closed-loop relationship between variability in work performance and that of economic and nation-state performance underscores the second reason for the analysis in this section; namely, that variability observed in economic and nation-state performance recapitulates themes related to the nature and sources of variability articulated earlier in this chapter. Specifically, observed variability in economic and nation-state performance rests ultimately on variability in the performance of human work, is highly context-specific, displays dynamic, complex, nonlinear, and deterministic variability patterns, and can be understood from a control systems perspective.

The latter three of these issues are addressed in the following subsections. Because this book is concerned with performance variability and not with the theory and practice of economics, the analysis below necessarily is somewhat abbreviated and superficial. As just noted, however, there is an evident continuity in the nature and sources of economic and nation-state variability relative to influences observed at less complex levels, suggesting that the common denominator across all levels of complexity is variability in work performance.

9.5.1   A CONTROL SYSTEMS PERSPECTIVE ON VARIABILITY IN ECONOMIC PERFORMANCE

The idea introduced above of a closed-loop relationship between economic and/or nation-state performance and work performance suggests a control systems interpretation of this relationship. K.U. Smith (1962b, p. 1–4) has proposed such an interpretation for economic systems, illustrated in Figure 9.13. The figure contrasts the conventional view of the market process as an open-loop system (Figure 9.13a) versus a feedback concept of the market process as closed-loop control system (Figure 9.13b). The control systems model suggests that a market economy creates a series of market systems (five are depicted in the model) whose functioning and operation collectively generates a market environment that feedbacks to influence the entire market economy. In this manner, the economy reciprocally directs its own variability and evolution.

Much more recently, Cooper (2014) argues for rethinking economic theory from a competitive perspective as an alternative to neoclassical economic theory, a view that also lends itself to a control systems interpretation. Neoclassical economic theory understands an individual in an economic system to be a rational decision-maker who tends to make economic decisions based on self-interest in order to maximize his or her welfare

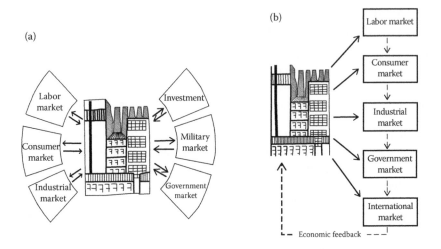

FIGURE 9.13 The conventional view of the market process compared with a closed-loop model of market control. (a) The conventional view describes the market as an open-loop, stabilized environment controlling factory operation and organization (division) of labor. (b) The closed-loop model describes industrial organization as a dynamic mechanism under feedback control, with different market systems generated by the industrial economy collectively establishing a market environment that provides feedback to enable the economy to reciprocally direct its own variability and evolution. (Reproduced from Smith, K.U., *Work theory and economic behavior* [Indiana Business Paper No. 5]. Bloomington, IN: Indiana University Foundation for Economic and Business Studies, pp. 1–4, 1962b.)

(Buttonwood 2014b). Cooper's view rather is that there is a constant struggle among economic agents (hence the allusion to Darwin in the title of his book), resulting in continuous striving for personal wealth and wellbeing. The competitive synergism of democracy and capitalism results in a feedback relationship between wealth creation and wealth distribution, a linkage that is depicted by the control systems model in Figure 9.14.

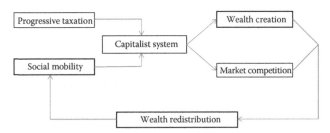

FIGURE 9.14 Control systems model of a competitive model of economic performance that links two key features of democracy (social mobility and progressive taxation) to wealth creation and wealth redistribution, mediated by capitalism, in a feedback manner that ensures a circulatory pattern of economic performance. (From Buttonwood, 2014b. Revolutionary fervor. In *The Economist* [March 8], p. 74; Cooper, G., 2014. *Money, blood and revolution: how Darwin and the doctor of King Charles I could turn economics into a science*. Harriman House: Petersfield, Hampshire, UK.)

In this model, functioning of the capitalist system (the controller) is influenced by two key design (input) factors, namely, social mobility and progressive taxation (both factors are features of democratic systems). The latter factor tends to favor the flow of wealth to the less economically advantaged. Some of these individuals, riding the wave of social mobility, are able to ascend the economic ladder and create wealth, but of course at some point, some of this is redistributed due to progressive taxation. Hence, as suggested by the control systems model in Figure 9.14, output factors of wealth creation and wealth redistribution are feedback-linked to the input factors in an economic circulatory pattern and hence the allusion in the title of Mr. Cooper's book to the discovery of William Harvey (the doctor of Charles I) of the circulation of the blood.

9.5.2 WORK PERFORMANCE VARIABILITY AND VARIABILITY IN ECONOMIC PERFORMANCE

The great historian Gibbon (1788, Chapter LXXI, p. 1070), a contemporary of Smith, encapsulates the essence of economic functioning with a somewhat comparable summary:

> The value of any object that supplies the wants or pleasures of mankind, is compounded of its substance and its form, of the materials and the manufacture. Its price must depend on the number of persons by whom it may be acquired and used; on the extent of the market; and consequently on the ease or difficulty of remote exportation, according to the nature of the commodity, its local situation, and the temporary circumstances of the world.

As Free Exchange (2013d) points out, there is an intimate link between variability in economic performance and that of the performance of firms. This analysis finds that 48 percent of the volatility of the GDP in the United States can be attributed to the performance of individual big firms. The analysis also finds that trade is another contributor to this link—foreign trade exposes economies more to the fortunes of large firms since they trade disproportionately.

To summarize, direct links have been documented between observed variability in work performance, that of organizations and institutions, and that of economic and nation-state performance. The nature and sources of variability in the latter two classes of complex sociotechnical systems are considered in this and the next section.

Both nativist and empiricist perspectives (Goldhaber, 2012) have been offered to explain variability in economic performance, recapitulating a comparable observation regarding the performance of firms cited in Section 9.4.1. Brooks (2010) broaches what appears to represent a nativist perspective by arguing that conventional models and analyses of economic performance ignore the likely role of psychology, emotion, and morality in influencing the behavior of agents (consumers and investors) whose decisions are likely to have an economic impact. Benjamin et al. (2012) is more explicit in supporting a nativist point of view. These authors analyzed genotype data from 9836 Swedish twins that had responded to a survey about their economical and political inclinations. Standard twin-based estimates of heritability indicated moderate (30–40 percent) heritability for these traits.

However, as addressed in more detail below, the preponderance of observations regarding factors likely to influence economic performance cite design factors. Indeed, Lotterman (2012b) argues that the traditional view that workers succeed largely because of their abilities and efforts is contradicted by observations that circumstances and environment—that is context—play a role in many outcomes. He offers a number of examples, but one is worth noting. A study of how a worker's year of birth affected lifetime earnings found that cohorts of workers born such that their prime earning years occurred during substantial downturns (such as the Depression) had substantially lower average lifetime earnings relative to those that came before or after them.

The views of McCloskey (2010), who asserts that economics cannot explain the modern world, may also be construed as colored by a nativist perspective, in this case human psychology. This author confronts the question of how and why the biggest story in economic history occurred; namely, the explosion in economic growth—first in northern Europe and eventually the rest of the world—that transpired between the eighteenth century and today. McCloskey (2010) attributes this dramatic transformation in economic system variability to changes in the way people thought about economic activity. In her words:

> In the eighteenth and nineteenth centuries, a great shift occurred in what Alexis de Tocqueville called "habits of the mind"—or more exactly, habits of the lip. People stopped sneering at market innovativeness and other bourgeois virtues.

She concludes that as attitudes changed, so did behavior, leading to more than two centuries of constant innovation and rising living standards.

As for the *nature* of variability in economic performance, the long history of commerce (Watson, 2005) has featured recurrent episodes of upturns and downturns in economic prosperity. Figure 9.2 illustrates one such episode, using the recent Great Recession as an example (Section 9.4.1). A much broader perspective on this question is offered by Hubbard and Kane (2013), who point out that

> One of the underappreciated realities of history is that most of it has been lived in relative misery. From the Paleolithic era 2.5 million years ago up to the early 19th century, average life expectancy topped out at about 35. And for much of this period there was no such thing as economic growth—humans subsisted on what they could kill or scratch from the ground or on the proceeds of a minimal barter economy. While some civilizations outperformed others, sooner or later the standouts fell into decline. These declines tend to follow a template or sequence of error: denying the internal nature of stagnation, centralizing power, and shortchanging the future to overspend on the present. When the inability to corral fiscal profligacy coincides with a breakdown of political institutions, a toxic imbalance ensues and decline follows.

A more recent synopsis of Buttonwood (2014a) suggests that a propensity for "toxic imbalance" (i.e., financial instability) may be inherent to economic systems. In explaining this conclusion, the account first cites a quote from J.K. Galbraith (a prominent Harvard economist) that summarizes the problem: "All crises have involved debt that, in one fashion or another, has become dangerously out of scale in relation to the underlying means of payment." In more prosaic terms, the death spiral of economic instability features the following key stages: (1) individuals use credit to finance consumption in excess of income, (2) easier credit drives up asset prices, which encourages more bank lending, (3) increased credit also leads to increased volumes and excess trading in the financial markets, and (4) this likely does not benefit the economy, but it does lead to a bubble in asset prices that ultimately bursts. These stages also are implicated in many of the major U.S. economic slumps between 1792 and 2007–2009 (The Economist [2014], p. 54).

Table 9.7 represents a selective survey of design factors that have been shown to contribute to variability in economic and nation-state performance. Two features of the table merit emphasis. First, the choice of factors listed in the table is selective—there are other factors not listed that also have been implicated as influencing variability in economic or nation-state performance. Second, all of the listed factors are design factors. In other words, context specificity—design—has a major influence on such performance.

An observation by Shermer (2013) suggests that the survey in Table 9.7 may be superficial. His summary conclusion, recapitulating that originally set forth by Adam Smith (1776, 1977) is that, "Free trade and the division labor constitute the greatest generator of wealth in history..." Why then proceed with the analysis offered in Table 9.7? The answer offered here is that trade and division of labor may be necessary, but not sufficient, context-specific contributors to variability observed in economic and nation-state performance, and that the contributions of other design factors, which are delineated in Table 9.7, also have been clearly documented.

Table 9.7 is divided into two sections—the top listing factors shown to influence economic performance variability and the bottom listing factors shown to influence nation-state performance variability. These factors are specified in the leftmost column, explanatory comments are specified in the middle column, and references are provided in the rightmost column. The first entry in both sections deals with factors specified by Adam Smith (1776, 1977), whose analysis pertains to sources of both economic and nation-state variability.

A key resource that I relied on with analysis of economic variability is the Great Courses lecture series on "Why Economies Rise or Fall" by Rodriguez (2010), the second entry in the first section of Table 9.7. There are a series of features of the Rodriguez (2010) analysis that merit emphasis:

- With one exception (high consumer savings rates), factors contributing to economic success specified by Rodriguez are identical to those specified over 23 decades earlier by Adam Smith (1776, 1977).
- What distinguishes the perspective of Rodriguez is that unlike Smith, who is credited with founding the field of economics, the former has benefitted from well over two centuries of economic experience in developing his perspective. This enables Rodriguez to compare and contrast more with less successful systems. Here is a synopsis of his analysis of keys to economic success and growth (Rodriguez, 2010, Lecture 14):

TABLE 9.7
Sources of Variability in Economic and Nation-State Performance: A Selective Survey

Factors Implicated in Effective Performance	Comments	Reference
Observations on Sources of Variability in Economic System Performance		
Incentives to worker productivity	Productivity incentives identified include competition, division of labor, free markets, free trade, political stability and the rule of law, stable income (especially for the poor), and technological innovation.	Adam Smith (1776, 1977)
Incentives to worker productivity	Productivity incentives identified include competition, free markets, high consumer savings rates (encourages investment), openness to trade, political stability and trust in the political system, stable income, and technological innovation.	Rodriguez (2010)
Disincentives to worker productivity	One of the underappreciated realities of history is that most of it has been lived in relative misery. From the Paleolithic era 2.5 million years ago up to the early nineteenth century, average life expectancy topped out at about age 35. For much of this period there was no such thing as economic growth—humans subsisted on what they could kill or scratch from the ground or on the proceeds of a minimal barter economy. While some civilizations outperformed others, sooner or later the standouts fell into decline. These declines tend to follow a template or sequence of error: denying the internal nature of stagnation, centralizing power, and shortchanging the future to overspend on the present. When the inability to corral fiscal profligacy coincides with a breakdown of political institutions, a toxic imbalance ensues and decline follows.	Hubbard and Kane (2013)
Debt	Unsustainable levels of private debt precipitated the recent Great Recession. Governments responded in two different ways to mitigate the effects, austerity (mainly southern European countries), and stimulus (the United States and the United Kingdom are two prime examples). Austerity initially inflicted economic damage, but signs of delayed recovery in selected countries are emerging. Stimulus exacerbated public debt; there is controversy over what level of public debt is detrimental to economic growth. Runaway debt also is implicated in other major U.S. economic slumps	The Economist (2013b, c, d; 2014); Free Exchange (2012, 2013b)

(continued)

TABLE 9.7 (Continued)
Sources of Variability in Economic and Nation-State Performance: A Selective Survey

Factors Implicated in Effective Performance	Comments	Reference
	Observations on Sources of Variability in Economic System Performance	
Demographics	The productivity challenge of the rich world's demography.	Buttonwood (2013a)
Human health	Fogel attempted to quantify how much economic growth came from improvements in nutrition and health. The key finding is that one-third of the growth in per capita incomes in Britain over the two centuries ending in 1980 came from improvements in nutrition alone.	Floud et al. (2011)
Income inequality	Growing income inequality is one of the biggest social, economic, and political challenges of our time, but it may not be inevitable.	Beddoes (2012) Schafer (2014)
	Design factors that are key to aiding low-income regional economies in overcoming income immobility include housing integration, lack of sprawl, high social capital (such as church and community volunteerism), good schools, adequate school funding, and a relatively low percentage of single-parent families.	
System complexity	Milton Friedman believed that a modern economy's complexities almost always thwart even the best-intentioned efforts by government officials to intervene into markets.	Boudreaux (2012) Buttonwood (2010) Pentland (2013)
	Two noted Austrian economists (von Mises and Hayek) theorized that financial lubrication has adverse economic consequences (low interest rates create a credit boom, which creates a project investment boom, which in turn creates misallocation of capital). Collectively, because of their complexity, these effects take years to clear up. The recent Great Recession featured all of these elements.	
	The era of big data exacerbates complexity of modern economic systems, with unpredictable consequences.	
Technology	Delivered computing power in the U.S. economy has increased by more than two orders of magnitude since 1970, yet productivity, especially in the service sector, seems to have stagnated.	Brynjolfsson (1993) Buttonwood (2013b)
	The Internet has not yet produced the hoped-for productivity miracle.	

(continued)

TABLE 9.7 (Continued)

Sources of Variability in Economic and Nation-State Performance: A Selective Survey

Factors Implicated in Effective Performance	Comments	Reference
Observations on Sources of Variability in Nation-State Performance		
Incentives to worker productivity	Productivity incentives identified include competition, division of labor, free markets, free trade, political stability and the rule of law, high wages for poor, and technological innovation.	Adam Smith (1776, 1977)
Balance of supply and demand	Successful growth of the economy of a nation-state entails expanding supply as quickly as possible without allowing demand to grow even faster.	Free Exchange (2013f)
Culture	Culture—the sense of inner values and attitudes that guide a population—is held to make all the difference in the wealth or poverty of nations, yet cultural effects are not consistent. By facilitating the movement of jobs to where labor is cheaper, globalization has benefited some poor countries	Landes (1998)
Crony capitalism	Business people have become too influential in government.	Schumpeter (2013e)
Entitlements	Entitlements in the United States and other countries—such as health care, education, retirement income, and so forth—have both positive and negative consequences.	Young (2012)
Middle class decline	A rich array of indicators points to the hollowing out of the middle class in the United States with profoundly negative consequences for nation-state governance and wealth.	Young (2013)
Military prowess and trade	Military prowess, international trade, and the relative strength of neighbors in these attributes are key determinants of the rise or fall of the great powers over the last five centuries.	Kennedy (1987)
Poor governance	The basic problem among democratic nation-states is not more government or less government, but better government.	Ringen (2013)
Power	There are four sources of nation-state power: military, economic vitality, values and ideals, and primacy, equated with hegemony. Of these, primacy may be the most important.	Carr (2013)

Broadly speaking the successful economies we've studied have all managed to get people to behave more productively by rewarding them for doing so… Investments are the key to productivity growth, and productivity growth is what ultimately raises living standards.

Successful governments have found ways to make local savings attractive rather than seeing people send their money abroad. This…requires stability and trust. People need to believe that the current government will remain intact long enough for the promised future to become a reality. To retain stability, governments have to make productivity possible and profitable—to make what is profitable productive and what is productive profitable.

Another aspect of a strong economy is competition, which brings the forces of innovation and survival to the commercial environment. Still another quality is openness to trade, which ties in with the idea of being competitive.

Over time, good economies invest in education and in research and development...

Growth is not an easy formula. It's about productivity, incentives, stability, confidence, competition, and many other things. Successful economies must also continue to adapt to new challenges and needs.

His wrap-up summary to this analysis also is worth noting (Rodriguez, 2010, Lecture 24):

In the end, what seems to matter is a complex combination of incentives for productivity, but also critical to productivity is the advance of technology.

In contrast to these keys to economic success, struggles by less successful economies have been documented over past decades to enable Rodriguez (2010, Lecture 15), using the former Soviet Union as an example, to identify some keys to economic failure:

The Soviet Union is a classic example of failure. This was a full command economy, with the state making every decision about what was produced, bought, and sold. In such an economy, people's jobs are decided for them, as are their wants and needs. The result is a disaster of personal incentives. One of the good things about a capitalist system or a free-market system is that it has built-in rewards for good behavior. It's almost impossible to have those rewards in an economy in which everyone's incomes and consumption patterns are equilibrated. Further, the Soviet Union had effectively no competition and no trade except with other economies in the Soviet bloc. Without competition, there was no pressure to become more productive. Finally, the Soviets had no personal property. The land, the resources, and the labor that could be used to make the economy grow were all owned by people with political power. They, in turn, had no incentives to make the entities they owned more productive.

Rodriguez (2010, Course Scope) refers to the development and establishment of a given economic system as a "natural experiment." The implication of this term is that an economy represents a complex sociotechnical system whose design emerges as the consequence of collective decisions and policies on the part of individual workers, groups of workers, organizations, institutions, and governments. Consequently, as noted earlier in this chapter (Section 9.3.3, point 3) and in Chapter 1 (Section 1.1.3, principle 4), based on evidence from studies of context specificity (Chapter 3), the emergence of new designs of economic systems will evoke new patterns and modes of individual, organizational, institutional, and governmental behavior that are not necessarily predictable in advance.

A quote by the noted Austrian economist, Friedrich Hayek, alludes to the same idea (Buttonwood, 2010): "the curious task of economics is to demonstrate to men how little they really know about what they imagine they can design."

A selected sample of design factors that have been shown by different observers to influence variability in the performance of economic systems, with consequent effects on work performance variability, is offered in the first section of Table 9.7. Paragraphs below briefly expand on these factors, starting with the fourth entry in the table's first section.

Debt. In their historical perspective, Hubbard and Kane (2013) note that one of the recurrent precursors to economic decline is shortchanging the future to overspend on the present. Debt is widely understood to represent the major contributor to the onset of the recent Great Recession in the 2007–2009 period, and of a number of earlier major U.S. economic slumps as well (The Economist, 2013b, c, d; 2014; Free Exchange, 2012, 2013b; Buttonwood, 2014a). This last slump was preceded by a housing boom in the United States that inflated housing prices, accompanied by a surge of subprime mortgages to marginally qualified buyers. During 2007–2009, a record number of homeowners defaulted on their mortgage payments, securities backed by these mortgages collapsed, causing a solvency crisis among financial institutions holding these securities, and the crisis was exacerbated by the creation of exotic assets, the failure of rating agencies to accurately assess the risk of such assets, and lax regulation. As noted in Section 9.4.1, it was the workers, not the bankers, who suffered the most severe consequences of these events (Figure 9.2).

Demographics. Every developed country in the world is confronting the challenge of an aging workforce—one indicator is the "grad-to-granny" ratio (the ratio of those aged 20–29 to those aged 55–75), and this ratio is declining almost everywhere in the world. Buttonwood (2013a) reviews this challenge and highlights the following observations: (1) productivity gains seem harder to generate with an aging workforce, (2) one assumption is that worker productivity peaks somewhere between the ages of 30 and 50 and declines more quickly after age 55, (3) economic growth in developed countries during the first decade of the twenty-first century has been modest compared with previous decades, (4) manual dexterity and willingness to adapt to new technologies decline with age but cognitive ability does not, and (5) the best evidence for a link between an aging workforce and slow economic growth is Japan, where labor productivity has been steadily declining over the past 30 years as the number of workers per pensioner has dropped from about 10 to about 8. However, Buttonwood (2013a) also points out that technological change may have an even greater adverse impact on productivity growth (see below) than demographics.

In the decades to come, the demographic crunch in developed countries is only likely to get worse, primarily because of declines in fertility rates and replacement ratios (births to deaths) that consequently no longer are adequate. According to the United Nations, in 83 countries and territories around the world, women will not have enough daughters to replace themselves unless their fertility rates rise (The Economist, 2011b). This report goes on to offer a rather amusing extrapolation of this situation by projecting (based on dubious assumptions) when populations will drop to zero in different countries or territories. In Hong Kong, the population is predicted to drop to zero in the year 2798. In Germany, Italy, Japan, Portugal, Russia, and Spain, this will occur between the years 3000 and about 3600. Even in China, the most populous country in the world for centuries, the population is predicted to drop to zero by about the year 3500.

For many countries in the shorter term (say the next five to ten decades), the implications of the demographic crunch for variability in both economic and work performance are sobering, given that fewer and fewer workers at peak productive capacity will be available to support the remainder of the population. Although a number of strategies have been proposed to mitigate these effects, the outcome is unpredictable. Given that declines in fertility rates around the world represent a design factor (an outcome of conscious decisions by millions of women not to become pregnant), the unpredictability of the consequences arguably represents yet another illustration of a basic principle of context specificity—that new designs evoke new patterns of behavior that are not necessarily predictable in advance.

Human health. Floud et al. (2011) argue, with ample empirical evidence, that human health and nutrition have made a far greater contribution to labor productivity and economic growth over the past few centuries than is commonly recognized. Their thesis is as follows:

> The health and nutrition of one generation contributes, through mothers and through infant and childhood experience, to the strength, health, and longevity of the next generation; at the same time, increased health and longevity enable the members of that next generation to work harder and longer and create the resources which can then, in their turn, be used to assist the next, and succeeding, generations to prosper.

The authors point out that much of the evidence for this thesis, such as growth in childhood mortality, adult living standards, an adequate food supply, manufacturing output, or labor productivity, typically has been studied as discrete topics, whereas the goal of their analysis is to link them. One specific observation cited in support of the health and nutrition thesis is that one-third of the growth in per capita incomes in Britain over the two centuries ending in 1980 came from improvements in nutrition alone.

Income inequality. In his review of the nature and extent of income inequality around the world, Beddoes (2012) points out that growing inequality represents one of the biggest social economic and political challenges of our time. In the United States, the share of national income going to the richest 1 percent has doubled since 1980. A similar pattern of growing inequality is observed in the United Kingdom, Canada, China, India, Russia, and Sweden. Beddoes (2012) points out that many economists now worry that widening income inequality may have damaging economic side effects related to such consequences as barring talented poor people from access to education, and/or feeding resentment that may result in populist government policies antithetical to growth. These concerns are supported by recent International Monetary Fund (IMF) research showing that income inequality slows economic growth, causes financial crises, and weakens demand (also see Lowrey, 2012). Some studies link health problems (obesity, suicide) among those on the less privileged side of income gaps. A 2012 meeting of the World Economic Forum identified inequality as one of the most pressing economic problems of the coming decade.

System complexity. A number of observers have commented on the complexity inherent to economic system performance and to the consequent intractability in making predictions about how such systems will vary in the future:

- The noted economist Milton Friedman believed that a modern economy's complexities almost always thwart even the best-intentioned efforts by government officials to intervene into markets (Boudreaux, 2012).
- Two noted Austrian economists (von Mises and Hayek) theorized that financial lubrication has adverse economic consequences (i.e., low interest rates creates a credit boom, which creates a project investment boom, which creates misallocation of capital). Collectively, because of their complexity, these effects take years to clear up (Buttonwood, 2010). Specifically, in his 1974 Nobel speech, Hayek noted that (http://www.nobelprize.org/nobel_prizes/economic-sciences/laureates/1974/hayek-lecture.html) (retrieved 12/18/13):

To act on the belief that we possess the knowledge and the power which enable us to shape the processes of society entirely to our liking, knowledge which in fact we do not possess, is likely to make us do much harm... If man is not to do more harm than good in his efforts to improve the social order, he will have to learn that in this, as in all other fields where essential complexity of an organized kind prevails, he cannot acquire the full knowledge which would make mastery of the events possible. He will therefore have to use what knowledge he can achieve, not to shape the results as the craftsman shapes his handiwork, but rather to cultivate a growth by providing the appropriate environment...

- The era of big data exacerbates complexity of modern economic systems, with unpredictable consequences (Pentland, 2013).

These observations extend those offered earlier for organizations and institutions (Section 9.4.1) in supporting the conclusions of Karwowski (2012) and Silver (2012) that predicting the performance of complex ST systems is inherently difficult.

Technology. The benefits of technology to support human work have been recognized for millennia, long before the idea of an economic system emerged (Section 9.2, point 3). For example, as Floud et al. (2011) point out, the application during the medieval period of water and fossil-fuel power to tasks formerly done with human labor certainly was important. Indeed, it probably was the most important factor in growth of per capita output over the past two or three centuries. Not surprisingly, since the onset of the industrial revolution, a major focus of economic analysis has been on documenting the role of technology, industrialization, and innovation in supporting economic growth.

However, as noted in Section 9.4.1, experience over recent decades with the computerization of society suggests that the bloom is off newer technology as a key driver of economic growth. Thus, as Brynjolfsson (1993) points out, delivered computing power in the U.S. economy has increased by more than two orders of magnitude since 1970, yet productivity, especially in the service sector, seems to have stagnated. Buttonwood (2013b) updates this story in noting that the Internet, a worldwide technological phenomenon, has not yet produced the hoped-for productivity miracle.

9.5.3   WORK PERFORMANCE VARIABILITY AND VARIABILITY
IN NATION-STATE PERFORMANCE

The inspiration and a key resource for addressing nation-state performance in this chapter is a work by Landes (1998), with a title—*The Wealth and Poverty of Nations*—almost identical to that of Adam Smith (1776, 1977). In providing this updated perspective, Landes enjoys the luxury of over two centuries of additional observation and experience with economic and nation-state performance. Yet his take on why nations and states differ in wealth and prosperity is much more nuanced than that of his predecessor for reasons that will be addressed below.

This section may strike some readers as simply a rehash of the analyses of organizational and institutional performance and of economic performance provided in the preceding sections. For example, unpredictability is an earmark of performance variability at the nation-state level, as is also the case with organizational and economic performance. For example, Landes (1998, p. 525) closes his analysis with a *mea culpa*, noting that, "No sooner had I sent in the final text for this volume when the countries of East Asia, featured as the big winners in the context for growth and development, fell into crisis and contraction" (he is referring to South Korea, Taiwan, Singapore, Hong Kong, and Japan). A quote from Paul Samuelson (1976, p. 107), one of the foremost economists of the modern era, appears to get at the same point: "No new light has been thrown on the reason why poor countries are poor and rich countries are rich."

Another example is that, as with work behavior and job preferences as well as economic performance, both nativist and empirical perspectives (Goldhaber, 2012) have been advanced to explain variability in political behavior underlying the governance of nations and states. As Buchen (2012) and The Economist (2012) both point out, studies of paired preferences of fraternal versus identical twins indicate that biology—from genes to hormones—can make a considerable contribution to attitudes and beliefs linked to political behavior. Data presented in the latter report indicates that this contribution accounts for anywhere from almost 60 percent (for political knowledge) to under 5 percent (for political party identification) in total observed variance.

Landes (1998, Chapter 1) begins his analysis by pointing out that the natural environment can have a major influence on the wealth or poverty of nations. He cites one example that illustrates this point; namely, that most underdeveloped countries lie in the tropical and semitropical zones between the Tropic of Cancer and the Tropic of Capricorn. Diamond (1997) advances the same theme, focusing on the role of environmental factors—favorable conditions for domestication of plants and animals in the Fertile Crescent, China, and Mesoamerica and for latitudinal dispersion of these innovations—in accounting for advances in Eurasian relative to African, Australasian, and Native American civilizations.

However, as was the case also for organizational and economic performance addressed above, the remainder of this section will focus on the contribution of context specificity—design factors—to variability in nation-state performance. One indication of the scope of this contribution is that all of the design factors listed in the first section of Table 9.7 implicated as influencing variability in economic performance also are implicated in influencing variability in nation-state performance.

However, a set of eight additional selected design factors, listed in the second section of Table 9.7, also have a distinctive influence on variability in nation-state performance. Each of these is addressed below.

Balance of supply and demand. Free Exchange (2013f) points out that, "To become rich, poor countries must enlarge their productive powers, mobilizing workers, absorbing new technology and accumulating capital." Success with this prescription is proposed to rely on nation-state expansion of what is called the supply side of the economy, based on the premise that economic growth of a country can be effectively created by lowering barriers for people to produce (supply) goods and services through a combined strategy of lower taxes and reduced regulation. The supply side of a country's economy determines how much a country can produce, and therefore how much it can earn and spend.

The other side of the equation is demand, related to the level of spending directed at consumption of available supply. Growth of a country's economy, and therefore its wealth, depends on effective economic management of the balance between supply and demand. Excessive demand, by definition, may result in excessive spending, with a possible consequence of inflation, trade deficits, and overreliance on foreign borrowing, which may result in financial crisis. Regarding the goal of balancing supply and demand, Free Exchange (2013f) concludes that, "Simply put, successful development entails expanding supply as quickly as possible without allowing demand to grow even faster."

What appears to represent a reasonably successful effort to balance supply and demand is the strategy adopted by the Federal Reserve Bank (the Fed) of the United States, confronted by the recent Great Recession, to stimulate the national economy (demand) by keeping interest rates close to zero and increasing the money supply by buying U.S. bonds. Indeed, on the very date these lines are being written (December 19, 2013), the Fed announced that it was tapering purchase of U.S. bonds given evidence that the U.S. economy was exhibiting a pattern of solid and consistent growth.

Culture. Landes (1998) suggests that differences in culture—the sense of inner values and attitudes that guide a population—make a substantive contribution to observed differences in the wealth or poverty of nation-states. These cultural distinctions relate to differences in history, in traditions, in social norms, in political and economic frameworks, and in both private and public institutions across different nation-states. A key distinction is that some nation-states are more willing to entrust their economic resources to markets, whereas others resort more instinctively to government control. As Lotterman (2012a) points out, the idea of a cultural influence on nation-state performance as encountered some resistance. However, this author underlines a basic point that if such a cultural influence does in fact exist, it essentially is impossible to deal with in mathematical models of economic performance.

Crony capitalism. As Schumpeter (2013e) suggests, business, in the form of "public-private partnerships," has become too influential in governments around the world. Firms—and the capital they provide—have a dominant influence on nation-state performance in many countries, from state-owned enterprises in China, India, and Russia, to the dramatic influence of business lobbying and election spending on politics and government decision-making in the United States.

Entitlements. Young (2012) points out that entitlements conferred by government—such as education, health care, a living wage, retirement income, vacations, parental leave, and so forth—have benefited the health and well-being of millions of workers worldwide. This author provides a historical perspective and traces the entitlement system back to the ideas of Rousseau. However, he also observes that the entitlement vision of Rousseau has negative consequences as well, in the form of personal freeloading and irresponsible behavior that increase both social and fiscal costs for many nation-states.

Middle class decline. By my count, Young (2013) cites at least 15 statistical indicators pointing to a dramatic hollowing out of the middle class in the United States. These include declines in middle class neighborhoods, wages, and jobs, increasing challenges to traditional middle class values (such as marriage), and a rise in health and welfare problems associated with socioeconomic instability. This author identifies a series of factors contributing to this state of affairs, including globalization (transfer of jobs overseas), labor-saving technologies (such as robotics), growth of financial services (antithetical to economic productivity), greater reliance on debt instead of savings as a source of funds, and higher government taxes and regulations (compromising job creation linked to economic activity). Collectively, these effects have clearly negative consequences for work performance variability in the United States.

Young (2013) points out that there also are a series of negative consequences for U.S. nation-state performance. These include:

- Undermining the middle class foundation dating back to origins of the country for constitutional government and democracy and justice under the rule of law
- A loss of cultural moderation that tends to favor governmental stasis linked to ideological, uncompromising gridlock
- Deterioration of middle class values and behaviors, including importantly a dissipation of work ethic among the U.S. workforce

Young cites a historical precedent for these cautionary warnings. Cicero, the famous Roman historian, claims that Rome's slide toward dictatorship was exacerbated by a decline in middle class behaviors and expectations.

The argument of Young (2013) may perhaps be viewed as overstated and over-dramatized. His central point is indisputable however. The decline of the middle class—in the United States or in any other nation-state—represents a fundamental threat to the health and prosperity of the nation-state and its workforce.

Military prowess and trade. In his survey of the rise and fall of the great powers over the past five centuries, Kennedy (1987) concludes that there are two key determinants of the relative rise or decline of the great powers over this period; namely, competitive advantage in international trade and in military prowess. The importance of trade to economic and nation-state performance has been recognized for many decades (Table 9.1; Rodriguez, 2010; Adam Smith, 1776, 1977). Relative to great power success, Kennedy (p. xxii) observes:

...there is detectable a causal relationship between the shifts which have occurred over time in the general economic and productive balances and the position occupied by the individual Powers in the international system. The move in trade flows from the Mediterranean to the Atlantic and northwestern Europe from the sixteenth century onward, or the redistribution in the shares of world manufacturing output away from western Europe in the decades after 1890, are good examples here. In both cases, the economic shifts heralded the rise of new Great Powers which would one day have a decisive impact upon the military/territorial order.

Kennedy (1987, p. xxii) also emphasizes the importance of military prowess to great power success:

...the historical record suggests that there is a very clear connection *in the long run* between an individual Great Power's economic rise and fall and its growth and decline as an important military power (or world empire). This, too is hardly surprising... economic resources are necessary to support a large-scale military establishment.

Finally, it is pointed out that the power and riches of a given nation-state are relative, dependent on whether its neighbors have more or less of the same.

Poor governance. Ringen (2013) argues that the major challenge to democratic states is lack of good governance. Despite idealistic notions that democracy equates with self-rule, this author points out this notion is a fantasy. Democratic governments still give orders (in the form of laws and regulations), and citizens residing in democracies are obliged to obey. However, if government rule-making is compromised by poor decision-making, lobbying bias, corruption, or other disruptive influences, citizen trust in government is undermined and democratic rule corroded. Hence the need for good government, which shows leadership, wins trust, facilitates the governmental decision-making process, and overall, supports nation-state performance.

Power. Carr (2013, p. 5) provides a nice summary of the critical contribution of power to nation-state performance that focuses on the United States, currently the world's most powerful nation-state, as an exemplar:

Americans would be entirely sensible to ask themselves whether taking on the job of being the biggest power in an ungrateful world is worth the effort... Walter Russell Mead, a writer and academic defines four sorts [of power]. The sharp power of military force [see above] serves as a foundation; the sticky power of economic vitality rewards others for joining the system and makes it expensive for them to pull out; and the sweet power of values attracts and inspires them. The fourth kind, drawing on the work of Joseph Nye, a political scientist, is hegemonic power, which is sometimes called primacy.

Primacy is to geopolitics what a full card is to a game of bingo. As a prize for scoring in all the other sorts of power, a country may get the chance to set the agenda. Primacy makes a state attractive. Other states want to win its favour and to benefit from its goodwill. Their support is a form of consent which gives the system legitimacy. On the global stage, the hegemonic country becomes what Colonel Edward House, President Woodrow Wilson's friend and adviser, called 'the gyroscope of world order'.

Carr (2013, p. 5) goes on to identify three advantages that the United States enjoys in the power game:

1. *Geography.* The United States has friends to the north and south (and oceans to the east and west), whereas all other great powers have had to defend themselves against their neighbors.
2. *History.* Following World War II, the United States prospered whereas other powers either were exhausted or physically and economically devastated. The lessons of World War I prompted President Roosevelt and then President Truman to promote democracy and peace in the defeated countries, which in turn promoted the primacy of the United States.
3. *Values and ideals.* The federal system of U.S. governance means that governmental power is distributed among the states and the federal government. Consequently, democratic decision-making in peacetime is in a constant state of flux and turmoil. One result is that, in exercising its hegemonic power, the United States consistently has declined the temptation, for reasons of both desire and ability, to conquer and administer other countries.

9.5.4 A Control Systems Perspective on Variability in Nation-State Performance

A variety of governance designs for nation-states have emerged over the millennia, but it is democratic government that has achieved steadily growing popularity since World War II. In 1941, there were only a dozen democracies worldwide (Carr, 2013, p. 6). In 2006, the Economist Intelligence Unit started compiling the Democracy Index, which measures the state of democracy in 167 countries (www.eiu.com/public/topical_report.aspx?campaignid=DemocracyIndex12). In 2012, seventy-one years after 1941, out of these 167 countries there were 116 countries (69.5 percent) with full, flawed, or hybrid regime democracies, and these countries contained 62.9 percent of the world's population. In his book on democratic leadership, Ringen (2013) observes that, "In democracy, we control our governors and they rule us." The stirring close to Abraham Lincoln's Gettysburg address—"...government of the people, by the people, for the people, shall not perish from the earth"—essentially gets at the same idea (I write this on November 19, 2013, the 150th anniversary of this address).

These concepts of the fundamental closed-loop nature of the design of democratic systems inspire the control systems perspective on democratic forms of government offered here. In particular, Table 9.8 equates key features of democratic governments with essential elements of a behavioral control system (Chapter 1, Section 1.3.2) to underscore the point that there is a one-to-one relationship between the key elements of the two systems. Specifically, for the governmental system: (1) control goals and objectives are embodied in the constitution and the chief executive; (2) sensory feedback to system is provided by elected representatives; (3) learning and memory is contained in the body of laws and regulations, which also serve as effectors to provide operational output for the system; (4) sensory feedback control of the system

TABLE 9.8

Behavioral Control Systems Elements of a Democratic Government

Democratic Governmental Feature	Behavioral Control System Element
Constitution; chief executive	Control goals and objectives
Elected representatives	Sensory feedback
Body of law	Learning and memory
Laws and regulations	Effectors
Elections; bill of rights; balance of power	Sensory feedback control
Lobbying by special interests; entrenched incumbency; partisan gridlock	Sources of perturbed feedback

if provided by elections, by a bill of rights, and by the balance of power; (5) perturbations to the system, which may compromise its effective functioning, originate with disturbing influences such as lobbying by special interests, partisan gridlock, and/ or entrenched incumbency (the 2013–2014 situation in Washington, at the time of preparation of this book, exhibits all of these features).

9.6 CONCLUSIONS

This chapter deals with variability in three classes of complex sociotechnical systems—organizations and institutions, economies, and nation-states—whose performance, individually and collectively, exerts the primary influence on variability in human work performance. In turn, human work performance arguably represents the most significant mode of performance variability addressed in this book, in that human work defines the human condition and has guided the course of human evolution and the emergence and elaboration of human civilization. From this perspective then, analysis of human performance variability cannot be dissected from analysis of variability in the complex systems addressed in this chapter.

Five common themes prevail across the different classes of complex systems variability addressed in preceding sections; these themes also recapitulate themes discussed in previous chapters. First, variability is inherent to the performance of organizations and institutions, economies, and nation-states, and consequently to human work performance as well. As Kennedy (1987, p. xv) puts it in his analysis of the differential performance of prominent nation-states, the most complex class of performance variability considered in this book, "The relative strengths of the leading nations in world affairs never remain constant...".

Second, just as with individual behavior, patterns of variability observed with the different classes of complex systems are inherently unpredictable. Third, both nativist and empiricist interpretations have been advanced to account for sources of variability observed with the different classes of complex systems. Fourth, control systems perspectives can be applied to interpret the underlying origins of variability observed for each class of complex systems variability considered in this chapter (Figures 9.1, 9.7, 9.10, 9.11, 9.12, 9.13).

Finally, as Tables 9.1 and 9.7 illustrate, a variety of design factors can be identified that contribute to the variability observed with organizations and institutions, economies, and nation-states. This suggests (but does not definitively prove) that context specificity contributes prominently to the performance variability observed with these different classes of complex systems, thereby holding out the promise that such variability can be directed in positive directions beneficial to human work by appropriate attention to, and application of, suitable design interventions.

However, an open question remains. With individual behavior, at least at the motor level, it has been established with reasonable certainty that variability is essential for optimal performance (Stergiou et al., 2006; Chapter 1, Section 1.2.1; Chapter 2). Can the same be said for variability in the performance of complex sociotechnical systems? Juran's theory of breakthrough management (Section 9.4.3.2) suggests an affirmative answer to this question. On the other hand, the Great Depression, the Great Recession, and numerous other episodes of economic and nation-state downturns (The Economist, 2014), with consequent adverse effects on work and organizational performance, suggest that variability is an unwelcome feature of complex systems performance. I will close this chapter by simply noting that this question represents an intriguing and important issue for further inquiry.

10 Variability in Fracture-Critical Systems

Thomas Fisher

10.1 INTRODUCTION

This chapter will consider one aspect of how complex infrastructure systems fail, often in sudden and catastrophic ways. The first part of this chapter will attempt to explain this phenomenon, called "fracture-critical" failure, the second part will examine the various reasons for its occurrence, and the third part will explore ways of addressing it in the design of systems. This chapter will draw from a book that I have written on the topic (Fisher, 2012) and will expand on it in ways that reflect further research on the topic since then.

By their nature, infrastructure systems involve multiple, complex networks and because of that complexity, those responsible for such systems prior to World War II often overdesigned them with a lot of redundancy, in part because prewar calculation and construction methods lacked the precision of what we have available to us today and in part because governments, utilities, and private corporations viewed infrastructure as a more-or-less permanent investment. After World War II, however, we became better at both calculating the stresses in systems and constructing them to handle these loads. At the same time, the governments and utilities largely responsible for our infrastructure began to put greater emphasis on reducing expenditures and not spending any more taxpayer or ratepayer money than absolutely necessary. Both of those trends led to the design of infrastructure with as much efficiency and as little redundancy as possible. This also left little room for error, creating systems that we now recognize as fracture-critical, in which the failure of any one part of the system can lead to an unanticipated and total collapse of the whole.

10.2 FRACTURE-CRITICAL SYSTEM FAILURES

The first fracture-critical failures started to show up in bridges designed and built in the 1950s and 1960s. As one state's Department of Transportation defines it, "A fracture-critical bridge is one that does not contain redundant supporting elements. This means that if those key supports fail, the bridge would be in danger of collapse. This *does not* mean the bridge is inherently unsafe, only that there is a lack of redundancy in its design" (Iowa DOT, 2013).

Failures of fracture-critical bridges continue to occur and often receive a lot of media attention because of the sudden and sometimes spectacular nature of their collapses. The fracture-critical I-35W bridge in Minneapolis, completed in the early

1960s, collapsed in 2007 when one of the steel gusset plates holding its steel trusses together cracked and the entire bridge fell unexpectedly into the Mississippi River (Fisher, 2012). Another example of this occurred with the 1955-era fracture-critical I-5 bridge north of Seattle, when one of its spans fell into the Skagit River immediately after a truck struck one of its truss members (Rafferty & Kirschner, 2013). In both cases, these bridges had highly integrated structures, with each member of their steel trusses as minimal in size as necessary to meet its design load and with each element dependent on all of the others working exactly as designed, with very little redundancy in the system as a whole.

Fracture-critical designs might not present a problem in an ideal world, in which such structures receive frequent maintenance, carry no extra loading, and encounter no outside force that puts an unanticipated stress on them. But such ideal situations rarely occur: maintenance budgets get cut and loads get increased, as happened with the I-35W bridge, or unexpected accidents happen in a vulnerable location, as happened with the I-5 bridge. And when fracture-critical structures fail, they give no warning; one moment they appear strong and stable and the next moment, they collapse.

Many hundreds of such bridges exist in the United States alone, but if this problem pertained only to bridge design, it might not warrant much attention. Unfortunately this is not the case; fracture-critical systems pervade America's infrastructure and they extend far beyond bridges. We can spot fracture-critical systems by their failure pattern. Had we put strain gauges on the I-35W or the I-5 bridges, we would have seen, just prior to their collapse, a sudden spike in the stress curve on the bridge as a result of the cascading collapse that characterizes such systems. When one piece gives way or gets damaged, the rest of the structure does not have enough resilience to take up the additional load and so the next piece fails and then the next one until, like a line of dominoes, the entire structure collapses.

We have witnessed such rapid, cascading failures in other systems. We have a fracture-critical electrical grid system in this country, in which the failure of a few key transmission stations can knock out power for large sections of the United States, as happened in 2003 when a local power failure near Cleveland set off a cascade of power outages that left 50 million people without electricity for days and caused an estimated $10 billion in losses (Barron, 2003). We have a fracture-critical transportation system in America so heavily dependent on fossil fuel that any major disruption of the oil supply immediately affects our economy, as has happened during the Yom Kippur War in 1973 and the Iranian Revolution in 1979 (Dr. Econ, 2007). And we have a fracture-critical road system across the nation, where federal investments in roads after World War II have left local communities unable to repair or replace them without major tax increases or unsustainable levels of growth in their tax base (Marohn, 2012). The list goes on, but these examples show how the creation of fracture-critical infrastructure vulnerable to collapse has characterized post-war America and has become a systemic problem.

Nor does it only apply to the work of governments and utilities: investment banks have become a fracture-critical system. Because of the highly complex and highly leveraged nature of some investment products and the highly efficient and interconnected nature of the banks themselves, when Bear Stearns and Lehman Brothers failed suddenly and without warning (at least to their customers) in 2008, their fall led to a cascading set of failures across the entire global banking industry (Dash et al., 2008; Sorkin et al., 2008). A quick and massive response by the U.S. government staved off that collapse, but few noticed how the liabilities of these banks just prior to their closing followed the same pattern as other fracture-critical systems, with an exponential spike in their debt load just prior to collapse.

A similar fracture-critical pattern characterized the collapse of our housing and home mortgage system. Here too, we had a highly efficient and highly integrated system that worked well as long as prices continued to rise. Like a fracture-critical bridge, our housing market needed to work exactly as planned in order to keep going, but once enough defaults hit, it set off a cascading set of failures through the system (Andrews, 2007). And once enough houses of a certain type went into foreclosure and their prices fell, all of the other comparable houses in an area experienced the same precipitous drop in value, causing many homeowners to find themselves suddenly underwater with their mortgage.

Fracture-critical failures can harm a lot of people and damage entire industries and sometimes the entire global economy. What starts as a localized failure of one part or in one area rapidly expands and becomes quickly magnified to encompass the whole. Most worrisome of all, some of the systems on which we depend not just as a society but also as a civilization and as a species have a fracture-critical nature, which means that their collapse could ultimately harm everyone, even those who may think of themselves as immune to catastrophes. Like the people who happened to be on the I-35W bridge when it fell, fracture-critical failures can affect anyone at any time, and because of that apparent randomness combined with their magnifying effect, this phenomenon deserves immediate attention.

10.3 WHY FRACTURE-CRITICAL FAILURES HAPPEN

Engineers have learned their lessons with fracture-critical bridges and now design them to have greater redundancy and more resilience. Bridge designers have also found that such precautions do not have to cost much more if designed that way from the beginning, knowing that it costs much more to replace a structure after it has prematurely failed. Because we rarely recognize the fracture-critical nature of other types of systems, however, we continue to suffer from bad design and inadequate oversight. Had the lessons engineers have learned with bridges been applied to the design of financial products or mortgage rules, we might have prevented the economic meltdown of 2008. And had we understood the nature of fracture-critical failures, we might have responded to the economic collapse of 2008 in a more robust way than we have, leaving open the possibility of it happening again.

As the ecological historian Jared Diamond has shown, civilizations have experienced collapses before (Diamond, 2005). Whether it involved the collapse of the Easter Island population because of deforestation or the collapse of the Cahokia community because of pollution or the collapse of the Mayan civilization because of resource exhaustion, humans have proven adept at one thing: deluding ourselves into believing that we can escape the consequences of our most self-destructive behavior and that we will find a way to keep going as we have even when signs begin to suggest otherwise. This is one area in which we have had little variability in our performance; when something threatens what we have come to see as normal, we seem almost invariably to ignore, downplay, or deny the peril, at least until circumstances no longer allow it.

In this, we may not differ much from other species. As the ecologists C.S. Holling and Lance Gunderson have observed, there exists across multiple ecosystems a tendency for one species to become so dominant and productive that it crowds out the very diversity of resources or other species on which it depends (Gunderson & Holling, 2002). This almost inevitably leads to a collapse and reorganization of the ecosystem in which no one species dominates or becomes so productive that it absorbs all of the resources needed by others. Holling and Gunderson call this cycle of dominance, collapse, and reorganization "panarchy."

The same cycle seems to apply to human ecosystems. We have seen an exponential growth in the human population as well as in our economic productivity and technological prowess to the point where we have become so dominant that scientists now call this the Anthropocene: the human-dominated era (Seielstad, 2013). But according to panarchy, our very dominance of the planet makes human civilization exceedingly vulnerable to collapse and reorganization. The panarchy cycle also follows a fracture-critical pattern; when an ecosystem collapses, the dominant species tends to have an exponential spike in its numbers, which precipitates its collapse and which then allows other species and the resources they depend on to recover and renew the ecological diversity that leads to a more resilient environment.

Unlike other species, humanity has the capacity to comprehend and communicate our situation, which holds the promise that we can alter our behavior and avert a panarchic collapse. There does exist some human variability here. Some communities of people have begun to respond to the prospect of collapse of the human ecosystem, ranging from the voluntary simplicity movement to the transition town movement to the sustainability and resiliency movements (Alexander, 2009). But the resistance to change among those with the most to lose, be it their power, their business, or their job, suggests that we may not change our thinking or alter our behavior fast enough to avoid what seems likely to lie ahead for us.

How then can we accelerate our understanding of the situation we face? To answer that, I will use a framework developed by the political economist Albert Hirschman (Hirschman, 1991). He identified three ways in which people react to change, which he labels as perversity, futility, and jeopardy, and uses that framework to explain why political groups respond so negatively to progressive movements ranging from the eighteenth century French Revolution to the Social Welfare programs of twentieth-century America. As he did in so much of his work, Hirschman neither condemns nor condones this "rhetoric of reaction," but his framework does provide a useful way of understanding the overly optimistic beliefs that can lead to fracture-critical design as well as the skeptical responses that can serve as correctives to it.

10.4   JEOPARDIZING THE FUTURE

Let's look at Hirschman's framework in relation to three fracture-critical situations. The first involves the failure of fracture-critical infrastructure systems such as the levee system around New Orleans after Hurricane Katrina, the subway system in

New York after Hurricane Sandy, or the nuclear power plant in Fukushima, Japan, after the earthquake and tsunami there (Fisher, 2012). In all of these cases, their designers created highly efficient, interconnected systems that, when working as designed, did their job effectively. But in each case, the system had a key weakness that made it vulnerable to catastrophic failure.

In New Orleans after Hurricane Katrina, the levee system got breeched and because no second line of defense or backup series of levees existed, large sections of the city flooded, resulting in what remains the most expensive disaster in human history. In New York City after Hurricane Sandy, the subway system, sitting well below sea level had no system in place to stop floodwaters once they overtopped the streets of Manhattan, and because of the interconnection of the system, a large segment of it became inundated, causing huge disruptions in New York City's transportation network and significant damage to equipment because of the saltwater. And in Fukushima, the nuclear power plant not only did not have a barrier high enough to stop the tsunami wave that hit it, but the plant also had no backup to the electrical power, diesel engines, and batteries, all of which stopped working in the wake of the tsunami, eventually leading to the release of radioactive material into the air and water.

While all different in many ways, these three examples share a common failing: the inability of those who put these systems in place to imagine events that had happened before and/or the unwillingness of those funding this infrastructure to put in place sufficient backup or enough redundancy to prevent the systems' failure. This does not mean those responsible for these systems showed negligence. Their designers and funders no doubt did cost-benefit analyses that probably showed that the likelihood of events like a category 5 hurricane or a major earthquake and tsunami did not justify the added expense of additional backups or higher barriers.

Hirschman called this "the hiding hand" principle (Hirschman, 1967). He saw humans in risky enterprises frequently "misjudging the nature of the task, by presenting it to ourselves as more routine, simple, undemanding of genuine creativity than it will turn out to be. Or put differently: since we necessarily underestimate our creativity it is desirable that we underestimate to a roughly similar extent the difficulties of the tasks we face, so as to be tricked by these two offsetting underestimates into undertaking tasks which we can, but otherwise would not dare, tackle."

In other words, we continually underestimate risks as well as underestimate our ability to address them creatively. Such underestimation can force people to develop new, innovative ideas, in part because they may have no other option when in the midst of a task that has proven more difficult than initially assumed. Underestimation, though, can also lead to catastrophic failures when we respond with routine or overly simplistic solutions to nonroutine, complex situations.

That certainly seemed true of the three examples here. Why would we put the same kind of levee—a single line of defense—around a major city that stood largely below sea level as we would around higher ground or a rural area? Why would we put so much of the infrastructure of Manhattan—not just the subway system, but also much of the mechanical rooms of major buildings—below water level on an island open at one end to the sea? And why would we build a nuclear power plant in a seismically active area, protected by a barrier lower than the tsunami levels that history showed to have happened there? These actions represent not just errors of judgment, but also a profound lack of imagination in envisioning situations likely to happen within the life of the system or structure.

The widespread nature of fracture-critical systems and the prevalence of our underestimating the potential hazard they present suggests that this occurs not because of one person's or one group's incompetence or malfeasance, but because we have set up disincentives that inhibit our ability to question such short-sighted and easily subverted designs and that prevent us from seeing the full costs and consequences of our decisions.

To understand the nature of those disincentives, we might look to the work of another political economist, Adam Smith (Smith, 1776, 1977). Two of his ideas seem particularly relevant here. One involves his realization that increases in the productivity and profitability of enterprises requires a degree of specialization never before imagined. Workers as well as those who own the means of production have benefited from such specialization, but with it has come a specialization in the way we account for costs, leading to the disaggregation of sectors and industries across an economy.

In each of the cases mentioned above, the people making the decision about how much to spend on building a levee, a subway, or a nuclear plant no doubt had to work within a budget and with the resources allocated to them to accomplish this task. But the cost of the system's failure, beyond perhaps the replacement cost of the system itself, rarely gets factored into the equation because the true costs of a failure gets borne elsewhere: by the property owners of New Orleans, the commuters in New York, or the people in northern Japan. As a result of this disconnection between design decisions and the full consequences of their failure, those making the decisions have little or no incentive to do anything beyond the minimum required to meet the project's goals, even though the extra cost of additional protection or backup systems, at the beginning of a project, pales in comparison to the enormous cost of a key system's failure.

If the disaggregation of our cost accounting, and with it, our sense of accountability, arises from one aspect of Smith's work, a paradox of value emerges from another. Smith identified this paradox early on in *The Wealth of Nations*. "The word value, it is to be observed, has two different meanings, and sometimes expresses the utility of some particular object, and sometimes the power of purchasing other goods which the possession of that object conveys. The one may be called 'valueinuse;' the other, 'valueinexchange.' The things which have the greatest value in use have frequently little or no value in exchange; on the contrary, those which have the greatest value in exchange have frequently little or no value in use. Nothing is more useful than water: but it will purchase scarce anything; scarce anything can be had in exchange for it. A diamond, on the contrary, has scarce any use-value; but a very great quantity of other goods may frequently be had in exchange for it" (Adam Smith, 1776, 1977).

While later economists have criticized Smith for comparing incomparable commodities like diamonds and water, his paradox speaks to the difficulty we have in measuring the true value of things. As Oscar Wilde remarked, "Nowadays people know the price of everything and the value of nothing" (Wilde, 1890). While we typically equate the price of something with its value, what price really measures, as Smith observes, is something's exchange value, having to do with the demand for it and its scarcity. That which we can't live without—water—has tremendous use value, but little exchange value, because we see it as plentiful and readily available. The problem arises when that is no longer the case, as has happened now in many parts of the world and, indeed, in parts of the United States, where water scarcity has become a real issue, limiting the ability of farmers to grow crops, cities to serve their populations, and people to have access to enough water in which to cook or to bathe.

All of this relates to Hirschman's idea of jeopardy. He argued that frequently the first opposition to progressive political ideas comes from those who focus on what we will lose as a result; the good work that has gone on in the past and the best qualities of a situation that will be jeopardized because of the innovation or new idea. But we can read Hirschman's idea of jeopardy in another way as well. A system designed with overly optimistic assumptions about how it will perform, be maintained, or handle additional stress can jeopardize the lives of people and the well-being of communities.

This partly comes from the pressures mentioned above of having to design a system based on the constraints and resources available at the time of its creation, a present-mindedness that often leads to a discounting of future effects we don't yet know. The design process, however, offers a rigorous way of exploring alternative scenarios about what doesn't yet exist and what we don't yet know. Likewise historical data can give us some sense of what might happen in the future. We know that certain coastlines remain vulnerable to hurricanes and tsunamis, certain structures like levees and bridges often suffer from deferred maintenance, and certain conditions such as climate change can greatly disrupt the food systems on which we all depend.

The real jeopardy, though, happens because of a disconnection between the costs and their consequences over time. We have designed a political and economic system in which, on one hand, we have externalized consequences that we have difficulty quantifying, such as the effect of a decision on long-term environmental and social factors, and, on the other hand, we have distanced decision makers, who have every incentive to focus on the next quarterly return or political campaign, from those who will bear the brunt of their decisions years later. The jeopardy here inverts the reactionary rhetoric that Hirschman writes about. While he dwelt on those who dislike progress because of what it forsakes from the past, the jeopardy here involves those who make irresponsible decisions that forsake future generations.

10.5   THE FUTILITY OF MASSIVE SCALE

A second type of fracture-critical system involves the question of scale. The natural world, from one point of view, operates at a global scale, and we have certainly seen that with a changing climate, in which human activities in one location, such as the emission of air pollutants, have dramatic impacts on ecosystems far away. But, from another point of view, nature's ecosystems exist as a series of patches that remain relatively discreet and largely unaffected by each other. A healthy ecosystem typically has patches in various states of transition, with some thriving, others crashing, and others in the process of reorganizing.

   The scale of these patches matters. Invasive species from a distant ecosystem can run riot in a patch that has no predators or way of defending itself from the invader, although the dominance of an alien species also sets up a scenario in which it and the ecosystem it has overtaken become vulnerable to collapse either because of the exhaustion of key resources internal to the system or as a result of an external shock to the system that alters its balance in some way. The natural world involves massive die-offs of species and major crashes of ecosystems, and we humans cannot count ourselves immune from that happening to us as well.

   This becomes especially urgent because we have, through modern technology, tried to do something nature never does: organize the human ecosystem into a single, global patch, with networks of transportation, communication, distribution, and energy generations, among others systems, crossing the entire planet rapidly. Never has a species on the planet attempted this level of interconnectedness at such a massive scale, with such efficiency and such an impact on every part of the globe. The end of nature, as Bill McKibben calls it, really means the end of nature not somehow affected by human nature and our drive to become the absolutely dominant species on earth (McKibben, 1989).

   The problem comes in both managing such a complex, global system and in preventing it from becoming fracture-critical, in which the failure of one part would bring down the whole. Technology has enabled us to accomplish the former: managing global systems with a high degree of confidence. Everything from the global financial system to the global airline system to the global telecommunications system all operate with amazing efficiency and with relatively few breakdowns, in part because of the capacity of computer-aided operations to enhance our a ability to visualize and grasp the larger patterns that help us make sense of complexity.

   The challenge arises when we don't see the potential for cascading failures, when something that passes a tipping point in one part of a complex system alters the rest of it. No amount of monitoring can predict such fracture-critical failures because they often happen so quickly that we only see them after the fact; even if there had been stain gauges on the I-35W bridge, it would not have mattered since the entire bridge fell almost instantly after the stress on the one gusset plate pushed it into failure mode. The rapidity with which fracture-critical systems fail demands a different approach. We cannot manage them or monitor them, since neither our computers nor their human operators can react that quickly. We can only design systems initially in ways that make them less vulnerable to fracture-critical collapse.

   This brings up the second of Hirschman's categories: futility. In his analysis of political systems, Hirschman looks at how efforts to improve something often lead others to see such efforts as futile out of a belief that, in the end, they will not work. In many cases, these claims of futility come in response to attempts to devise large-scale or one-size-fits-all solutions to seemingly intractable problems. The futility argument then arises not out the belief that nothing will ever change, but rather that the scale of the change matters and that solutions that seem too sweeping or overly simplistic may not address the often local causes of a problem.

   We can make a similar argument against fracture-critical systems. They tend to be so efficient, so interconnected, and so large in scale that it becomes futile to try to manage their vulnerability to failure. They collapse before we know what happened. Instead, we need to design systems more along the lines of natural eco-systems, as patches that remain connected and semiautonomous, dynamic and eco-logically diverse. A system made up not of one singular and interconnected entity that spans the globe (or at least a large territory or realm of activity), but instead of multiple, varied, and mostly self-sufficient units becomes a system that has much greater resiliency and must greater resistance to complete failure. Even when one or more units collapse—as happens with ecosystems all the time—the entire system does not crash any more than the natural world, as a whole, ever does. An ecosystem approach to human-designed systems counters the futility argument with facility, with the ease with which any system, as a whole, can accommodate an outside shock, internal malfunction, or human error.

Our bias against such thinking emerges in the connotation of the word "patch-work." We associate it with words like "makeshift" or "jerry-rigged," as if something made up of patches suggests a lack of care, foresight, or craft. In contrast, the more unified, consistent, and efficient a system, the better we think it will work and the more we seem to trust its performance. As we have seen, such systems do perform well when everything operates as planned. But when the unexpected, the unusual, or the unplanned-for event happens, the very unity that we admire in highly integrated, efficient systems becomes their Achilles' heel. The only resilient systems are those characterized by what we have for too long dismissed: patchworks of diverse parts in dynamic and semi-independent relationships with each other.

This also represents a shift away from hierarchical command-and-control ways of managing systems toward networked, inclusive, and participatory forms of self-governance. The idea of the "holon" has some value here. Arthur Koestler coined that term to describe nonhierarchical, flattened systems in which every part is both a part of a larger whole and a composite of smaller parts (Koestler, 1967). Thinking about systems as holons can help us conceive of them not as all elements of a single hierarchy but instead as a set of nested entities, each with its own identity and yet each a part of and also comprising parts other entities. Ecosystems, which contain multiple species and which remain parts of larger biomes, have a holon-like struc-ture, and human systems designed in the same way will avert their fracture-critical failure.

10.6 THE PERVERSITY OF UNINTENDED CONSEQUENCES

A falling bridge, a failing housing market, and faltering banking system may injure or harm many people, but the number of people who die as a result remains relatively small. But one fracture-critical system, in particular, threatens every one of us: our rapidly growing human population. Human population has witnessed an explosive growth from 3 billion in 1960 to 7 billion 52 years later, with an expectation of our hitting 9 billion by or before 2050 (United Nations, 2013). The question we face is: will we be able support such a large population without exhausting critical resources like fresh water or altering climatic conditions in ways that disrupt our food systems? Like other species whose numbers outstrip the carrying capacity of their environ-ment, we, too, stand poised for a collapse unless we change how we now occupy the planet.

The physicists Geoffrey West and Luis Bettencourt (West & Bettencourt, 2013) have shown how innovation can forestall the collapse of systems in the midst of rapid growth, such as the human population. But that innovation has to happen at an ever-greater speed, argue West and Bettencourt, to compensate for the rapid multiplying effect of exponential growth in our numbers and in our demand for resources. So far, the rate of innovation, especially with the rise of global networks of research and development teams in companies and at universities, has managed to keep pace, although carried to its logical conclusion this argument suggests that we will reach a point where innovation will have to happen almost instantaneously and where its adoption by ever-greater numbers of people cannot occur fast enough.

That brings to mind Hirschman's perversity argument. He points to a long line of thinkers who have claimed that our overly optimistic views of progress can blind us to unintended consequences that are often opposite of those we had hoped and planned for. In the case of human population, letting our numbers grow at an almost exponential rate in hopes that innovations will enable us to sustain that many people on a finite planet represents the ultimate perversity. An unintended consequence—say, our inability to come anywhere near supporting that size of a population—involves human deaths at a potentially massive scale, including those who believed otherwise. Human beings have always taken risks. But a risk that involves potentially the entire human species seems foolhardy and, to use Hirschman's term, perverse.

The human population grows, however, because it can. Without any global family planning in place, people have too many children for all sorts of personal reasons, while the global economy sees an ever-growing population as an expanded marketplace in which to sell products and services. And because of the disconnect between local and global effects, it often doesn't appear that our childbearing decisions have any discernible effect on the entire population any more than our consumption patterns of vital resources like water have any effect on the worldwide supply. This characterizes all fracture-critical systems. They often appear fine until the moment an internal weakness or outside force brings the entire system down. We cannot discount the possibility of that happening with the human species and the more we plan for unintended consequences and perverse effects, the more resilient we will be in the face of such a prospect.

It isn't the absolute number of human beings that makes our population fracture-critical, but the way in which we have linked ourselves together in a highly efficient, interconnected way. This interconnectedness, for example, leaves us open to one of the greatest threats we face: a zoonotic disease against which we have neither immunity nor any vaccine with which to treat it. There remain three essential components of our designed environment that make us vulnerable to that happening. First, massive numbers of people continue to move into cities; over 50% of humanity now lives in urban areas for the first time in human history, and projections place it at 75% by 2050. Some have even argued that as we approach the next century, we will have reached over 90% (West & Bettencourt, 2013). At the same time, a lot of that urbanization has occurred in informal settlements, mostly in the developing world, with people eager for economic opportunity but also living in unsanitary and unsafe conditions.

Those conditions present themselves as not just inhumane, but also the potential sites for the transfer of zoonotic disease from animals to humans. In the developing world, some of the worst informal settlements exist in large cities near airports, which harbor the most efficient disease-carrying mechanism ever invented: the jet airplane. It takes just one person infected by an easily transferred, zoonotic virus to board a plane and infect others on the plane, who will in turn infect others still, starting a pandemic that will move through the human population much faster than we can inoculate ourselves against it. We tracked how the H1N1 virus moved around the planet, going to major urban centers and infecting, first, the wealthiest populations that travel by air most often, and we should see that not-very-virulent viral outbreak as a warning of what will happen when a virulent strain begins to spread.

The sheer density, mobility, and inequality of the human population, in other words, make us exceedingly vulnerable to a collapse, leading to the ironic conclusion that we have never been so dominant and so vulnerable as now. No one intended this as a consequence of modern technology. But Hirschman's perversity argument makes clear, whenever we try to overcome the limits in which humanity has evolved as a species—living much longer, dwelling more densely, and moving much farther and faster—we set ourselves up for unintended results, in this case, creating environments that literally threaten our survival as a species.

10.7   CREATING A MORE RESILIENT FUTURE FOR OURSELVES

Going forward, how can we avoid fracture-critical failures, either of systems we depend on or of our species itself? Such a question has specific, technical answers. We need to steward the resources we most need to survive and stop acting as if we can control systems at a scale far beyond our ability to comprehend. And we need to start moving bytes and not bodies around the world to address the terrible living conditions in global slums that give birth to zoonotic diseases, and to change simple behaviors that spread infection ranging from shaking hands with strangers to going to work when ill.

Such recommendations, though, miss the larger issue. Unless we think differently and behave in new ways, we will continue to set ourselves up for the fracture-critical failures of the sort mentioned above. So let's start with the thought process that has brought us to this point. Much of it involves what Daniel Kahneman calls "thinking fast" as opposed to "thinking slow" (Kahneman, 2011). Although the creation of systems takes a long time and involves many people and decision-points, too often their design arises out of "fast" thinking about the nature of the need the system seeks to address. Kahneman and his colleague Amos Tversky identified three ways of thinking that I would argue lead us toward fracture-critical failures (Kahneman & Tversky, 1979).

They call the first one "anchoring," in which people faced with an unknown tend to start from what they know and then adjust it to fit, often insufficiently, a new situation. As we have seen in some of the examples above, this can lead to incremental change that obscures a phase change in the nature of what we have created: a supersonic jet airline may be anchored in early aircraft design, but the impact of our flying rapidly around the world differs in kind and not just degree from what the Wright

Brothers achieved. Because our thinking remains anchored in what we know, we often miss the unintended or perverse consequences of our continual adjustments.

The second rule of thumb they call "availability," which involves our tendency to overlook the risk involved in the unfamiliar. This echoes Hirschman's "hiding hand" principle and, as we have seen, it can lead us to take risks far greater than more careful reflection might warrant. As we have expanded systems globally and, via new technology, made them faster and more powerful than ever before, we have also made them into a kind of abstraction that distances us from them and makes them, in that sense, unfamiliar to us. Our telecommunications systems, which once involved wires on telephone poles, now involves satellite communications and distant cell phone towers; our banking system, which once involved going to a bank and depositing or withdrawing cash, now entails digital information and automatic deposits done online. While this advance in the efficiency and speed of services make them more convenient, it also encourages us to engage in riskier behavior with regard to them because of their relative invisibility.

A third rule of thumb—similarity—involves our thinking that because two things look alike that they are alike. This assumption creates the problem of our making one thing like another even if the context or conditions around it differ. We tend to typecast not just people, but problems and their solutions, so that a situation we think we have encountered before will lead us to the same resolution as happened before, which can lead not only to inappropriate solutions but also to solutions that repeat errors from the past. We see this a lot in fracture-critical systems. The hundreds of fracture-critical bridges in the United States, for example, shows how engineers kept replicating poor design decisions in similar situations, resulting in a massive problem that need not have happened.

Social psychologists have identified other characteristics among groups of people that can lead to poor design decisions. We tend to have overconfidence that things will go as we assume they will, to have an aversion to changing what has been done before out of a fear of loss, and to feel peer pressure in conforming to the status quo (Thaler & Sunstein, 2009). Those human traits help create the conditions in which fracture-critical systems have emerged, bred by their designers' overconfidence and nurtured by their unwillingness to challenge assumptions or their bowing to peer pressure.

Given the incentives that exist in modern economies and cultures to push systems to the point of failure and the pressure on governments and utilities to minimize the expenditure of taxpayer or ratepayer money, the best strategy in reducing the likelihood of fracture-critical systems might involve what Thaler and Sunstein call "libertarian paternalism," in which we nudge people and organizations to build in more resiliency into systems. Their idea of choice architecture suggests that we can avoid a fracture-critical solution by framing a problem correctly, shifting the attention from the lowest first cost to looking at not just the life-cycle cost, but also the replacement cost of a system that fails prematurely. Over the life of a system, designing it to have some redundancy and conceiving of it as a series of semiautonomous patches costs much less than the extraordinary cost of its collapse.

The challenge then becomes one of structuring the choices so that there are a few easily understood choices, a clear map on how to achieve the goal, a default mode to make the desired choice that path of least resistance, and a set of incentives that encourage us to take that path. To avoid creating a fracture-critical situation, this might involve showing how systems that lack redundancy make us all vulnerable, developing simple rules of thumb for more resilient design, making such resiliency a part of the review and approval process, and rewarding those who factor this into their planning.

In addition to these nudges, Thaler and Sunstein also recommend that the design process include feedback loops to ensure that a system does not end up with a weak link that can bring it down unexpectedly, and that the final system have built in to it a tolerance for error or localized failure. Thinking of systems as semiautonomous patches allows for both of these conditions. Feedback can ensure that later patches incorporate lessons learned in earlier ones and the semiautonomy of each patch enables some parts of a system to fail without preventing the whole from functioning.

Ultimately, our ability to move away from creating fracture-critical systems will depend on our understanding that size matters. We have long had a system of planned obsolescence in consumer products, designing them so that they either have a limited lifespan as some key component wears out or fails or tempting consumers to get rid of their old purchase in favor of a new one because of the improved performance or added features of the latter. While this rapid cycling through of products often has negative environmental effects, its economic impact has led to a consumer economy that has kept people employed and spurred innovation.

At the scale of infrastructure, however, planned (or unplanned) obsolescence has a much different effect. The premature failure of a system upon which many people—and indeed, the entire economy—depend can cause great harm and have tremendously negative effects for a very long time. The larger the size and impact of a system and more we depend on it, the more resiliency it needs to have. Infrastructure is not just another product writ large; it differs in fundamental ways and until we recognize that difference and design such systems accordingly, we will continue to suffer the consequences of unnecessary and increasingly devastating failures.

11 Human Performance Variability
An Evolutionary Perspective

Thomas J. Smith

11.1 INTRODUCTION

There is no more dramatic illustration of the role and ubiquity of performance variability in nature than the panorama of evolution. As noted in the introduction to Chapter 1, this conclusion is explicit in Darwin's original account of the role of performance variability in the emergence of species, "descent with modification." With regard to the field of human factors/ergonomics (HF/E) targeted by this book—primarily concerned with the interaction of human performance and design of the performance environment—both human performance and designs created by such performance are products of evolution acting over millions of years.

In the following discussion, the term "human evolution" is used to refer to the span of evolution encompassing the appearance of premodern humans—hominins—some five to six million years ago, to the present day, with modern humans—*homo sapiens*—emerging about 200,000 years ago (Ehrlich, 2000).

It is fair to say that the evolution of human performance and human designs, in the sense of their emergence over the eons of time, have not been a central concern of the (HF/E) field. For example, based on information available on the Human Factors and Ergonomics Society (HFES) website:

- A November 2013 search of the HFES publication *Proceedings of the Human Factors and Ergonomics Society Annual Meeting* between the years 1974 and 2013 revealed 22 papers with the term "evolution" in the title, but without exception these papers deal with relatively recent trends in performance or design, the longest perspective dating back to the beginning of the twentieth century.
- A similar search of the HFES publication *Human Factors* between the years 1958 and 2013, revealed three papers concerned exclusively with the evolution of the HF/E field over the past 50–60 years and one paper that takes the longest term view of evolution among all those surveyed here, namely that of Van Cott (1984). This author argues that as technologic designs have evolved over the past two million years or so (from early stone tools to the complex

systems of today), human interaction with technology also has evolved from the role of a mechanical controller to that of a supervisor. This transition, it is argued, requires a greater emphasis on the cognitive demands of supervisory relative to direct mechanical actuation of technology.

- A similar search of the HFES publication *Reviews of Human Factors and Ergonomics* between the years 2005 and 2013 revealed no reviews with the term "evolution" in the title.
- A similar search of the HFES publication *Ergonomics in Design* between the years 1993 and 2013 revealed three papers concerned with the evolution of different designs over the twentieth century. Among these papers, the paper by Hancock (1996) is the only one to introduce a long-term evolutionary perspective. Specifically, this author compares the convergent evolution of human-computer interface designs over the course of the twentieth century with the convergent evolution of biological designs in nature over millions of years.

Recapitulating the foregoing observations, there is no entry for the term "evolution" in the index for the first edition of the *Handbook of Human Factors* (Salvendy, 1987).

This admittedly cursory analysis suggests that the evolutionary origins of human performance and of human designs have not represented a central concern of the HF/E community.

The fact remains, however, that a comprehensive account of the emergence and elaboration of the myriad patterns of human performance and designs, as they exist today, is incomplete without an evolutionary perspective.

11.1.1 MECHANISMS OF EVOLUTION

I have no intention of delving into the vast literature on evolution and its underlying mechanisms. However, it is worthwhile briefly reviewing the major explanations that have been advanced for how evolution occurs, anticipating an analysis later in this chapter of their relevance for interpreting evolutionary changes in human performance and design that have occurred over the millennia. From a scientific perspective, the general assumption is that the evolutionary process is mediated through some sort of selective mechanism acting upon adaptive changes in biological structure and function. Debate arises as to how exactly such selection occurs.

One view is that we really do not understand how selection works (Margulis, 1998).

Generally speaking, however, either nativist or empiricist mechanisms (Goldhaber, 2012) are invoked to explain why evolution occurs, recapitulating a theme noted in previous chapters. The nativist account is essentially that of Darwinian natural selection acting at the biological level—adaptive changes through genetic mutation occur at one or more levels of biological organization of an organism, and if these changes are favorable for survival, differential reproductive success ensures that the organism's genes are passed on to the next generation.

Yet in recent decades, this view has been complemented by empiricist interpretations that assume that environmental design factors—context—also influence evolution. Some selected examples are

- Harcourt (2012, Chapter 8) notes that better nutrition for those who move from more to less economically disadvantaged regions or countries results in both increased body mass and increased body size—and human evolution continues even today (Hawks, 2011, Lecture 24). The steady increase in human populations after the agricultural revolution some 10,000 years ago is attributed to the same effect.
- Livermore (2013, Lecture 13) points out that a series of cultural design factors—availability of food, economic productivity (at whatever level), customs, traditions and/or laws, and security—have been essential for survival throughout human evolution. As noted in Chapter 9, cultural norms among different peoples also are implicated in accounting, at least in part, for differences in the wealth of nations.
- Ehrlich (2000) devotes an entire book to articulating the argument that much of our biology makes sense only when considered in the context of culture, and that our nongenetic cultural evolution has had a potent influence in shaping our past and present.
- In her characterization of the concerns and scope of the field of biological anthropology, King (2001, Lecture 1) echoes Ehrlich (2000) in asserting that humans are best understood as biocultural beings, influenced by both biology (i.e., anatomy, genetics, and physiology) and culture (i.e., environmental influences on choices made by individuals and groups, on the nature and extent of learning, and on ideas developed over time); it is difficult to separate biology and culture as two different spheres of influence on human performance and evolution; and to understand the adaptation of humans (as one type of anthropoid ape) in evolution, biological anthropology assumes that both biology and culture must be considered together in terms of their combined influences.
- In line with the views of King (2001, preceding bullet), cultural differences among different troops of both monkeys and nonhuman anthropoid apes arising out of distinct environmental influences also have been observed in Vervet monkeys (van der Waal et al., 2013), bathing, diving, mountain climbing, and beachcombing behavior in Japanese macaques (research described in a Science Channel broadcast), and tool-using behavior in African chimpanzees residing near Lake Tanganyika (research described in a National Geographic Wild broadcast).

At least three interpretations have been advanced to explain the influence of environmental design factors on human adaptation and consequent selection.

1. The simplest is environmental determinism; that is, a direct, controlling influence of the environment on the selection of human biological and/or cultural adaptations (Diamond, 1997; Harcourt, 2012, Chapter 5; Hawks, 2011, Lecture 24).
2. A second idea is that the influences of environmental design factors on selection are secondary to primary biological influences and may in fact represent just another manifestation of these latter influences (Ridley, 2003).

3. A third idea is that humans (and possibly other organisms as well) have self-selected themselves in evolution, extending a concept first introduced in Chapter 1 (Section 1.4).

This control systems perspective on evolutionary selection, focusing on the synergism between adaptive changes in human structure and function and human technological and cultural innovations will be explored in more detail in Section 11.3. First, however, a topic that for many decades has charged the debate over how selection works will be addressed, namely that of nature versus nurture.

11.2   NATURE VERSUS NURTURE

Arguably, no other issue germane to the analysis of performance variability, and to the evolution of human performance and the designs created thereby, has attracted more attention, more analysis, and more vituperative debate than discussions over the relative roles of nature versus nurture in influencing variability in human behavior, performance, and evolution. Nature is a code word for genetic influences; nurture is a code word for environmental influences, or in the vernacular employed in this book, context specificity. As Goldhaber (2012, p. 1) points out, this debate dates back to the nineteenth century, yet it continues with hurricane force up to today. Thus, in the past 15 years alone, at least 10 books directly concerned with this topic, nine in the twenty-first century, have been published: Ceci and Williams (1999), Dowling (2004), Gander (2003), Goldhaber (2012), Harris (2006), Hernandez and Blazer (2006), Keller (2010), Moore (2003), Ridley (2003), and Rutter (2006). The last author cited introduces his book (p. 1) with an account of why the topic of genes and behavior is controversial.

It is axiomatic that ascertaining the influence of both nature and nurture, in terms of both effect and degree, is essential for understanding the properties and evolutionary sources of human performance variability. A common theme, expressed in somewhat different ways by each of the authors cited above, is that the idea of a strict dichotomy between nature versus nurture in terms of influencing human variability is outmoded. Instead, the genesis of variability manifest in the diverse modes and patterns of overt behavior and performance is assumed to originate with genetic and environmental design factors working in concert.

Goldhaber (2012) provides an up-to-date review of this debate, with a balanced treatment of what he terms the relative empiricist (nature) versus nativist (nurture) influences on behavioral variability. As applied to sources of variability in both psychological traits and mental disorders, Rutter (2006, pp. 3–4) makes the following relevant points: (1) genetic factors play a significant role, but rarely predominate, contributing to a lesser degree (20–60 percent variance in general populations), (2) both genetics and environmental factors are influential for the great majority of traits or disorders, and (3) except in rare circumstances, genes are not determinative of either psychological traits or mental disorders. An even more pointed and cogent perspective on the debate is provided by Ridley (2003, pp. 3–4), as follows:

I believe human behavior has to be explained by both nature and nurture. I intend to make the case that the genome has indeed changed everything, not by closing the argument or winning the battle for one side or the other, but by enriching the argument from both ends till they meet in the middle. The discovery of how genes actually influence human behavior, and how human behavior influences genes, is about to recast the debate entirely. No longer is it nature versus nurture, but nature via nurture. Genes are designed to take their cues from nurture.

Ridley goes on to conclude (p. 280) that, "Nature versus nurture is dead. Long live nature via nurture."

Further discussion of the nature–nurture debate here will be limited to two topics that continue today to attract spirited controversy: heritability of intelligence (typically equated with IQ), and epigenetics.

11.2.1 HERITABILITY OF INTELLIGENCE

As Devlin et al. (1997) point out, study of IQ variability dates back nearly a century. From an evolutionary perspective however, hypotheses about the evolution of hominid and human knowledge and skill capabilities in relation to such factors as brain size, social and cultural adaptations, and technological innovation are directly connected to the IQ debate.

It is only in the last four decades that studies of fraternal (dizygotic) and identical (monozygotic) twins reared apart have provided more quantitative insight into the relative contributions of nature versus nurture to IQ. Nevertheless, these results have not resolved the controversy. At one extreme of this debate are the views of Jensen (1998) and Herrnstein and Murray (1994), who claim that IQ can be reduced to a single number (g), and that IQ is immutable and almost entirely heritable. More rigorous results are provided by the seminal contribution of the Minnesota Twin Family Study (Bouchard et al., 1990), which found (based on analysis of 100 monozygotic twins reared apart) that about 70 percent of the variance in IQ is associated with genetic variation. However, analysis of 13 other psychological variables—ranging from information processing, to special mental abilities, to personality, to interest attributes—found lower indices of heritability. For example, the heritability of spatial processing speed is only 36 percent. Additionally, with regard to their contribution to sources of variability in intelligence, Rutter (2006, p. 6) summarizes various types of scientific criticism leveled against twin studies of this sort, including concerns over research design, sampling issues, and bias in subject participation.

The IQ heritability study of Bouchard et al. (1990) and related studies do not identify the possible sources of *non*heritable contributions to IQ variance, although environmental design factors represent a tempting, yet entirely speculative, candidate. However, this shortcoming is somewhat resolved by the meta-analysis of Devlin et al. (1997) of 212 previous IQ heritability studies that yielded a provocative model for environmental influence. The analysis grouped the subjects of these previous studies into two maternal womb environment categories, one for twins, the other for siblings. The shared maternal environment effects, often assumed to be negligible, account for 20 percent of covariance between twins and 5 percent between siblings, and the effects of genes are correspondingly reduced, with two measures of heritability being less than 50 percent. The authors argue that the shared maternal environment may explain the striking correlation between the IQs of twins, especially those of adult twins reared apart (Bouchard et al., 1990). The authors suggest further that their results have two implications: (1) a new model may be required regarding the influence of genes and environment on cognitive function, and (2) interventions aimed at improving the prenatal environment could lead to a significant boost in postnatal IQs.

Regarding the second implication above, and in line with the context specificity emphasis of this book, many mothers likely would not be surprised at the suggestion that different wombs feature different environmental design conditions throughout pregnancy.

Finally, the assumption that IQ is largely immutable has rather convincingly been refuted by Flynn (2012), who has compiled extensive and consistent evidence collected in a number of different countries, that IQ test scores have significantly increased from generation to generation over the past century (termed the eponymous Flynn effect). Flynn himself attributes this effect (p. 27) largely to the advent of a wider range of cognitive problems (and presumably the new environmental designs that have emerged to present such problems) than previous generations encountered that in turn have required new cognitive skills and the kind of brains that can deal with these problems. In other words, the discoverer of the Flynn effect provides what essentially amounts to a context-specific interpretation of its origins.

11.2.2 EPIGENETICS

Epigenetics is defined as, "an essential mechanism for pruning down the wide range of possible behaviors permitted by genes, selecting those that fit an individual's environment" (Berreby, 2011). In recent decades, its analysis has attracted growing interest as part of broader scientific attention to gene-environment interactions (Thomas, 2010) and behavioral genetics (Berreby, 2011; Harris, 2006). The motivation for this interest is succinctly stated by Thomas (2010): "Despite the yield of recent genome-wide association studies, the identified variants explain only a small proportion of the heritability of most complex diseases. This unexplained heritability could be partly due to gene–environment interactions."

The two earmarks of epigenetic effects are (1) changes in traits apparently governed by genes that appear to be attributable to responses to (often traumatic) life experiences, and (2) transmission of these changes across generations. These effects are attributable to chemical modifications to DNA itself (DNA methylation) or to histones, the proteins around which DNA is wound (Kubicek, 2011). This latter author also points out that, "The idea of trans-generational inheritance of acquired characteristics goes back to Lamarck in the early 19th century, but still only correlative evidence exists in humans." One interesting example of such correlative evidence is described by Morris (2012): "Adults conceived during the [Nazi-induced] Dutch famine of 1944–1945 had distinct epigenetic marks that reduced the production of insulin-like growth factor 2, and that were retained for several decades in the afflicted individuals. These marks were not evident in siblings born before or after the famine." Spector (2012) advocates study of distinct health or personality differences in monozygotic twins as a possible avenue for investigating possible epigenetic effects in humans.

11.2.3   IMPLICATIONS FOR CONTEXT SPECIFICITY

In summary, this brief survey of nature–nurture observations and evidence consistently points to the conclusion that context specificity makes some level of contribution to variability across all modes of performance so far considered by this research. This conclusion appears to apply even to heritability of intelligence, a question that has attracted the most ardent champions of a purely genetic influence on variability. As for epigenetics, the evidence strongly suggests that, at least in some cases, genes no longer have a unidirectional effect on variability, as has long been assumed, but that at least in some cases there is a closed-loop relationship between gene action and the environments in which these genes are acting. This evidence has caused some to rethink the original ideas of Lamarck over two centuries ago, whose ideas about inheritance of acquired characteristics have been equated with Kipling's "just so stories" (inspired by Lamarck's ideas) and which have long been dismissed by modern biological science. As Morris (2012) puts it, "Although biologists have generally considered Lamarck's ideas to contain as much truth as Kipling's fables, the burgeoning field of epigenetics has made some of us reconsider our ridicule." More generally, the implications of the nature–nurture debate for understanding human evolution arguably points to support for the adage of Ridley (2003), "nature via nurture."

11.3   PHYLOGENETIC ORIGINS OF HUMAN PERFORMANCE VARIABILITY

There is compelling evidence that the role of context specificity as a prominent influence on human performance variability, which was addressed in preceding chapters and in the sections above from an evolutionary perspective, has phylogenetic origins in subhuman species, particularly anthropoid apes (i.e., bonobos, chimpanzees, and orangutans). King (2001, Lectures 6–10) reviews this evidence, pointing out that between-individual and between-group differences in what is broadly termed cultural behavior have been documented for such modes of behavioral expression as tool using, feeding habits, and vocalization. Given the widely accepted idea that anthropoid apes represent the direct evolutionary ancestors of humans, this evidence suggests that many if not most of the patterns of human performance variability that we observe today have ancient phylogenetic origins and have emerged over millions of years in a process involving gradual adaptive change. As proposed in the next section, the major distinguishing feature of adaptation and evolutionary progress among humans, in contrast to their subhuman ancestors, has been their unparalleled ability to create and innovate organized patterns of work on a comprehensive tribal and societal scale and to thereby self-select themselves in evolution.

11.4   HUMAN SELF-SELECTION THROUGH WORK

The analytical approach in Chapter 9 explored the closed-loop linkages between variability in human work performance and that of organizations and institutions, economies, and nation-states whose performance both depends on, and reciprocally influences, human work in fundamental and pervasive ways. This section extends this approach by introducing the idea that the trajectory of human evolution itself has been linked in a closed-loop fashion to human work performance, such that through engaging in work humans have self-selected the biological adaptations that emerged from time to time to ensure their survival and the consequent retention of these adaptations in evolution.

The theory underlying this idea is a control systems concept of behavioral control of environmental conditions. That is, the almost universal metaphor deployed in discussions of evolution is that the evolutionary imperative is for an organism to survive, through favorable adaptations, to pass on its genes. However, no subhuman organism has any idea of what a gene is, much less why it exists (the same can be said for many humans as well). What every organism does understand, subconsciously or consciously, is that they must control their environment in order to survive. This is the behavioral imperative. Behavioral control begins at conception, persists until death (an event precipitated by loss of such control), and is ubiquitous across all levels of biological organization, encompassing molecular to cellular to organ physiology to whole organism modes of functional activity.

The possibility of a closed-loop link between human work performance and human evolution has been noted by a few observers. Thus, Van Cott (1984) suggests that during much of human history and prehistory humans relied on manual work based on use of simple tools for survival, and human intellectual development was linked to the rudimentary skills learned during work with these tools. During more recent human history, this author extends this idea by suggesting that: "Each new invention was accompanied by a parallel genesis in the structure of human thought: from simple to more complex skills, from simple to more complex rules, and from simple to more complex knowledge."

Hawks (2011, Lecture 24) asserts that modern humans are still evolving and points to new human environments created by human work in recent millennia as a principal driving force for the selection of genetic changes underlying this evolution. His summary statement of this linkage is

> When we think about how human evolution has gone in the past few thousand years, we have to see that even though cultures have massively adapted us to the environment in new ways, creating new opportunities for us, they also have created new selection on us. They've given us the environment that has prompted us to evolve even more rapidly.

The concept of a closed-loop linkage between human work and human evolution was briefly addressed by T.J. Smith and Smith in 1974 and 1978, following an extensive elaboration of this idea from both theoretical and observational perspectives by K.U. Smith (1965a, Chapters 1 and 2). In Chapter 9 (Section 9.3.2), a general control systems theory of human work (K.U. Smith, 1965a; T.J. Smith & Smith, 1987a) was introduced, and is illustrated in Figure 9.1. The theory views work as a closed-loop process in which the worker employs various behavioral strategies to control sensory feedback from the physical, social, and organizational design attributes of the work environment. One assumption of this theory, and of the model in Figure 9.1, is that the behavioral feedback effects of work represent the principal mechanism of human self-selection in evolution.

Figure 11.1 further elaborates on this concept. The premise in Figure 11.1 is that the evolutionary emergence of the major and distinctive biological features of species *homo*—upright posture, relatively hairless body, apposed thumb and fingers mediating manipulative skill, verbal communication, distinctive modifications in neurohumoral regulation, and specialization in brain structure and function—were influenced and guided in a major way as a self-selective process through the planning, organization and execution of work, and through the fabrication of tools and other artifacts to facilitate the performance of work. That is, most of the behavioral and technological functions of modern machines employed to facilitate work performance (e.g., those of shaping, striking, cutting, smashing, forming, lifting, or turning) are derivative and can be traced back to human use of hand tools in remote prehistory for similar purposes (Oakley, 1957).

Information gathered from present-day primitive tribal and village groups suggests that transitional processes in the social, cultural, and technological development of

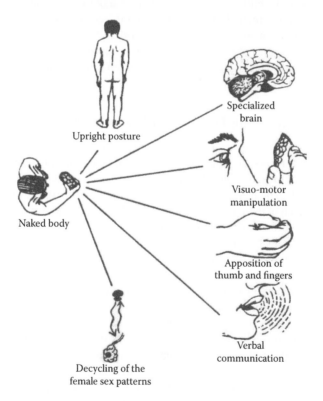

FIGURE 11.1 Work theory of human evolutionary self-selection. Through the use of work to control the behavioral environment, humans tend to self-select anatomical and physiological attributes of behavioral organization and personality that tend to enhance such control as a consequence of differential survival and reproductive success. Work behavior and natural selection thereby become coupled in a feedback manner. (Reproduced from Smith, T.J., in Karwowski, W. (Ed.), *International encyclopedia of ergonomics and human factors*. New York: Taylor & Francis, 2001.)

human societies and civilizations are related more prominently to distinctive adaptive changes in the organization and implementation of work than to those in other behavioral factors such as changes in patterns of plant or animal husbandry, consumption, and so forth (Wolf, 1954; Woolley, 1946). Indeed, the latter depend on the former. The relatively permanent nature of tools (some stone tools have been dated back to 2.6 million years [Hawks, 2011, Lecture 6]), machines, and other technological artifacts of work suggests a dominant role of behavioral feedback mechanisms of work activity and technology in defining both the direction and pattern of human development.

Finally, using parallels from present-day observations of human behavior, it can be argued that all of the mechanisms of natural selection invoked by classical evolutionary theory to explain the evolution of the modern humans—transformation of social and family structures, exploitation of new sources of nutrition, advantages in intra- and interspecies competition, dispersion into new ecosystems, heightened motor behavioral capabilities, improved exploitation of external sources of environmental energy—are directly linked in a behavioral feedback manner to advances in the organization, technology, and conduct of work to human factor and control the environment and to thereby facilitate differential reproductive success. From this perspective, human evolution may be viewed as a self-selective process largely influenced and guided by work behavior.

11.4.1 Landmarks in the Evolution of Human Work

Landmarks in the evolutionary progression of human work can be characterized in terms of the use of work to guide the emergence of new forms of societal organization (Figure 11.2), of technology (Figure 11.3), or of modes of symbolic communication (Figure 11.4). The heavy lines in the figures suggest that the development and elaboration of each of these manifestations of work has been more or less continuous and progressive throughout evolution (K.U. Smith, 1965a), in patterns that largely parallel the nonlinear growth of human population itself. However, at certain critical periods in prehistory and history, innovative patterns of adaptive organization of work emerged to mediate the creation of new types of architecture and modes of societal integration, of new tools, machines and other technologies, and of new methods of symbolic communication.

In particular, Figure 11.2 indicates that forms of societal organization have progressed from primitive shelters to fixed village and temple systems, to ramified mercantile and industrial centers, and finally to the national and global commercial systems of today. The figure also suggests that the emergence of each of these new levels of societal organizational sophistication has been accompanied and guided, in a feedback manner, by the emergence of new behavioral sophistication in the use of work to control time. Figure 11.3 indicates that technological innovation has progressed from primitive hand tools to megalithic temples used as architectural timekeeping machines, to human-powered and thence to human-controlled machines powered by environmental sources of energy, and finally to the semiautomated, automated, and cybernetic technologic systems of today. Finally, Figure 11.4 indicates that modes of symbolic communication have progressed from spoken language

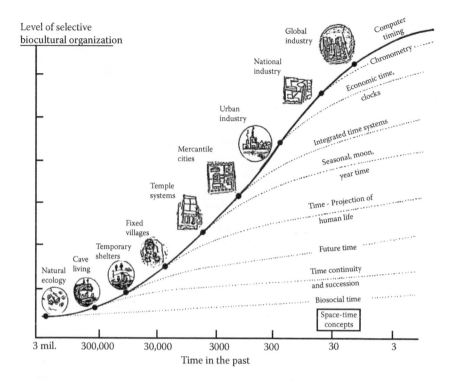

FIGURE 11.2 Progressive emergence in evolution of innovative forms of societal organization feedback-coupled to advances in work behavior and in time perception and control. As new forms of organization emerge, old forms persist and are conserved. (Reproduced from Smith, T.J., in Karwowski, W. (Ed.), *International encyclopedia of ergonomics and human factors.* New York: Taylor & Francis, 2001.)

and nonverbal communication to graphic expression, handwriting and printing, to various modes of electronic communication, and finally to the computer-mediated communication systems of today.

The diagrams in Figures 11.2 through 11.4 are meant to suggest that the innovative advances depicted are both integrated and cumulative. That is, work-related elaboration of new forms of societal organization, technology, and symbolic communication has occurred in an interdependent manner. In addition, all of the innovations indicated in the figures, plus related skills, patterns, and systems of work on which they depend, can be found today. Legacies in work innovation accumulate because of the unique human capacity for conserving knowledge and information about new technology and work methods over time.

Thus, many of the so-called "modern" features of work in fact have ancient origins. For example, many observers agree that a key prehistoric landmark in the organization of work was the establishment of relatively fixed settlements facilitated in a feedback manner by innovations in agricultural and animal husbandry work. A likely behavioral feedback effect of this trend was establishment of more elaborate

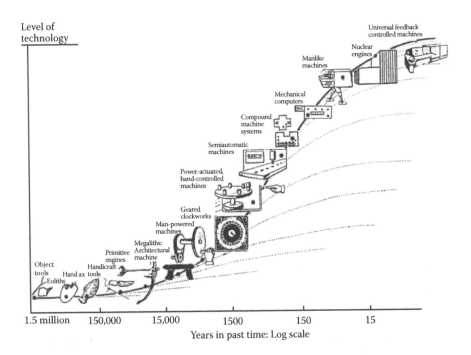

Level of technology

Years in past time: Log scale

1.5 million   150,000   15,000   1500   150   15

FIGURE 11.3 Progressive emergence in evolution of innovative forms of technology feedback-coupled to advances in work behavior. As new forms of technology emerge, old forms persist and are conserved. (Reproduced from Smith, T.J., in Karwowski, W. (Ed.), *International encyclopedia of ergonomics and human factors.* New York: Taylor & Francis, 2001.)

systems of community decision-makers, today termed administrators, managers, authorities, bureaucrats, and government. This was necessary to ensure that the work of residents directed at expanding the scope of community control over critical environmental conditions—food supply, security, design, and fabrication of technology, commerce, and so forth—was effectively organized and managed. Division of labor certainly exists in mobile hunting and gathering societies and also is apparent in nonhuman species. However, emergence of larger populations in more static settlements dramatically reduced the ability of a given resident to effectively control all key environmental conditions on an individual basis. Hence the need arose for better defined systems of organizers and managers to organize and guide community work processes.

One of the earliest historical records of this macroergonomic approach (Hendrick, 2007) to work organization is that of the ancient Sumerians starting about 7000 years ago. These peoples—precursors of the Hebrews, the Babylonians, the Greeks, and the Roman—elaborated over the next three millennia most if not all of the basic concepts of work organization that we observe and practice today (Kramer, 1963). These include writing and phonetic notation, laws of work and commerce, ethical principles and mechanisms of justice, organization of crafts, professions, and trades,

Level of symbolic
control

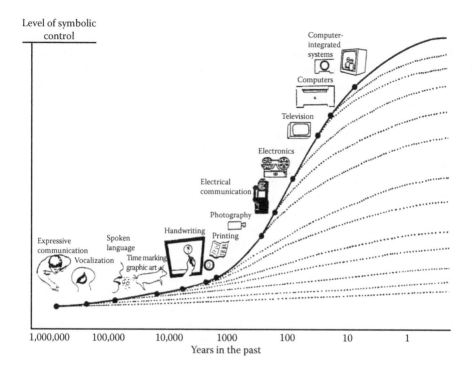

1,000,000   100,000   10,000   1000   100   10   1

Years in the past

FIGURE 11.4 Progressive emergence in evolution of innovative modes of symbolic communication feedback-coupled to advances in work behavior. As new forms of communication emerge, old forms persist and are conserved. (Reproduced from Smith, T.J., in Karwowski, W. (Ed.), *International encyclopedia of ergonomics and human factors.* New York: Taylor & Francis, 2001.)

rules of fair play, concepts of freedom, and valuation of qualities of ambition and success, all hallmarks of current work practices. Anticipating subsequent Greek and Roman developments by some 2000 years or more, Sumer was the first ancient civilization to effectively combine and organize labor, crafts, professions, and civil and religious authority as an integrated societal system of work.

As another example, it is likely that methods of mass production, based on what we today call factory technology and work, originated with Greek, Persian, and Asian civilizations to deal with the demands of providing for burgeoning populations and of equipping large armies numbering in the hundreds of thousands (K.U. Smith, 1965a). For example, the largest naval battle in recorded history between the Greeks and Persians over 2000 years ago clearly points to a well-defined system of work for fabricating and equipping thousands of ships. A third example is the use of massive hydraulic, geared, and levered machines for mining work in northern Europe in the fifteenth and sixteenth centuries, documented by Agricola in 1556, which anticipated by some three centuries the application of similar machines to mechanize work during the industrial revolution.

What factors prompted the emergence of successive innovation in forms of societal organization, technology, and symbolic communication, depicted in Figures 11.2 through 11.4? Answers to this question remain a matter of speculation. Diamond (1997) points to favorable environmental circumstances. He postulates that Eurasian success in expeditions of conquest over other peoples may be traced ultimately to availability of native plant and animal species suitable for domestication in the Fertile Crescent area. This factor, coupled with suitable climatic conditions and lack of major geographic barriers, allowed early dispersal of domesticated crops and livestock from the Mediterranean basin to the Asian subcontinent and perhaps beyond, thereby facilitating early transformation of hunting and gathering modes of work in these areas to activities and systems of work associated with agriculture, mining, and development of settled communities. A self-selective impetus for the innovation of new technologies plus approaches to work organization was thereby generated to address associated needs of large, established communities, such as better communication (writing), better weapons for defense or conquest (mining), and better coordination and integration of work on a societal basis (government and religion). Early success in work innovation thus is assumed to anticipate later success in subjugating other peoples. More broadly, we may postulate that there are reciprocal feedback relationships between the availability of favorable environmental conditions readily controlled through work, the development of more stable and organizationally sophisticated settlements and societies, and the innovation of new modes and forms of work technologies and practices (Figure 9.1), a pattern that undoubtedly has prevailed throughout human evolution.

A more explicit answer to this question that supports hypotheses introduced in the preceding paragraph is advanced by Dean et al. (2012) in a report dealing with the accumulation of cultural changes and innovations in human evolution. These authors introduce their analysis with the following observations:

> The remarkable ecological and demographic success of humanity is largely attributed to our capacity for cumulative culture, with knowledge and technology accumulating over time, yet the social and cognitive capabilities that have enabled cumulative culture remain unclear.
>
> The success of humanity in colonizing virtually every terrestrial habitat on the planet and resolving countless ecological, social, and technological challenges is widely attributed to our species' unique capability for "cumulative culture"—the extensive improvements in technology, over time. Although many animals—especially mammals, birds, and fishes—acquire knowledge and skills from others (often manifest in behavioral traditions), in no instance have these unambiguously exhibited "ratcheting" in complexity. Given that the adaptive value of cumulative learning is well established, the question as to why social learning is so much more widespread [in subhuman species] than cumulative culture [in humans] constitutes a major evolutionary puzzle.

As an approach to answering this puzzle, Dean et al. (2012) compared the abilities of capuchin monkeys, chimpanzees, and children in carrying out high-level problem solving. The children, but not the monkey or ape subjects, were successful in reaching higher-level solutions, which the authors attribute to distinctive sociocognitive capabilities—including verbal instruction, imitation, and prosociality—among children. The implication of the study is that these capabilities are associated with the ability to accumulate culture over time, an ability that is unique to human evolution and whose trajectories are schematically illustrated in Figures 11.2 through 11.4.

Kurzban and Barrett (2012) provide a nuanced perspective on this study in drawing attention to the difficulty of drawing strong causal conclusions based on differences between humans and other primates and of inferring causality from correlation. These authors do not dismiss the implications of the Dean et al. (2012) study, but they point out that other factors may have contributed to the origins of cumulative culture in humans, such as additional cognitive capacities related to abilities to form complex concepts, to draw causal inferences and/or to engage in social tracking.

With reference to the human accumulation of technology over time cited in the observations by Dean et al. (2012) above and depicted in Figure 11.3, it is worthwhile noting that Van Cott (1984, Figure 1) documents the same pattern of accumulation of technology over historic time, linked to the steady incremental improvement and refinement in human cognitive capabilities.

11.5 CONCLUSIONS

This chapter is concerned with the evolutionary origins of human performance variability prompted by the consideration that both human performance, as well as designs created by such performance, are products of evolution acting over millions of years. Two themes are emphasized to address this topic. The first, recapitulating a recurrent focus of earlier chapters, is that environmental design—context—plays a prominent role in influencing evolutionary change in addition to the role of biological factors. The second theme is that through the creation of cultural and technologic work designs, context specificity in evolution is closely linked to human work performance.

In other words, the trajectory of human evolution has been linked in a closed-loop fashion to human work performance, such that through engaging in work humans have self-selected the biological adaptations that emerged from time to time to ensure their survival and the consequent retention of these adaptations in evolution. The theory underlying this idea is a control systems concept of behavioral control of environmental conditions at all levels of biological organization. This focus on the evolutionary significance of human work extends the perspective introduced in Chapter 9 that variability in human work performance is the common denominator underlying variability observed in organizational, institutional, economic, and nation-state performance.

Figure 11.1 illustrates this closed-loop relationship between human work performance and the evolutionary emergence of the distinctive biological features of species *homo*. The assumption in the figure is that these features were influenced and guided in a major way during the course of human evolution as a self-selective process through the planning, organization, and execution of work and through the fabrication of tools and other artifacts to facilitate the performance of work.

Figures 11.2 through 11.4 illustrate the evolutionary trajectories of societal organization, technological innovation, and symbolic communication, respectively, made possible by the recurrent intersection of human work performance, biological adaptation, cultural transformation, and technological development. The conceptual validity of these schemes is supported by Dean et al. (2012) and Kurzban and Barrett (2012), whose analyses point to the factors underlying the phenomenon of cumulative culture distinctive to human evolution.

However, for purposes of the focus of this book, the major implication of the illustrations in Figures 11.2 through 11.4 is that they epitomize, in most major respects, the saga of human performance variability spanning the prehistory and history of the species. That is, each of the stages of societal organization, technological innovation, and symbolic communication shown in these figures, and the transitions from one stage to the next, represent a product of cognitive, physical, and social behavioral expression collectively realized through coordinated variability in human work performance.

Finally, it was pointed out in the introduction to this chapter that that the HF/E community has largely ignored the evolutionary significance of performance-design interaction that represents the central focus of the field. This is unfortunate. Practically every other domain of human science includes a defined evolutionary perspective. Adopting a comparable perspective represents a major intellectual and epistemological challenge for HF/E.

12 Summary and Conclusions

Thomas J. Smith

12.1 SUMMARY

Variability in performance is ubiquitous to the human condition, and more broadly, to all life. It represents both the basis and the consequence of evolution—our understanding of how and why evolution occurs dates back to Darwin's original account of the role of performance variability in the emergence of species: descent with modification. Performance variability is manifest in the functioning of all levels of human organization, from the molecular to complex sociotechnical systems. All domains of human science are grounded in the investigation of performance variability. In practical terms, analysis of performance variability leads to our understanding of the distribution of human populations and cultures across the planet and of the relative success or failure of learning and education, of social relationships, of health status and care, of athletic competition, of hiring and promotion, of organizational and institutional effectiveness and productivity, of community and societal functioning, of financial markets, and of entire state, nation, or economic systems.

In a number of respects, the perspective on human performance variability adopted in this book is broader in scope than that seen in previous books on the topic. Specifically, there are separate chapters devoted to the kinesiology of performance variability (the nature and sources of variability in motor performance) (Chapter 2), variability in cognitive and psychomotor performance (Chapter 3), the role of design factors—educational ergonomics—in student learning (Chapter 4), the effects of displaced sensory feedback on performance variability (Chapter 5), human error and performance variability (Chapter 6), variability in affective performance (Chapter 7), variability in social and team performance (Chapter 8), the nature and sources of performance variability in complex sociotechnical systems; namely, organizations and institutions, economies, and nation-states (Chapter 9), adverse consequences of performance variability linked to fracture-critical systems (Chapter 10), and an evolutionary perspective on performance variability (Chapter 11). The rationale for this approach is that the ubiquity of human performance variability merits a multidimensional treatment that considers the myriad patterns of variability and their underlying influences and causes.

Another consideration is that human performance variability represents one of the most critical scientific and practical concerns for the field of HF/E, the target audience for this book. However, it is fair to say that at least three topics addressed in this book—variability in economic and nation-state systems, performance variability of fracture-critical systems, and evolutionary implications—heretofore have received only modest attention by the field. It is to be hoped that this book will arouse awareness and interest regarding the potential for additional opportunities for HF/E research and practice in the area of performance variability.

Four themes are emphasized across the different variability topics addressed in this book:

1. *Variability in motor behavior as a foundation for other modes of variability.* The first theme pertains to the assumption that the variability properties of motor performance analyzed in Chapter 2 underlie those of all of the other modes of variability considered in subsequent chapters. The premise for this assumption is that, in the last analysis, despite the complexity of a particular system, its functioning and therefore its performance variability ultimately relies on the collective contributions of individual and group behaviors.

   Three key observations are documented in Chapter 2 regarding the nature of performance variability in human movement behavior: (1) variability in human movement is characterized by a nonlinear, deterministic, chaotic temporal structure, a pattern that is neither regular and predictable (such as a sine wave) nor completely random, (2) with healthy movement behavior there is an optimal amount of temporal variability characterized by a complex, chaotic structure; a reduction in such variability equates with increased behavioral rigidity, leading to an increase in noise and instability in movement patterns, and (3) because of its temporal structure (point 1), the future trajectory of human movement variability is inherently unpredictable.

   The degree to which these observations apply to the other modes of variability addressed in this book is less well established. Complexity in the variability pattern very likely applies to variability in cognitive and psychomotor performance, performance under displaced feedback, performance under conditions that predispose to human error, variability in affective performance, social and team performance, performance of complex sociotechnical systems (organizations, institutions, economies, and nation-states), and fracture-critical system performance. Consequently, for all these modes of performance, the future trajectory of the variability pattern likely is unpredictable. Moreover, the degree to which these other modes of variability exhibit nonlinear, deterministic, chaotic temporal structures is uncertain, although some observers believe that such is the case for variability in complex sociotechnical system performance. Additionally, it remains to be established whether there is an optimal amount of temporal variability for these other modes of variability.

2. *Context specificity in performance variability.* The second theme is the concept of context specificity, the idea that environmental design factors—context—make a prominent contribution to variability observed in different modes of performance. Throughout this book, emphasis is placed on this idea, motivated by the insight this analytical approach provides into how and why such variability occurs.

In simplest terms, there are two major classes of influences on human performance variability: innate biological factors (gender, anthropometric differences, etc., that are ultimately attributable to genetic effects), and design (context-specific) factors in the performance environment. In Chapter 1, the terms "nativist" and "empiricist" were introduced to refer respectively to these two sources of influence.

As also noted in Chapter 1 (Section 1.1), Webster's definition of variability within and between organisms—"the power possessed by living organisms...of adapting themselves to modifications or changes in their environment, thus possibly giving rise to ultimate variation of structure or function"—carries with it the implication that an intrinsic property of performance variability is context specificity; that is, susceptibility to influence by environmental design factors. The rationale for this emphasis is as follows. First, compared with the relatively gradual pace of change in innate biological factors, variability patterns can undergo essentially instantaneous change under the influence of design factors, which implies that the most dynamic properties of performance variability are those affected by the context of the performance environment. Second, the basic premise of context specificity is that new designs (of products, technology, built environments, complex sociotechnical systems, etc.) inevitably will evoke new modes and patterns of performance variability.

There thus are two implications of context specificity for behavioral expression: (1) the introduction of a new design may have unintended consequences for performance variability (Chapter 10, Section 10.6), and (2) these consequences thus are not necessarily predictable. These considerations imbue the phenomenon of context specificity as a major influence on performance variability with a scientific and practical meaning and significance that far exceeds that of innate biological factors, at least in the judgment of this author.

3. *Control systems interpretations.* The third theme emphasized throughout the book is the application of control systems analyses to different modes of performance variability. The impetus for this approach is the goal of providing a behavioral interpretation of the role of context specificity in performance variability. That is, two alternative explanations have been advanced to account for evidence pertaining to context specificity (above). One is the idea that environmental design factors have a direct, open-loop influence on performance variability; this is an environmental determinism interpretation.

The alternative explanation assumes a closed-loop feedback relationship between environmental design conditions (input or sensory feedback) and behavioral control mechanisms (output, mediated through movement) directed at feedback control of this input (Chapter 1, Section 1.3). This is the interpretation favored in this book, and control systems analyses of different modes of performance variability are introduced in Chapters 3 and 5 through 11. The strongest evidence for this control systems interpretation of context specificity comes from studies of behavioral performance effects of displaced feedback. As noted in Chapter 5, with either spatial displacement of visual feedback or temporal delay of either visual or auditory feedback, these effects can manifest themselves in essentially an instantaneous manner. Under displaced feedback, effective control of behavior is abruptly degraded, accompanied by loss of body sense and signs of sympathetic arousal and stress. From the perspective of an open-loop model of context specificity, these effects should not occur—delay or spatial displacement would not be expected to have direct environmental effects on behavior. From the perspective of a closed-loop model however, behavioral control of sensory feedback is closely referenced to the spatiotemporal properties of that feedback (functionally validated in the course of development), and displacement of sensory feedback imposes control demands on behavior that naïve subjects find difficult to initially adapt to.

4. *Variability in human work performance.* The fourth theme introduced in this book, emphasized explicitly in Chapters 9 and 11, is variability in work performance. However, it can be argued that this theme represents a common denominator for all of the other modes of performance variability addressed in previous chapters. That is, whether we are talking about motor behavior, cognition, learning, performance under displaced feedback, human error, affective performance, social and team performance, performance of complex sociotechnical systems, and/or performance of fracture-critical systems, in every case these different modes of performance and their distinctive patterns of variability are based on human work. The ubiquity with which work mediates all modes of human performance prompts the conclusions in Chapter 9 (Section 9.3.1) that work is the most ubiquitous form of human activity on the planet, that human engagement in work is synonymous with life itself, that all conscious human activity represents work directed at some sort of goal or another, that in the occupational sense all humans routinely engage in purposeful task or job work activity throughout most of their lives, and that, in the evolutionary sense, since emergence of the species work activity has served both as the engine of the human condition and as the means of human self-selection.

12.2 CONCLUSIONS: THE PURPOSE OF HUMAN PERFORMANCE VARIABILITY

Let me close this book by addressing the general question of why variability in human performance exists. In relation to both motor behavior and human evolution, the answer to this question seems quite clear. Scientific evidence regarding motor behavior, reviewed in Chapter 2, makes a strong case for the conclusions (Section 2.6) that chaotic temporal patterns of movement variability are associated with healthy movement behaviors, reduction or deterioration of such patterns is associated with a decline in healthy flexibility of movement behaviors coupled with the onset of behavioral rigidity and inability to adapt, and for healthy movement behavior there is an optimal amount of temporal variability characterized by a complex chaotic structure, whereas a reduction in such variability equates with increased behavioral rigidity leading to an increase in noise and instability.

As for evolution, observations dating back to Darwin, and a number of controlled studies since then, leave little doubt that performance variability, modulated by emergent biological adaptations, is essential for the process of evolution. Chapter 11 addresses this point.

However, as far as the objectives and properties underlying other modes of performance variability are concerned, and more broadly, as far as scientific perspectives on variability itself are concerned, a number of open questions remain. Regarding the latter point, dealing with individual differences in terms of rigorous scientific inquiry has posed a challenge and even a conundrum for behavioral scientists for over a century. As discussed in Chapter 1 (Section 1.2.1), HF/E research often treats individual differences as a nuisance variable that either is controlled in the study or covaried out in the statistical analyses of the results. However, another point of view is that individual differences contribute a significant amount of variance to many human factors related situations (yet relatively few human factors studies make an attempt to systematically investigate these variables).

Beyond motor behavior and evolution, is there an underlying goal to the variability evident in the other modes of performance considered in this book? Is there a purpose to the volatility observed in human error performance? In affective behavior? In social and team performance? In the performance of complex sociotechnical systems? It strikes this observer that in general, accounts of these phenomena treat them something like a force of nature comparable to gravity or the sun rising in the east. In other words, is performance variability in these various domains both necessary and inevitable?

Consider first the idea introduced in Chapter 2 and cited above, that there is an optimal level of performance variability essential for the health of the performing system. In the case of human error, affective behavior, social and team performance, and economic and nation-state performance, I am not aware of any definitive evidence regarding this question. To be sure, it was argued in Chapter 6 that human science generally, and certainly HF/E science in particular, treat human performance variability and human error as one and the same (i.e., "error" measures pervade analytical approaches to performance variability assessment). From this perspective, it may be argued that the purpose of variability intrinsic to motor behavior may be to introduce a degree of human error in system performance essential for effective guidance of behavior. The HF/E community may object that this is not what is meant by human error. As far as I am concerned, error is error is error.

Another example of the likely essentiality of performance variability is introduced in Chapter 9 in relation to organizational performance. The proposition of Juran, described in Section 9.4.3.2, is that an organizational system that relies exclusively on feedback control to minimize error in system output, and to thereby achieve system stability, is incapable of effectively dealing with and controlling new challenges in the organizational performance environment that inevitably arise from time to time. Juran's solution is to combine feedback with feedforward control of organizational performance so that variability in system performance is built into system design as an essential feature.

Beyond these examples, whether variability in the other modes of performance cited above is essential remains unanswered. Arguably, volatility in emotional behavior, in social or team behavior, in economic behavior, in nation-state behavior, and/or in fracture-critical design, that leads to breakdown, collapse, or failure, cannot be considered as desirable. Similarly, the dramatic breakdown in behavioral control evoked by displaced feedback conditions can also be viewed as an unwelcome mode of performance variability. Perhaps these phenomena represent extreme and malignant distortions of normal, healthy variability that characterizes the basic modes of performance.

Another open question is the degree to which the manifestations of performance variability, documented for motor behavior, obtain for other modes performance addressed in other chapters of this book. One such manifestation is the property of self-organized behavior. Some observers cited in Chapter 9 (Section 9.4.2) suggest that self-organized behavior may represent a general property of complex technological and organizational systems and that, in particular, self-organization may represent a strategy for managing organizational complexity. However, no definitive empirical evidence is provided to support these suggestions.

Another manifestation of performance variability documented for motor behavior, but with uncertain applicability for other modes of performance, is a dynamical systems perspective of variability, best understood as dynamic, complex, nonlinear, and deterministic. Some observers assume (Chapter 9, Section 9.4.6) that dynamical systems behavior also represents an intrinsic property of variability of complex sociotechnical systems performance. Suggested examples include population dynamics, ecological systems, economic systems, business cycles, learning, social interaction, business management, health care management, human service systems, information and software engineering, and manufacturing operations and design. Here again, however, no definitive empirical evidence is provided to support any of these suggestions.

I will close by offering a hypothesis and a suggestion. The hypothesis is that performance variability itself is context-specific in that the meaning and properties of variability observed are critically influenced by both the mode of performance as well as the performance environment in which each different mode takes place. The suggestion is that exploring the meaning and properties of different modes of performance variability that remain unresolved and/or relatively unexplored at present offers fertile ground for further HF/E research.

Note

- *Televised nondisplaced visual feedback* with assembly task. (From Smith, K.U. & Smith, W.M., *Perception and motion: an analysis of space-structured behavior.* Philadelphia: Saunders, p. 163, 1962.)
- *Reversed, inverted-reversed, and inverted visual feedback* with the following tasks: (1) contact times in writing, tracing, and drawing tasks (From Smith, K.U. & Smith, W.M., *Perception and motion: an analysis of space-structured behavior.* Philadelphia: Saunders, pp. 170–171, 1962.); (2) travel times in writing and drawing tasks (From Smith, K.U. & Smith, W.M., *Perception and motion: an analysis of space-structured behavior.* Philadelphia: Saunders, pp. 170–171, 1962.); and (3) percent time off path in maze tracking task. (From Smith, K.U. & Smith, W.M., *Perception and motion: an analysis of space-structured behavior.* Philadelphia: Saunders, p. 172, 1962.)
- *Inverted-reversed visual feedback* with the following additional task: duration of component assembly movement in complex assembly task. (From Smith, K.U. & Smith, W.M., *Perception and motion: an analysis of space-structured behavior.* Philadelphia: Saunders, p. 163, 1962.)
- *Inverted visual feedback* with the following additional task: manipulation and travel times in dial setting, knob turning, button pushing, and switch pressing tasks. (From Smith, K.U. & Smith, W.M., *Perception and motion: an analysis of space-structured behavior.* Philadelphia: Saunders, p. 239, 1962.)

The teleoperation results are based on two studies of remote telemanipulation under spatial displacement: Smith and Stuart (1989) and Stuart et al. (1991).

References

Aardex Corporation (2004). *User effective buildings*. Denver, CO: Aardex Corporation.

Abbs, J.H. & Smith, K.U. (1970). Laterality differences in the auditory feedback control of speech. *Journal of Speech and Hearing Research, 13*, 298–303.

Ackerman, P.L. (1986). Individual differences in information processing: An investigation of intellectual abilities and task performance during practice. *Intelligence, 10*, 101–139.

Ackerman, P.L. (1987). Individual differences in skill learning: An integration of psychometric and information processing perspectives. *Psychological Bulletin, 102*(1), 3–27.

Ackerman, P.L. (1988). Determinants of individual differences during skill acquisition: Cognitive abilities and information processing. *Journal of Experimental Psychology: General, 117*, 288–318.

Ackerman, P.L. & Lohman, D.F. (2006). Individual differences in cognitive functions. In P.A. Alexander & P.H. Winne (Eds.), *Handbook of educational psychology* (2nd edition) (Chapter 7, pp. 139–161). Mahwah, NJ: Lawrence Erlbaum Associates.

Active Living Research (2007). *Active education. Physical education, physical activity and academic performance*. San Diego, CA: Active Living Research, San Diego State University. Downloaded from: http://www.activelivingresearch.org/files/Active_Ed.pdf.

Adams, J.A. (1953). *The prediction of performance at advanced stages of training on a complex psychomotor task* (USAF Human Resources Center Research Bulletin No. 53-49). Dayton, OH: Wright-Patterson USAF Base.

Adams, J.A. (1987). Historical review and appraisal of research on the learning, retention, and transfer of human motor skills. *Psychological Bulletin, 101*(1), 41–74.

Adams, N.L., Barlow, A., & Hiddlestone, J. (1981). Obtaining ergonomics information about industrial injuries: A five-year analysis. *Applied Ergonomics, 12*, 71–81.

Adelman, J. (2013). *Worldly philosopher. The odyssey of Albert O. Hirschman*. Princeton, NJ: Princeton University Press.

Adolph, K.E., Robinson, S.R., Young, J.W., & Gill-Alvarez, F. (2008). What is the shape of developmental change? *Psychological Review, 115*, 527–543.

Agricola, G. (1556). *De re metallica* (1950 translation of the first Latin edition of 1556, by H.C. Hoover and L.H. Hoover). New York: Dover.

Albin, T.J. (1988). Relative contribution of behavior to slip and fall accidents in mining maintenance. In *Proceedings of the Human Factors Society 32nd Annual Meeting* (pp. 511–514). Santa Monica, CA: Human Factors and Ergonomics Society.

Alexander, P.A. & Winne, P.H. (Eds.) (2006). *Handbook of educational psychology* (2nd edition). Mahwah, NJ: Lawrence Erlbaum Associates.

Alexander, S. (Ed.) (2009). *Voluntary simplicity: The poetic alternative to consumer culture*. Whanganui, New Zealand: Stead & Daughters.

Allan, N. (2012). A national report card. In *The Atlantic* (October, Volume 310, Number 3), pp. 84–87.

Allen, I.E. & Seaman, J. (2013). *Changing course. Ten years of tracking online education in the United States*. Babson College, FL: Babson Survey Research Group.

Allen, I.E. & Seaman, J. (2014). *Grade change. Tracking online education in the United States*. Babson Park, MA & San Francisco: Babson Survey Research Group & Quahog Research Group, LLC.

Allport, G.W. & Odbert, H.S. (1936). Trait names: A psycholexical study. *Psychological Monographs, 47*(1), i–171.

American Society of Interior Designers (ASID) (2005). *Sound solutions. Increasing office productivity through integrated acoustic planning and noise reduction strategies.* Washington, DC: ASID.

Anderson, S. (2013). The CEO "performance pay" charade. In *St. Paul Pioneer Press* (September 2), p. 7A.

Andrews, E. (2007). Fed shrugged as subprime crisis spread. In *New York Times* (December 18), http://www.nytimes.com/2007/12/18/business/18subprime.html?pagewanted= all&_r = 0 (retrieved 12/1/13).

Annett, J. (1968). *Feedback and human behaviour.* Baltimore: Penguin.

Ansell, S. & Smith, K.U. (1966). Application of a hybrid computer system to human factors research. In *Human Factors Conference. ASME Publication 66-HUF-9* (pp. 1–8). Washington, DC: American Society of Mechanical Engineers.

Antonovsky, A., Pollock, C., & Straker, L. (2014). Identification of the human factors contributing to maintenance failures in a petroleum operation. *Human Factors, 55*(2), 306–321.

Armstead, A.G. & Henning, R.A. (2007). Effects of long audio communication delays on team performance. *Proceedings of the Human Factors and Ergonomics Society 51st Annual Meeting* (pp. 136–140). Santa Monica, CA: Human Factors and Ergonomics Society.

Arutyunyan, G.H., Gurfinkel, V.S., & Mirskii, M.I. (1968). Investigation of aiming at a target. *Biophysics, 14*, 1162–1167.

Ashton, T.S. (1948). *The industrial revolution (1760–1830).* Oxford: Oxford University Press.

Bailey, A. & Puckett, J. (2011). Don't automate the status quo; redesign the process. In *St. Paul Pioneer Press* (October 7), p. 8B.

Ballantyne, G.H. (2002). Robotic surgery, telerobotic surgery, telepresence, and telementoring. *Surgical Endoscopy, 16*, 1389–1402.

Bandura, A. (1991). Social cognitive theory of self-regulation. *Organizational Behavior and Human Decision Processes, 50*, 248–287.

Bandura A. (2001). Social cognitive theory: An agentic perspective. *Annual Reviews of Psychology, 52*, 1–26.

Barnett, W.S. (2011). Effectiveness of early educational intervention. *Science, 333*, 975–978.

Barron, J. (2003). The blackout of 2003: The overview; power surge blacks out Northeast, hitting cities in 8 states and Canada; midday shutdowns disrupt millions. In *New York Times* (August 15), http://www.nytimes.com/2003/08/15/nyregion/blackout-2003-overview-power-surge-blacks-northeast-hitting-cities-8-states.html?pagewanted=all&src=pm.

Bartlett, F.C. (1932). *Remembering: A study in experimental and social psychology.* Cambridge: Cambridge University Press.

Barton, S. (1994). Chaos, self-organization, and psychology. *American Psychologist, 49*, 5–14.

Bauer, H.U. & Schöellhorn, W. (1997). Self-organizing maps for the analysis of complex movement patterns. *Neural Processing Letters, 5*(3), 193–199.

Bauer, S. (2011). Private-school voucher students trail public-school peers on state tests. In *St. Paul Pioneer Press* (March 29), p. 6A.

Bayle, P. (1647–1706) (1985). *Historical and critical dictionary* (Translated and edited by R.H. Popkin). Indianapolis: Hackett.

Beamish, R. (2011). Back to school for billionaires. In *Newsweek* (May 9), pp. 38–43.

Beare, A.N., Donovan, M.D., Lassiter, D.L., & Gray, L.H. (1985). *The effects of supervisor experience and assistance of a shift technical advisor (STA) on crew performance in control room simulators.* Oak Ridge, TN: Oak Ridge National Laboratory Report, TM-9660, Oak Ridge National Laboratory.

Beckstrom, M. (2003). "Invest in kids" taken literally. In *St. Paul Pioneer Press* (October 24), pp. 1A, 6A, 7A.

Beddoes, Z.M. (2012). For richer, for poorer. Special report. World economy. In *The Economist* (October 13), pp. 1–24.

Bejczy, A.K. (2002). Teleoperation and telerobotics. In O.D.I. Nwokah & Y. Hurmuzlu (Eds.), *The mechanical systems design handbook. Modeling, measurement, and control.* Boca Raton, FL: CRC Press.

Belden, D. (2006). Novel seats keep students on the ball. In *St. Paul Pioneer Press* (October 16), pp. 1B, 5B.

Bendick, J. (2002). *Galen and the gateway to medicine.* Bathgate, ND: Bethlehem Books.

Benjamin, D.J., Cesarini, D., van der Loos, M.J.H.M., Dawes, C.T., Koellinger, P.D., Magnusson, P.K.E., Chabris, C.F., Conley, D., Laibson, D., Johannesson, M., & Visscher, P.M. (2012). The genetic architecture of economic and political preferences. *Proceedings of the National Academy of Sciences, 109*(21), 8026–8031.

Bennett, S. (2001). *Human error—By design?* Leicester, UK: Perpetuity Press.

Bennett, S. & Kalish, N. (2007). *The case against homework: How homework is hurting children and what parents can do about it.* New York: Three Rivers Press.

Ben-Porat, O., Shoham, M., & Meyer, J. (2000). Control design and task performance in endoscopic teleoperation. *Presence, 9*(3), 256–267.

Berliner, D.C. & Biddle, B. (1995). *The manufactured crisis—Myths, fraud, and the attack on America's public schools.* Columbia, MO: University of Missouri Center for Research in Social Behavior.

Berliner, D.C. & Calfee, R.C. (Eds.) (1996). *Handbook of educational psychology.* New York: Simon and Schuster Macmillan.

Bernard, C. (1865). *Introduction à l'étude de la médecine expérimentale.* Paris: Baillière. Trans. by H.C. Greene (1929). New York: Macmillan.

Bernstein, N.A. (1967). *The coordination and regulation of movements.* London: Pergamon Press.

Berreby, D. (2011). Environmental impact. *The Scientist, 25*(3), 40–44.

Biddle, B.J. & Berliner, D.C. (2002). *What research says about small classes and their effects.* San Francisco: WestEd. Downloaded from: http://www.WestEd.org/policyperspectives.

Bill and Melinda Gates Foundation (2010). *Primary sources: America's teachers on America's schools.* New York: Scholastic Inc. Downloaded from: http://www.scholastic.com/primarysources/pdfs/Scholastic_Gates_0310.pdf.

Bilodeau, E.A. (1953). Speed of acquiring a simple motor response as a function of the systematic transformation of knowledge of results. *American Journal of Psychology, 66,* 409–420.

Bilodeau, E.A. & Ryan, F.J. (1960). A test for interaction of delay of knowledge of results and two types of interpolated activity. *Journal of Experimental Psychology, 59,* 414–420.

Bizarro, A. (2013). *The distinct roles of first impressions and physiological compliance in establishing effective teamwork* (unpublished MA thesis). Storrs, CT: University of Connecticut.

Black, J.W. (1951). The effect of delayed side-tone upon vocal rate and intensity. *Journal of Speech and Hearing Disorders, 16,* 56–60.

Black, J.W. (1955). The persistence of the effects of delayed side-tone. *Journal of Speech and Hearing Disorders, 20,* 65–68.

Blakemore, S.-J., Frith, C.D., & Wolpert, D.M. (2001). The cerebellum is involved in predicting the sensory consequences of action. *Neuroreport, 12*(9), 1879–1884.

Blankinship, D.G. (2010). Worth of master's degree pay for teachers challenged. In *St. Paul Pioneer Press* (November 21), p. 3A.

Block, J. (2010). The five-factor framing of personality and beyond: Some ruminations. *Psychological Inquiry: An International Journal for the Advancement of Psychological Theory, 21*(1), 2–25.

Boashash, B. (Ed.) (2003). *Time-frequency signal analysis and processing: A comprehensive reference.* Oxford: Elsevier.

Böhlmark, A. & Lindahl, M. (2008). *Does school privatization improve educational achievement? Evidence from Sweden's voucher reform.* Bonn, Germany: Institute for the Study of Labor.

Boldt, M. (2010). Junior high to stop giving students, teachers a laptop. In *St. Paul Pioneer Press* (July 25), p. 4B.

Boldt, M. (2011). Schools do get better with open enrollment. In *St. Paul Pioneer Press* (August 21), pp. 1A, 7A, 8A.

Bouchard, T.J. Jr., Lykken, D.T., McGue, M., Segal, N.L., & Tellegen, A. (1990). Sources of human psychological differences: The Minnesota study of twins reared apart. *Science, 250*(4978), 223–228.

Boudreaux, D.J. (2012). A man of intellectual stature. In *Minneapolis Star Tribune* (July 31), p. A9.

Brand, J. (2008). Office ergonomics: A review of pertinent research and recent developments. In C.M. Carswell (Ed.), *Reviews of human factors and ergonomics, Volume 4* (pp. 245–282). Santa Monica, CA: Human Factors and Ergonomics Society.

Brandt, S. (2013). School engages parents to boost reading success. In *Minneapolis Star Tribune* (May 22), pp. B1, B8.

Bransford, J., Vye, N., Stevens, R., Kuhl, P., Schwartz, D., Bell, P., Meltzoff, A., Barron, B., Pea, R., Reeves, B., Roschelle, J., & Sabelli, N. (2006). Learning theories and education: Toward a decade of synergy. In P.A. Alexander & P.H. Winne (Eds.), *Handbook of educational psychology* (2nd edition) (Chapter 10, pp. 209–244). Mahwah, NJ: Lawrence Erlbaum Associates.

Bransford, J.D., Brown, A.L., Cocking, R.R., Suzanne, M.S., & Pellegrino, J.W. (Committee on Developments in the Science of Learning, and Committee on Learning Research and Educational Practice) (Eds.) (2000). *How people learn. Brain, mind, experience, and school.* Washington, DC: Commission on Behavioral and Social Sciences and Education, National Research Council, National Academy Press.

Bredo, E. (2006). Conceptual confusion and educational psychology. In P.A. Alexander & P.H. Winne (Eds.), *Handbook of educational psychology* (2nd edition) (Chapter 3, pp. 43–57). Mahwah, NJ: Lawrence Erlbaum Associates.

Brooks, D. (2005). Psst! Human capital. In *The New York Times* (November 13).

Brooks, D. (2009). Not so much genius as hard work. In *Minneapolis Star Tribune* (May 4), p. A11.

Brooks, D. (2010). Economic wizards forget human factor. In *Minneapolis Star Tribune* (November 17), p. A13.

Brooks, D. (2013). "Big data." What you'll do next. In *St. Paul Pioneer Press* (April 22), p. 9A.

Brown, J.W. & Braver, T.S. (2005). Learned predictions of error likelihood in the anterior cingulate cortex. *Science, 307*(5712), 1118–1121.

Brown, O. Jr. (2002). Macroergonomic methods: Participation. In H.W. Hendrick & B.M. Kleiner (Eds.), *Macroergonomics. Theory, methods, and applications* (pp. 25–44). Mahwah, NJ: Lawrence Erlbaum.

Bruening, J.C. (1989). Incentives strengthen safety awareness. *Occupational Hazards, 51*(11), 49–52.

Bruening, J.C. (1990a). Employee participation strengthens incentive programs. *Occupational Hazards, 52*(3), 45–47.

Bruening, J.C. (1990b). Shaping workers—Attitudes toward safety. *Occupational Hazards, 52*(3), 49–51.

Bryce, G.K. (1981). *Joint labour-management occupational health committees. An example of worker participation in work site health and safety programs* (unpublished master of health administration thesis). Ottawa, Ontario: University of Ottawa.

Brynjolfsson, E. (1993). The productivity paradox of information technology. *Communications of the ACM, 36*(12), 66–77.

Buchen, L. (2012). The anatomy of politics. *Nature, 490*, 466–468.

Bullinger, H.-J., Korndorfer, V., & Salvendy, G. (1987). Human aspects of robotic systems. In G. Salvendy (Ed.), *Handbook of human factors* (pp. 1657–1693). New York: Wiley.

Burch, D. (2013). Taken in vein. The joy of bloodletting. *Natural History, 121*(5), 10–13.

Buttonwood (2010). Taking von Mises to pieces. In *The Economist* (November 20), p. 87.

Buttonwood (2013a). Age shall weary them. In *The Economist* (May 11), p. 78.

Buttonwood (2013b). Net gains and losses. In *The Economist* (August 17), p. 61.

Buttonwood (2014a). The inevitability of instability. In *The Economist* (January 25), p. 60.

Buttonwood (2014b). Revolutionary fervor. In *The Economist* (March 8), p. 74.

Cacciabue, P.C. (2005). Human error risk management methodology for safety audit of a large railway organization. *Applied Ergonomics, 36*, 709–718.

Cacioppo, J.T. & Gardner, W.L. (1999). Emotion. *Annual Reviews of Psychology, 50*, 191–214.

Calabrese, C. & Henning, R.A. (2013). *Testing the job demands-control model of stress using teams as the unit of analysis.* Paper presented at Work Stress & Health 2013: The 10th International Conference on Occupational Stress and Health, May 16–19, Los Angeles.

Calder, J. (1899). *The prevention of accidents.* New York: Longmans, Green.

Caldwell, B.S. (1992). Human factors and educational quality. In *Proceedings of the Human Factors Society 36th Annual Meeting* (pp. 548–552). Santa Monica, CA: Human Factors Society.

Calvin, W.H. & Bickerton, D. (2000). *Lingua ex machina. Reconciling Darwin and Chomsky with the human brain.* Cambridge, MA: MIT Press.

Cannon, W.B. (1939). *The wisdom of the body.* New York: W.W. Norton.

Cao, C.G.L. & Rogers, G. (2007). Robotics in health care: HF issues in surgery. In P. Carayon (Ed.), *Handbook of Human Factors and Ergonomics in health care and patient safety* (Chapter 26, pp. 411–421). Boca Raton, FL: CRC Press.

Carayon, P. & Smith, M.J. (2000). Work organization and ergonomics. *Applied Ergonomics, 31*, 649–662.

Carayon, P., Karsh, B.-T., Gurses, A.P., Holden, R.J., Hoonakker, P., Hundt, A.S., Montague, E., Rodriguez, A.J., & Wetterneck, T.B. (2013). Macroergonomics in health care quality and patient safety. In D.G. Morrow (Ed.), *Reviews of human factors and ergonomics,* Volume 8, (pp. 4–54). Santa Monica, CA: Human Factors and Ergonomics Society.

Carlson, S.A., Fulton, J.E., Lee, S.M., Maynard, M., Brown, D.R., Kohl, H.W., III, & Dietz, W.H. (2008). Physical education and academic achievement in elementary school: Data from the early childhood longitudinal study. *American Journal of Public Health, 98,* 1–7.

Carmichael, M. (2007). Stronger, faster, smarter. In *Newsweek* (March 26), pp. 38–46.

Carnegie Foundation for the Advancement of Teaching (1995). *The basic school.* Princeton, NJ: Carnegie Foundation.

Carr, E. (2013). Time to cheer up. In *The Economist* (November 23, Special Report), pp. 1–16.

Carr, N. (2008). Is Google making us stupid? What the Internet is doing to our brains. In *The Atlantic* (July/August, Volume 302(1)), pp. 56–63.

Carver, C.S. & Scheier, M.F. (1998). *On the self-regulation of behavior.* Cambridge: Cambridge University Press.

Ceci, S.J. & Williams, W.M. (Eds.) (1999). *The nature-nurture debate. The essential readings.* Malden, MA: Blackwell.

CedarCrestone (2013). *The seven practices of top performing organizations* (taken from CedarCrestone 2012–2013 HR systems survey white paper). Alpharetta, GA: CedarCrestone.

Centers for Disease Control and Prevention (2010). *The association between school-based physical activity, including physical education, and academic performance.* Washington, DC: U.S. Department of Health and Human Services, Centers for Disease Control and Prevention, National Center for Chronic Disease Prevention and Health Promotion, Division of Adolescent and School Health. Downloaded from: http://www.cdc.gov/ healthyyouth/health_ and_academics/pdf/pa-pe_paper.pdf.

Chadwick, R.A. & Pazuchanics, S. (2007). Spatial disorientation in remote ground vehicle operations: Target localization errors. In *Proceedings of the Human Factors and Ergonomics Society 51st Annual Meeting* (pp. 161–165). Santa Monica, CA: Human Factors and Ergonomics Society.

Chandlee, G.O., Smith, R.L., & Wheelwright, C.D. (1988). Illumination requirements for operating a space remote manipulator. In W. Karwowski, H.R. Parsaei, & M.R. Wilhelm (Eds.), *Ergonomics of hybrid automated systems I* (pp. 241–248). Amsterdam: Elsevier.

Chanel, G., Kivikangas, J.M., & Ravaja, N. (2012). Physiological compliance for social gaming analysis: Cooperative versus competitive play. *Interacting with Computers, 24*, 306–316.

Chapanis, A. (1988). Some generalizations about generalization. *Human Factors, 30*(3), 253–267.

Chapanis, A. (1991, 1992). To communicate the human factors message, you have to know what the message is and how to communicate it. *Human Factors Society Bulletin, 34*(11), 1–4 (Part 1); *35*(1), 3–6 (Part 2).

Chapanis, A. (1996). *Human factors in systems engineering.* New York: Wiley.

Charlemagne (2006). Back to school. In *The Economist* (March 25), p. 58.

Chase, R.A. (1958). Effect of delayed auditory feedback on the repetition of speech sounds. *Journal of Speech and Hearing Disorders, 23*, 583–590.

Chen, J.Y.C. (2008). Effectiveness of concurrent performance of military and robotics tasks and effects of cueing in a simulated multi-tasking environment. In *Proceedings of the Human Factors and Ergonomics Society 52nd Annual Meeting* (pp. 237–241). Santa Monica, CA: Human Factors and Ergonomics Society.

Chen, J.Y.C. & Barnes, M.J. (2010). Supervisory control of robots using RoboLeader. In *Proceedings of the Human Factors and Ergonomics Society 54th Annual Meeting* (pp. 1483–1487). Santa Monica, CA: Human Factors and Ergonomics Society.

Chen, J.Y.C. & Joyner, C.T. (2006). Individual differences in concurrent performance of gunner's and robotic operator's tasks. In *Proceedings of the Human Factors and Ergonomics Society 50th Annual Meeting* (pp. 1759–1763). Santa Monica, CA: Human Factors and Ergonomics Society.

Cherry, C. & Sayers, B.McA. (1956). Experiments on total inhibition of stammering by external control and clinical results. *Journal of Psychosomatic Research, 1*, 233–246.

Chien, S.-Y., Wang, H., & Lewis, M. (2011). Effects of spatial ability on multi-robot control tasks. In *Proceedings of the Human Factors and Ergonomics Society 55th Annual Meeting* (pp. 894–898). Santa Monica, CA: Human Factors and Ergonomics Society.

Chieppo, C. (2012). The teachers we need (and the ones we don't). In *St. Paul Pioneer Press* (November 4), p. 12B.

Chomitz, V.R., Slining, M.M., McGowan, R.J., Mitchell, S.E., Dawson, G.F., & Hacker, K.A. (2009). Is there a relationship between physical fitness and academic achievement? Positive results from public school children in the northeastern United States. *Journal of School Health, 79*, 30–37. Downloaded from: http://extension.oregonstate.edu/physical-activity/sites/default/ files/Fit_kids_are_smart_kids.pdf.

Chomsky, N. (1965). *Aspects of the theory of syntax.* Cambridge, MA: MIT Press.

Christensen, J.M. (1987). The human factors profession. In G. Salvendy (Ed.), *Handbook of human factors* (Chapter 1.1, pp. 3–16). New York: Wiley.

Chua, A. (2011). *Battle hymn of the tiger mother.* New York: Penguin Press.

Cleveland, R.J. (1976). *Behavioral safety codes in select industries.* Madison, WI: Wisconsin Department of Industry, Labor and Human Relations.

Coase, R.H. (1937). The nature of the firm. *Economica, 4*(16), 386–405.

Coase, R.H. (1993). 1991 Nobel lecture: The institutional structure of production. In O.E. Williamson & S.G. Winter (Eds.), *The nature of the firm. Origins, evolution, and development* (pp. 227–235). New York: Oxford University Press.

Cohen, A.L. (1977). Factors in successful occupational safety programs. *Journal of Safety Research, 9*(4), 168–178.

Cohen, A.L., Gjessing, C.C., Fine, L.J., Bernard, B.P., & McGlothlin, J.D. (1997). *Elements of ergonomics programs. A primer based on workplace evaluations of musculoskeletal disorders.* Cincinnati, OH: National Institute for Occupational Safety and Health.

Cohen, A.L., Smith, M.J., & Anger, W.K. (1979). Self-protective measures against workplace hazards. *Journal of Safety Research, 11*(3), 121–131.

Cohen, W.J. & Smith, J.H. (1994). Community ergonomics: Past approaches and future prospects towards America's urban crisis. In *Proceedings of the Human Factors and Ergonomics Society 38th Annual Meeting* (pp. 734–738). Santa Monica, CA: Human Factors and Ergonomics Society.

Coleman, P.J. & Sauter, S.L. (1978). *The worker as a key control component in accident prevention systems.* Presentation to the 1978 Convention of the American Psychological Association, Toronto, Ontario, Canada.

Coleman, P.J. & Smith, K.U. (1976). *Hazard management: Preventive approaches to industrial injuries and illnesses.* Madison, WI: Wisconsin Department of Industry, Labor, and Human Relations.

Conklin, J.E. (1957). Effect of control lag on performance in a tracking task. *Journal of Experimental Psychology, 49*, 261–268.

Conklin, J.E. (1959). Linearity of the tracking performance function. *Perceptual and Motor Skills, 9*, 387–391.

Cook, V. & Newson, M. (2007). *Chomsky's universal grammar. An introduction* (3rd edition). Malden, MA: Blackwell.

Cooper, G. (2014). *Money, blood and revolution: How Darwin and the doctor of King Charles I could turn economics into a science.* Petersfield, Hampshire, UK: Harriman House.

Cronbach, L.J. (1957). The two disciplines of scientific psychology. *American Psychologist, 12*, 671–684.

Crosby, W.B. (1979). *Quality is free.* New York: McGraw-Hill.

Crossman, E.R.F.W. (1959). A theory of the acquisition of speed skill. *Ergonomics, 2*, 153–166.

Daileader, P. (2001). *The high middles ages* (Part I). Chantilly, VA: The Great Courses.

Dainoff, M.J. (1990). Ergonomic improvements in VDT workstations: Health and performance effects. In S.L. Sauter, M.J. Dainoff, & M.J. Smith (Eds.), *Promoting health and productivity in the computerized office: Models of successful ergonomic interventions* (pp. 49–67). London: Taylor and Francis.

Darwin, C. (1872). *The expression of the emotions in man and animals.* London: John Murray.

Dash, E., White, B., & de la Merced, M.J. (2008). Lehman expected to file for bankruptcy protection. In *New York Times* (September 18), dealbook.nytimes.com/2008/09/14/lehman-to-file-for-bankruptcy-protection/ (retrieved 12/2/13).

Davids, K., Glazier, P., Araújo, D., & Bartlett, R. (2003). Movement systems as dynamical systems. *Sports Medicine 33*(4), 245–260.

Davison, G.M. (2013). K-12 education is flawed to the core; we need a revolution. In *Minneapolis Star Tribune* (February 27), p. A11.

Dawson, J. (1996). Study suggests IQ starts forming in infancy. In *Minneapolis Star Tribune* (February 10), p. A11.

de Jong, T. & Pieters, J. (2006). The design of powerful learning environments, In P.A. Alexander & P.H. Winne (Eds.), *Handbook of educational psychology* (2nd edition) (Chapter 32, pp. 739–754). Mahwah, NJ: Lawrence Erlbaum Associates.

Dean, L.G., Kendal, R.L., Schapiro, S.J., Thierry, B., & Laland, K.N. (2012). Identification of the social cognitive processes underlying human cumulative culture. *Science, 335*(6072), 1114–1118.

Dekker, S. (2004). *Ten questions about human error. A new view of human factors and system safety.* Mahwah, NJ: Lawrence Erlbaum.

Dekker, S. (2006). *The field guide to understanding human error*. Burlington, VT: Ashgate.

Dekker, S. (2007). *Just culture: Balancing safety and accountability*. Burlington, VT: Ashgate.

Dekker, S. (2011). *Patient safety. A human factors approach*. Boca Raton, FL: CRC Press.

Deming, W.E. (1982). *Out of the crisis*. Cambridge, MA: Massachusetts Institute of Technology, Center for Advanced Engineering Study.

Demsetz, H. (1993). The theory of the firm revisited. In O.E. Williamson & S.G. Winter (Eds.), *The nature of the firm. Origins, evolution, and development* (pp. 159–178). New York: Oxford University Press.

Denby, D. (2010). School spirit (review of "Waiting for Superman"). In *The New Yorker* (October 11), pp. 122–123.

Descartes, R. (1596–1650) (1996). *Meditations on first philosophy* (edited and translated by J. Cottingham. Revised edition). New York: Cambridge University Press.

Devlin, B., Daniels, M., & Roeder, K. (1997). The heritability of IQ. *Nature, 388*(6641), 468–471.

Dewey, J. (1916). *Democracy and education*. New York: Macmillan.

Diamond, J. (1997). *Guns, germs, and steel. The fates of human societies*. New York: W.W. Norton.

Diamond, J. (2005). *Collapse. How societies choose to fail or succeed*. New York: Viking.

Dickman, A. & Kovach, M. (2008). *Toward high quality early childhood education: An imperative for the regional economy*. Atlanta, GA: Economic Justice Summit Draft Working Paper, NOW Foundation, Institute for Women's Policy Research, and National Council of Negro Women. Downloaded from: http://www.publicpolicyforum.org/pdfs/atlanta paper.pdf.

Digman, J.M. (1990). Personality structure: Emergence of the five-factor model. *Annual Review of Psychology, 41*, 417–440.

Dodge, R. (1927). *Elementary conditions of human variability*. New York: Columbia University Press.

Dodge, R. (1931). *Conditions and consequences of human variability*. New Haven: Yale University Press.

Dove-Steinkamp, M.L. & Henning, R.A. (2012). Training under imposed communication delays benefits performance effectiveness of distributed teams. *Proceedings of the Human Factors and Ergonomics Society 56th Annual Meeting* (pp. 2432–2436). Santa Monica, CA: Human Factors and Ergonomics Society.

Dowling, J.E. (2004). *The great brain debate. Nature or nurture*. Princeton, NJ: Princeton University Press.

Dr. Econ (2007). *What are the possible causes and consequences of higher oil prices on the overall economy*. San Francisco: Federal Reserve Board of San Francisco.

Draper, J.V., Kaber, D.B., & Usher, J.M. (1998). Telepresence. *Human Factors, 40*(3), 354–375.

Draper, N. (1996). Schools see boost from breakfast. In *Minneapolis Star Tribune* (February 7), pp. B1, B5.

Draper, N. (1998). Educators pondering pitfalls of computers. In *Minneapolis Star Tribune* (July 11), pp. A1, A12.

Draper, N. (2002). Math, reading scores up statewide. Schools still plagued by racial gap in performance. In *Minneapolis Star Tribune* (April 19), pp. A1, A20, A21.

Draper, N. (2005). Reading, writing, redecorating: School décor for better scores? In *Minneapolis Star Tribune* (December 18), pp. A1, A23.

Drucker, P.F. (1999). Beyond the information revolution. In *The Atlantic Monthly* (October, Volume 284, Number 4), pp. 47–59.

Du, W. & Milgram, P. (2002). Effect of history trail display on human spatial performance under normal and rotated spatial mappings. In *Proceedings of the Human Factors and Ergonomics Society 46th Annual Meeting* (pp. 1623–1627). Santa Monica, CA: Human Factors and Ergonomics Society.

Dul, J., Bruder, R., Buckle, P., Carayon, P., Falzon, P., Marras, W.S., Wilson, J.R., & van der Doelen, B. (2012). A strategy for human factors/ergonomics: Developing the discipline and profession. *Ergonomics, 55*(4), 377–395.

Durso, F.T. (2013). HFES leadership sets priorities at midyear meeting. *HFES Bulletin, 56*(5), 1–2.

Eberts, R.E. & Brock, J.F. (1984). Computer applications to instruction. In F.A. Muckler (Ed.), *Human factors review* (Chapter 7, pp. 239–284). Santa Monica, CA: Human Factors Society.

Eberts, R.E. and Brock, J.F. (1987). Computer-assisted and computer-managed instruction. In G. Salvendy (Ed.), *Handbook of human factors* (Chapter 8.4, pp. 976–1011). New York: Wiley.

The Economist (2004). The world's oldest companies. The business of survival. In *The Economist* (December 18).

The Economist (2009). Out the window. In *The Economist* (April 25), pp. 59–60.

The Economist (2010a). Homo administrans. The biology of business. In *The Economist* (September 25), pp. 99–101.

The Economist (2010b). How to get good grades. In *The Economist* (November 27), p. 68.

The Economist (2011a). Lessons learned. In *The Economist* (January 8), pp. 26–27.

The Economist (2011b). The last woman and the end of history. In *The Economist* (August 27), p. 52.

The Economist (2012). Body politic. The genetics of politics. In *The Economist* (October 6), pp. 93–94.

The Economist (2013a). Speed isn't everything. In *The Economist* (April 20), p. 12.

The Economist (2013b). Schools brief (first in a series). In *The Economist* (September 7), pp. 74–75.

The Economist (2013c). Schools brief (second in a series). In *The Economist* (September 14), pp. 74–75.

The Economist (2013d). Schools brief (fourth in a series). In *The Economist* (September 28), pp. 72–73.

The Economist (2014). The slumps that shaped modern finance. In *The Economist* (April 12), pp. 49–54.

Egan, D.E. (1988). Individual differences in human-computer interaction. In M. Helander (Ed.), *Handbook of human-computer interaction* (pp. 543–568). Amsterdam: North-Holland.

Ehrlich, P.R. (2000). *Human natures. Genes, cultures, and the human prospect.* Washington, DC: Island Press.

Eklund, J. (1997). Ergonomics, quality and continuous improvement—Some current issues. In P. Seppälä, T. Luopajärvi, C.-H. Nygård, & M. Mattila (Eds.), *Proceedings of the 13th Triennial Congress of the International Ergonomics Association, Tampere, Finland. 1997* (Vol. 1, pp. 10–12). Helsinki, Finland: Finnish Institute of Occupational Health.

Ekman, P. (1992). An argument for basic emotions. *Cognition and Emotion, 6*(3/4), 169–200.

Elkins, A.N., Muth, E.R., Hoover, A.W., Walker, A.D., Carpenter, T.L., & Switzer, F.S. (2009). Physiological compliance and team performance. *Applied Ergonomics, 40*, 997–1003.

Ellis, L. (1975). A review of research on efforts to promote occupational safety. *Journal of Safety Research, 7*(4), 180–189.

Ellis, S.R., Tyler, M., Kim, W.S., & Stark, L. (1991). *Three-dimensional tracking with mis-alignment between display and control axes* (SAE Technical Report Number 911390). Warrendale, PA: Society of Automotive Engineers International.

Emery, F.E. (1969). *Systems thinking.* London: Penguin.

Emery, F.E. & Trist, E. (1965). The causal nature of organizational environments. *Human Relations* (February), 21–32.

Endsley, M.R. (1988). Design and evaluation for situation awareness. In *Proceedings of the Human Factors Society 32nd Annual Meeting* (pp. 97–101). Santa Monica, CA: Human Factors and Ergonomics Society.

Endsley, M.R. (1995a). Toward a theory of situation awareness in dynamic systems. *Human Factors, 37*(1), 32–64.

Endsley, M.R. (1995b). Measurement of situation awareness in dynamic systems. *Human Factors, 37*(1), 65–84.

Endsley, M.R. & Garland, D.J. (2000). *Situation awareness analysis and measurement.* Mahwah, NJ: Lawrence Erlbaum.

Endsley, M.R. & Jones, D.G. (2011). *Designing for situation awareness: An approach to user-centered design* (2nd edition). Boca Raton, FL: CRC Press.

Enfield, N.J. (2010). Without social context? *Science, 329*, 1600–1601.

Estrada, H.M. (2013). Burnsville schools gauging technology's impact. In *Minneapolis Star Tribune* (March 16), pp. AA1, AA4.

Evans, G.W. & Maxwell, L. (1997). Chronic noise exposure and reading deficits: The mediating effects of language acquisition. *Environment and Behavior, 29*, 638–656.

Ewert, P.H. (1930). A study of the effect of inverted retinal stimulation upon spatially coordinated behavior. *Genetic Psychology Monographs, 7*, 177–363.

Fairbanks, G. (1955). Selective vocal effects of delayed auditory feedback. *Journal of Speech and Hearing Disorders, 20*, 333–346.

Fairbanks, G. & Guttman, N. (1958). Effects of delayed auditory feedback upon articulation. *Journal of Speech and Hearing Research, 1*, 12–22.

Falk, J.H. & Dierking, L.D. (2010). The 95 percent solution. School is not where most Americans learn most of their science. *American Scientist, 98*, 486–493.

Ferdig, B. (2013). Your full attention, please. In *Minneapolis Star Tribune* (May 5), pp. OP1, OP4.

Ferrell, W.R. (1966). Delayed force feedback. *Human Factors, 8*, 449–455.

Finn, C.E. Jr. (2009). *Reroute the preschool juggernaut.* Stanford, CA: Hoover Institution Press.

Fisher, T. (2012). *Designing to avoid disaster. The nature of fracture-critical design.* New York: Routledge.

Fitch, W.T. (2010). *The evolution of language.* Cambridge: Cambridge University Press.

Fitts, P.M. (1951). Engineering psychology and equipment design. In S.S. Stevens (Ed.), *Handbook of experimental psychology* (pp. 1287–1340). New York: Wiley.

Fitts, P.M. (1954). The information capacity of the human motor system in controlling the amplitude of movement. *Journal of Experimental Psychology, 47*(6), 381–391.

Flach, J.M. (1990). Control with an eye for perception: Precursors to an active psychophysics. *Ecological Psychology, 2*(2), 83–111.

Flach, J.M. (1994). Ruminations on mind, matter, and what matters. *Proceedings of the Human Factors and Ergonomics Society 38th Annual Meeting.* Santa Monica, CA: Human Factors and Ergonomics Society.

Flach, J.M. (1995). Situation awareness: Proceed with caution. *Human Factors, 37*(1), 149–157.

Flach, J.M. & Hancock, P.A. (1992). An ecological approach to human-machine systems. *Proceedings of the Human Factors and Ergonomics Society 36th Annual Meeting* (pp. 1056–1058). Santa Monica, CA: Human Factors and Ergonomics Society.

Flach, J.M., Tanabe, F., Monta, K., Vicente, K.J., & Rasmussen, J. (1994). An ecological approach to interface design. *Proceedings of the Human Factors and Ergonomics Society 38th Annual Meeting* (pp. 295–299). Santa Monica, CA: Human Factors and Ergonomics Society.

Fleishman, E.A. (1954). *A factorial study of psychomotor abilities* (USAF Personnel & Training Research Center Research Bulletin No. 54-15). Dayton, OH: Wright-Patterson USAF Base.

Fleishman, E.A. (1960). Abilities at different stages of practice in rotary pursuit performance. *Journal of Experimental Psychology, 60*, 162–172.

Fleishman, E.A. (1966). Human abilities and acquisition of skill. In E.A. Bilodeau (Ed.), *Acquisition of skill* (pp. 147–167). New York: Academic Press.

Fleishman, E.A. & Hempel, W.E. Jr. (1954). Changes in factor structure of a complex psychomotor test as a function of practice. *Psychometrika, 19,* 239–252.

Fleishman, E.A. & Hempel, W.E. Jr. (1955). The relation between abilities and improvement with practice in a visual discrimination reaction task. *Journal of Experimental Psychology, 49,* 301–312.

Fleishman, E.A. & Quaintance, M.K. (1984). *Taxonomies of human performance: The description of human tasks.* New York: Academic Press.

Floud, R., Fogel, R.W., Harris, B., & Hong, S.C. (2011). *The changing body: Health, nutrition, and human development in the western world since 1700.* Cambridge: Cambridge University Press.

Flynn, J.R. (2012). *Are we getting smarter? Rising IQ in the twenty-first century.* Cambridge: Cambridge University Press.

Fox, M. (2007). *Talking hands: What sign language reveals about the mind.* New York: Simon & Schuster.

Fredericq, L. (1887). Methode des gekrcutzen kreislaufs. *Bulletin de l'Académie r. de Belgique. Classe des Sciences, 13*(4), 417.

Free Exchange (2012). No short cuts. In *The Economist* (October 27), p. 76.

Free Exchange (2013a). Net benefits. In *The Economist* (March 9), p. 76.

Free Exchange (2013b). The 90% question. In *The Economist* (April 20), p. 82.

Free Exchange (2013c). Making pay work. Why bosses should be careful when using performance-related pay. In *The Economist* (May 25), p. 76.

Free Exchange (2013d). The goliaths. In *The Economist* (June 22), p. 79.

Free Exchange (2013e). One of the giants. In *The Economist* (September 7), p. 73.

Free Exchange (2013f). A reasonable supply. In *The Economist* (November 30), p. 76.

Funder, D.C. (2001). Personality. *Annual Reviews of Psychology, 52,* 197–221.

Furnham, A. (1990). Faking personality questionnaires: Fabricating different profiles for different purposes. *Current Psychology: Research & Reviews, 9*(1), 46–55.

Galbraith, J.K. (1958). *The affluent society.* Boston: Houghton Mifflin.

Galton, F. (1883). *Inquiries into human faculty and its development.* New York: Macmillan.

Gander, E.M. (2003). *On our minds: How evolutionary psychology is reshaping the nature-versus-nurture debate.* Baltimore: Johns Hopkins University Press.

Gerisch, H., Staude, G., Wolf, W., & Bauch, G. (2013). A three-component model of the control error in manual tracking of continuous random signals. *Human Factors, 55*(5), 985–1000.

Gibbon, E. (1788). *The history of the decline and fall of the Roman empire* (Volume the Sixth) (D. Womersley, Editor, 1994). London: Allen Lane The Penguin Press.

Gibson, E.J. & Pick, A.D. (2000). *An ecological approach to perceptual learning and development.* New York: Oxford University Press.

Gibson, J.J. (1966). *The senses considered as perceptual systems.* Boston: Houghton-Mifflin.

Gibson, J.J. (1979). *The ecological approach to visual perception.* Boston: Houghton Mifflin.

Gil, M.C. & Henning, R.A. (2000). Determinants of perceived teamwork: Examination of team performance and social psychophysiology. *Proceedings of the Human Factors and Ergonomics Society 44th Annual Meeting* (pp. 743–746). Santa Monica, CA: Human Factors and Ergonomics Society.

Gill, J. & Martin, K. (1976). Safety management: Reconciling rules with reality. *Personnel Management, 8*(6), 36–39.

Gladwell, M. (2008). *Outliers. The story of success.* New York: Back Bay Books.

Gladwell, M. (2013). The gift of doubt. In *The New Yorker* (June 24, Volume LXXXIX, Number 18), pp. 74–79.

Glendon, A.I., Clarke, S.G., & McKenna, E.F. (2006). *Human safety and risk management* (2nd edition). Boca Raton, FL: CRC Press.

Glynn, S., Fekieta, R., & Henning, R.A. (2001). Use of force-feedback joysticks to promote teamwork in virtual teleoperation. *Proceedings of the Human Factors and Ergonomics Society 45th Annual Meeting* (pp. 1911–1915). Santa Monica, CA: Human Factors and Ergonomics Society.

Goertz, R.C. & Thompson, R.C. (1954). Electronically controlled manipulator. *Nucleonics, 14*, 46–47.

Goldberg, H.S. (2006). *Hippocrates. Father of medicine*. Lincoln, NE: Authors Choice Press.

Goldhaber, D. (2012). *The nature-nurture debates. Bridging the gap*. Cambridge: Cambridge University Press.

Goldhaber, D. & Anthony, E. (2004). *Can teacher quality be effectively assessed? National board certification as a signal of effective teaching*. Seattle, WA and Washington, DC: University of Washington and the Urban Institute. Downloaded from: http://www.urban.org/UploadedPDF/410958_NBPTS_Outcomes.pdf.

Goldhaber, D.D. & Brewer, D.J. (1997). Evaluating the effect of teacher degree level on educational performance. In *Developments in school finance, 1996* (NCES State Data Conference). Washington, DC: National Center for Education Statistics, U.S. Department of Education Institute of Education Sciences. Downloaded from: http://nces.ed.gov/pubsearch/pubsinfo.asp?pubid=97535.

Gopnik, A. (2010). Market man. What did Adam Smith really believe? In *The New Yorker* (October 18, Volume LXXXVI, Number 32), pp. 82–87.

Gopnik, A. (2012). Scientific thinking in young children: Theoretical advances, empirical research, and policy implications. *Science, 337*(6102), 1623–1627.

Gorman, J.C., Cooke, N.J., & Amazeen, P.G. (2010). Training adaptive teams. *Human Factors, 52*(2), 295–307.

Gottlieb, M.S. (1976*). Worker's awareness of industrial hazards: An analysis of hazard survey results from the paper mill industry*. Madison, WI: Wisconsin Department of Industry, Labor and Human Relations.

Gottlieb, M.S. & Coleman, P.J. (1977). *Inspection impact on injury and illness totals*. Madison, WI: Wisconsin Department of Industry, Labor and Human Relations.

Gould, J.D. (1990). How to design usable systems. In M. Helander (Ed.), *Handbook of human-computer interaction* (pp. 757–789). Amsterdam: Elsevier.

Gould, J.D. & Smith, K.U. (1963). Angular displacement of visual feedback in motion and learning. *Perceptual and Motor Skills, 17*, 699–710.

Greene, J.P. & Winters, M.A. (2005). Five myths crying out for debunking. In *National Review* (October 24), pp. 49–51.

Greeno, J.G., Collins, A.M., & Resnick, L.B. (1996). Cognition and learning. In D.C. Berliner & R.C. Calfee (Eds.), *Handbook of educational psychology* (Chapter 2, pp. 15–46). New York: Simon and Schuster Macmillan.

Greenwood, J.D. (2009). *A conceptual history of psychology*. Boston: McGraw-Hill.

Grens, K. (2013). Human whiskers. *The Scientist, 27*(2), 19–20.

Grimaldi, J.V. & Simonds, R.H. (1989). *Safety management* (5th edition). Boston: Richard D. Irwin.

Grissmer, D.W., Flanagan, A., Kawata, J.H., & Williamson, S. (2000). *Improving student achievement. What state NAEP test scores tell us*. Washington, DC: Rand Corp.

Guarnieri, M. (1992). Landmarks in the history of safety. *Journal of Safety Research, 23*, 151–158.

Guastello, S.J. (1993). Do we really know how well our occupational accident prevention programs work? *Safety Science, 16*, 445–463.

Guastello, S.J. (2013). Occupational accidents and prevention. In S.J. Guastello, *Human factors engineering and ergonomics: A systems approach* (Chapter 10, pp. 269–322). Mahwah, NJ: Lawrence Erlbaum.

Gunderson, L. & Holling, C.S. (2002). *Panarchy. Understanding transformations in human and natural systems.* Washington, DC: Island Press.

Gustafsson, J.-E. & Undheim, J.O. (1996). Individual differences in cognitive functions. In D.C. Berliner & R.C. Calfee (Eds.), *Handbook of educational psychology* (Chapter 8, pp. 186–242). New York: Simon and Schuster Macmillan.

Haims, M.C. & Carayon, P. (1998). Theory and practice for the implementation of "in-house" continuous improvement participatory ergonomics programs. *Applied Ergonomics, 29,* 461–72.

Hale, R.L., Hoelscher, D.R., & Kowal, R.E. (1987). *Quest for quality. How one company put theory to work.* Minneapolis, MN: Tennant Company.

Hallinan, J.T. (2009). *Errornomics. Why we make mistakes and what we can do to avoid them.* London: Ebury Press.

Hamill, J., Haddad, J.M., & McDermott, W.J. (2000). Issues in quantifying variability from a dynamical systems perspective. *Journal of Applied Biomechanics, 16,* 407–418.

Hammer, W. (1972). *Handbook of systems and product safety.* Englewood Cliffs, NJ: Prentice-Hall.

Hammer, W. (1976). *Occupational safety management and engineering.* Englewood Cliffs, NJ: Prentice-Hall.

Hancock, P.A. (1996). On convergent technological evolution. *Ergonomics in Design, 4*(1), 22–29.

Hancock, P.A. (1997). On the future of work. *Ergonomics in Design, 5*(4), 25–29.

Hanley, C.N., Tiffany, W.R., & Brungard, J.M. (1958). Skin resistance changes accompanying the side-tone test for auditory malingering. *Journal of Speech and Hearing Research, 1,* 286–293.

Hanson, K.F. (2001). Deserting the classroom battleground. In *St. Paul Pioneer Press* (June 7), p. 23A.

Harbourne, R.T. & Stergiou, N. (2009). Movement variability and the use of nonlinear tools: Principles to guide physical therapist practice. *Physical Therapy, 89*(3), 267–282.

Harcourt, A.H. (2012). *Human biogeography.* Berkeley, CA: University of California Press.

Harper, R.F. (1904). *The code of Hammurabi, King of Babylon, about 2250 B.C.* (2nd edition). Chicago: University of Chicago Press.

Harris, C. & Straker, L. (2000). Survey of physical ergonomics issues associated with school childrens' use of laptop computers. *International Journal of Industrial Ergonomics, 26*(3), 337–346.

Harris, C.S. (1965). Perceptual adaptation to inverted, reversed, and displaced vision. *Psychological Review, 72*(6), 419–444.

Harris, J.R. (2006). *No two alike. Human nature and human individuality.* New York: W.W. Norton.

Hartnett, K. (2012). How to teach your children well. In *Minneapolis Star Tribune* (September 2), p. OP4.

Hathaway, S.R. & McKinley, J.C. (1940). A multiphasic personality schedule (Minnesota): I. Construction of the schedule. *The Journal of Psychology, 10*(2), 249–254.

Hauser, M.D., Chomsky, N., & Fitch, W.T. (2002). The faculty of language: What is it, who has it, and how did it evolve? *Science, 298*(5598), 1569–1579.

Havens, G. & Patterson, G. (2010). To fix schools, fix neighborhoods. In *Minneapolis Star Tribune* (February 14), pp. B1, B12.

Hawkins, J. & Blakeslee, S. (2004). *On intelligence.* New York: Henry Holt.

Hawks, J. (2011). *The rise of humans: Great scientific debates.* Chantilly, VA: The Great Courses.

Head, H. (1926). *Aphasia and kindred disorders of speech.* Cambridge: Cambridge University Press.

Healy, J. (1998). *Failure to connect: How computers affect our children's minds, for better and worse*. New York: Simon and Schuster.

Heckman, J.J. (2008). *Schools, skills, and synapses* (IZA Discussion Paper No. 3515). Bonn, Germany: Institute for the Study of Labor. Downloaded from: http://ftp.iza.org/dp3515.pdf.

Heiderscheit, B.C., Hamill, J., & van Emmerik, R.E.A. (2002). Variability of stride characteristics and joint coordination among individuals with unilateral patella-femoral pain. *Journal of Applied Biomechanics, 18*, 110–121.

Heinrich, H.W. (1931). *Industrial accident prevention. A scientific approach* (1st edition). New York: McGraw-Hill.

Heinrich, H.W. (1941). *Industrial accident prevention. A scientific approach* (2nd edition). New York: McGraw-Hill.

Heinrich, H.W. (1959). *Industrial accident prevention. A scientific approach* (4th edition). New York: McGraw-Hill.

Helander, M.G. & Khalid, H.M. (2006). Affective and pleasurable design. In G. Salvendy (Ed.), *Handbook of human factors and ergonomics* (3rd edition) (pp. 843–872). New York: Wiley.

Held, R. & Durlach, N. (1991). Telepresence, time delay and adaptation. In S.R. Ellis (Ed.), *Pictorial communication in virtual and real environments* (2nd edition, pp. 233–246). London: Taylor & Francis.

Heller, N. (2013). Laptop U. In *The New Yorker* (May 20, Volume LXXXIX, Number 14), pp. 80–91.

Helmholtz, H.L.F. von (1856–1866). *Handbuch der physiologischen Optic* (3 volumes). Hamburg & Leipzig: Voss.

Hendrick, H.W. (1984). Wagging the tail with the dog: Organizational design considerations in ergonomics. In *Proceedings of the Human Factors Society 28th Annual Meeting* (pp. 899–902). Santa Monica, CA: Human Factors and Ergonomics Society.

Hendrick, H.W. (1986). Macroergonomics: A conceptual model for integrating human factors with organizational design. In O. Brown and H. Hendrick (Eds.), *Human factors in organizational design and management II* (pp. 467–477). Amsterdam: North Holland.

Hendrick, H.W. (1991). Human factors in organizational design and management. *Ergonomics, 34*, 743–756.

Hendrick, H.W. (1996). The ergonomics of economics is the economics of ergonomics. In *Proceedings of the Human Factors and Ergonomics Society 40th Annual Meeting* (pp. 1–10). Santa Monica, CA: Human Factors and Ergonomics Society.

Hendrick, H.W. (1997). *Good ergonomics is good economics*. Santa Monica, CA: Human Factors and Ergonomics Society.

Hendrick, H.W. (2007). Macroergonomics: The analysis and design of work systems. In P.R. DeLucia (Ed.), *Reviews of human factors and ergonomics* (Volume 3), (pp. 44–78). Santa Monica, CA: Human Factors and Ergonomics Society.

Hendrick, H.W. & Kleiner, B.M. (Eds.) (2002). *Macroergonomics. Theory, methods, and applications*. Mahwah, NJ: Lawrence Erlbaum.

Henning, R.A. & Korbelak, K. (2005). Social-psychophysiological compliance as a predictor of future team performance. *Psychologia, 48*(2), 84–92.

Henning, R.A. & Reeves, D.W. (2013). An integrated health protection/promotion program supporting participatory ergonomics and salutogenic approaches in the design of workplace interventions. In G. Bauer & J. Gregor (Eds.), *Salutogenic organizations and change: The concepts behind organizational health intervention research* (pp. 307–325). The Netherlands: Springer.

Henning, R.A., Armstead, A.G., & Ferris, J.K. (2009a). Social psychophysiological compliance in a four-person research team. *Applied Ergonomics, 40*, 1004–1010.

Henning, R.A., Bopp, M.I., Tucker, K.M., Knoph, R.D., & Ahlgren, J. (1997). Team-managed rest breaks during computer-supported cooperative work. *International Journal of Industrial Ergonomics, 20*, 19–29.

Henning, R.A., Boucsein, W., & Gil, M.C. (2001). Social-physiological compliance as a determinant of team performance. *International Journal of Psychophysiology, 40*, 221–232.

Henning, R.A., Boucsein, W., Fekieta, R.E., Gil, M.O., & Pratt, J.H. (2000). *Team biocybernetics based on social psychophysiology.* Paper presented at the 3rd International Conference for Psychophysiology in Ergonomics, San Diego, CA.

Henning, R.A., Smith, T.J., & Korbelak, K. (2005). Social psychophysiological compliance as a gauge of the cognitive capacity of teams. In D.D. Schmorrow (Ed.), *Foundations of augmented cognition* (pp. 1228–1238). New York: CRC Press.

Henning, R.A., Smith, T.J., & Armstead, A.G. (2007). Use of communication delays to identify physiological indices for augmented team cognition. In D.D. Schmorrow, D.M. Nicholson, J.M. Drexler & L.M. Reeves (Eds.), *Foundations of augmented cognition* (4th edition) (pp. 238–247). Arlington, VA: Strategic Analysis, Inc. & Augmented Cognition International Society.

Henning, R.A., Warren, N.D., Robertson, M., Faghri, P., & Cherniack, M. (2009b). Workplace health protection and promotion through participatory ergonomics: An integrated approach. *Public Health Reports, 124*, S1, 26–35.

Henry, J.P., Junas, R., & Smith, K.U. (1967). Experimental cybernetic analysis of delayed feedback of breath pressure control. *American Journal of Physical Medicine, 46*, 1317–1331.

Hernandez, L.M. & Blazer, D.G. (Eds.) (2006). *Genes, behavior, and the social environment: Moving beyond the nature/nurture debate.* Washington, DC: Institute of Medicine of the National Academies.

Herrnstein, R.J. & Murray, C. (1994). *The bell curve. Intelligence and class structure in American life.* New York: Simon and Schuster.

Hersey, P. & Blanchard, K. (1977). *Management of organizational behavior: Utilizing human resources* (3rd edition). Englewood Cliffs, NJ: Prentice Hall.

Hick, W.E. (1952). On the rate of gain of information. *Quarterly Journal of Experimental Psychology, 4*, 11–26.

Hirschman, A.O. (1967). The principle of the hiding hand. *The Public Interest* (Issue 6, Winter), pp. 11–23.

Hirschman, A.O. (1991). *The rhetoric of reaction. Perversity, futility, and jeopardy.* Cambridge, MA: Harvard University Press.

Hobbes, T. (1588–1679) (2012). *Leviathan* (edited by N. Malcolm). Oxford: Oxford University Press.

Hofferth, S.L. (2010). Home media and children's achievement and behavior. *Child Development, 81*, 1598–1619.

Holden, C. (Ed.) (2005). Random samples. Signs support Chomsky. *Science, 310*, 1900.

Holden, J.G., Flach, J.M., & Donchin, Y. (1999). Perceptual-motor coordination in an endoscopic surgery simulation. *Surgical Endoscopy, 13*, 127–132.

Hollnagel, E. (2002). Understanding accidents—from root cause to performance variability. In *Proceedings of the IEEE 7th Human Factors Meeting* (pp. 1-1–1-6). New York: IEEE.

Hollnagel, E., Dekker, S., & Leveson, N. (Eds.) (2006). *Resilience engineering: Concepts and precepts.* Burlington, VT: Ashgate.

Hommel, B., Ridderinkhof, K.R., & Theeuwes, J. (2002). Cognitive control of attention and action: Issues and trends. *Psychological Research, 66*, 215–219.

Hornick, R.J. (1987). Dreams, design and destiny. *Human Factors, 29*(1), 111–121.

Horowitz, D. (1998). *Marx's Manifesto. 150 years of evil.* Los Angeles: Center for the Study of Popular Culture.

Horrell, P. (2009). *An exploration of how classroom ergonomics can contribute to a successful climate for teaching and learning*. Downloaded from: http://www.pirate-university.org/papers/AnExplorationofClassroomErgonomics.pdf.

Hough, L.M. & Schneider, R.J. (1996). Personality traits, taxonomies, and applications in organizations. In K.R. Murphy (Ed.), *Individual differences and behavior in organizations* (pp. 31–88). San Francisco: Jossey-Bass.

Hourcade, J.P. (2006). Design for children. In G. Salvendy (Ed.), *Handbook of human factors and ergonomics* (3rd edition) (Chapter 55, pp. 1446–1458). New York: Wiley.

Hubbard, A.W. (1976). Homeokinetics: Muscular function in human movement. In W.R. Johnson & E.R. Buskirk (Eds.), *Science and medicine of exercise and sport* (pp. 5–20). New York: Harper and Row.

Hubbard, G. & Kane, T. (2013). *Balance: The economics of great powers from ancient Rome to modern America*. New York: Simon & Schuster.

Huggins, C.Q., Malone, T.B., & Shields, N.L. (1973). Evaluation of human operator visual performance capability for teleoperator missions. In E. Heer (Ed.), *Remotely manned systems* (pp. 337–350). Pasadena, CA: California Institute of Technology.

Hunter, W.S. (1934). Learning: IV. Experimental studies of learning. In C. Murchison (Ed.), *Handbook of general experimental psychology* (pp. 497–570). Worcester, MA: Clark University Press.

Hurford, J.R. (2007). *The origins of meaning: Language in the light of evolution*. Oxford: Oxford University Press.

Hyman, R. (1953). Stimulus information as a determinant of reaction time. *Journal of Experimental Psychology, 45*, 188–196.

Imada, A.S. (2002). A macroergonomic approach to reducing work-related injuries. In H.W. Hendrick & B.M. Kleiner (Eds.), *Macroergonomics. Theory, methods, and applications* (pp. 151–171). Mahwah, NJ: Lawrence Erlbaum.

International Organization for Standardization (ISO) (1994). *International standard. Quality systems—model for quality assurance in design, development, production, installation and servicing* (Standard No. ISO 9001:1994, 2nd edition). Geneva, Switzerland: ISO.

Iowa Department of Transportation, Subcommittee on Public Affairs, Bridge Terms (2013), http://www.iowadot.gov/subcommittee/bridgeterms.aspx#f (retrieved 12/8/13).

Irwin, T. (1999) (Translator). *Aristotle. Nicomachean ethics* (2nd edition). Indianapolis, IN: Hackett Publishing.

Jackson, B. & Parry, K. (2011). *A very short, fairly interesting and reasonably cheap book about studying leadership* (2nd edition). Los Angeles: Sage.

Jackson, P.L. & Decety, J. (2004). Motor cognition: A new paradigm to study self-other interactions. *Current Opinion Neurobiology, 14*, 259–263.

Jagacinski, R.J. (1977). A qualitative look at feedback control theory as a style of describing behavior. *Human Factors, 19*(4), 331–347.

Jagacinski, R.J. & Flach, J.M. (2003). *Control theory for humans: Quantitative approaches to modeling performance*. Mahwah, NJ: Lawrence Erlbaum.

James, W. (1884). What is an emotion? *Mind, 9*(34), 188–205.

James, W. (1890). *Principles of psychology* (Volume 1). London: Holt.

Jennings, S., Craig, G., Reid, L., & Kruk, R. (2000). The effect of visual system time delay on helicopter control. In *Proceedings of the IEA 2000/HFES 2000 Congress* (3-69–3-72). Santa Monica, CA: Human Factors and Ergonomics Society.

Jensen, A.R. (1998). *The g factor: The science of mental ability*. Westport, CT: Praeger.

Johnson, D.W. & Johnson, R.T. (1987). *Joining together: Group theory and group skills* (3rd edition). Englewood Cliffs, NJ: Prentice-Hall.

Johnson, D.W. & Johnson, R.T. (1989). *Cooperation and competition. Theory and research*. Edina, MN: Interaction Book Company.

Johnson, D.W., Maruyama, G., Johnson, R., Nelson, D., & Skon, L. (1981). Effects of cooperative, competitive, and individualistic goal structures on achievement: A meta-analysis. *Psychological Bulletin, 89*, 47–62.

Jones, D.F. (1973). *Occupational safety programs—Are they worth it?* Toronto: Labour Safety Council of Ontario, Ontario Ministry of Labour.

Jones, M.B. (1966). Individual differences. In E.A. Bilodeau (Ed.), *Acquisition of skill* (pp. 109–146). New York: Academic Press.

Jones, M.B. (1969). Differential processes in acquisition. In E.A. Bilodeau and I.M. Bilodeau (Eds.), *Principles of skill acquisition* (pp. 141–170). New York: Academic Press.

Jubera, D., Paterik, S., & Lewis, K. (2010). Who should fund cyber schools? In *Parade Intelligence Report* (April 25), p. 12.

Jung, J.Y., Adelstein, B.D., & Ellis, S.R. (2000). Discriminability of prediction artifacts in a time-delayed virtual environment. In *Proceedings of the IEA 2000/HFES 2000 Congress* (1-499–1-502). Santa Monica, CA: Human Factors and Ergonomics Society.

Juran, J.M. (1954). Universals in management planning and control. *The Management Review* (November), 748–761.

Juran, J.M. (1964). *Managerial breakthrough. A new concept of the manager's job.* New York: McGraw-Hill.

Juran, J.M. (1992). *Juran on quality by design. The new steps for planning quality into goods and services.* New York: Free Press.

Juran, J.M. (1995). *Managerial breakthrough. The classic book on improving management for performance* (Revised edition). New York: McGraw-Hill.

Juran, J.M. & Gryna, F.M. Jr. (1980). *Quality planning and analysis. From product development through use.* New York: McGraw-Hill.

Jyoti, D.F., Frongillo, E.A., & Jones, S.J. (2005). Food insecurity affects school children's academic performance, weight gain, and social skills. *Journal of Nutrition, 135*, 2831–2839.

Kaber, D.B., Riley, J.M., Zhou, R., & Draper, J. (2000). Effects of visual interface design, and control mode and latency on performance, telepresence and workload in a teleoperation task. In *Proceedings of the IEA 2000/HFES 2000 Congress* (1-503–1-506). Santa Monica, CA: Human Factors and Ergonomics Society.

Kahneman, D. (2011). *Thinking, fast and slow.* New York: Farrar, Straus and Giroux.

Kahneman, D. & Tversky, A. (1979). Prospect theory: An analysis of decision under risk. *Econometrica, 47*(2), 263–292.

Kaiser, R.G. (2005). Best schools in the world. In *St. Paul Pioneer Press* (June 12), p. 13A.

Kalmus, H., Denes, P., & Fry, D.B. (1955). Effect of delayed acoustic feedback on some nonvocal activities. *Nature, 175*, 1078.

Kantowitz, H.H. & Sorkin, R.D. (1983). *Human factors: Understanding people-system relationships.* New York: Wiley.

Kao, H.S. & Smith, K.U. (1969). Cybernetic television methods applied to feedback analysis of automobile safety. *Nature, 222*, 299–300.

Kao, H.S. & Smith, K.U. (1971). Social feedback: Determination of social learning. *Journal of Nervous and Mental Disease, 152*, 289–297.

Kao, H.S.R. (1976). On educational ergonomics. *Ergonomics, 19*(6), 667–681.

Kaplan, M.C. & Coleman, P.J. (1976). *County highway department hazards: A comparative analysis of inspection and worker detected hazards.* Madison, WI: Wisconsin Department of Industry, Labor and Human Relations.

Kaplan, M.C., Knutson, S., & Coleman, P.J. (1976). *A new approach to hazard management in a highway department.* Madison, WI: Wisconsin Department of Industry, Labor and Human Relations.

Karasek, R.A. (1979). Job demands, job decision latitude, and mental strain: Implications for job redesign. *Administrative Science Quarterly, 24*(2), 285–308.

Karsh, B.-T., Moro, F.B.P., & Smith, M.J. (2001). The efficacy of workplace ergonomic interventions to control musculoskeletal disorders: A critical analysis of the peer-reviewed literature. *Theoretical Issues in Ergonomics Science, 2*(1), 23–96.

Karwowski, W. (2003). Considering the importance of individual differences in human factors research: No longer simply confounding noise. Panel summary. *Proceedings of the Human Factors and Ergonomics Society 47th Annual Meeting* (pp. 1082–1086). Santa Monica, CA: Human Factors and Ergonomics Society.

Karwowski, W. (2012). A review of human factors challenges of complex adaptive systems: Discovering and understanding chaos in human performance. *Human Factors, 54*(6), 983–995.

Kearns, S.K. & Sutton, J.E. (2013). Hangar talk survey: Using stories as a naturalistic method of informing threat and error management training. *Human Factors, 55*(2), 267–277.

Keller, E.F. (2010). *The mirage of a space between nature and nurture.* Durham, NC: Duke University Press.

Kelly, K. (2000). False promise. Parking your child in front of the computer may seem like a good idea, but think again. In *U.S. News and World Report* (September 25), pp. 48–55.

Kendall, R.M. (1986). Incentive programs with a competitive edge. *Occupational Hazards, 48*(3), 41–45.

Kendall, R.M. (1987). Recognition sparks safety excellence. *Occupational Hazards, 49*(3), 41–46.

Kennedy, P. (1987). *The rise and fall of the great powers.* New York: Random House.

Kennedy, P. (2013). Healthier companies provide superior results. In *Minneapolis Star Tribune* (June 16, Top Workplaces section), p. 25.

Kenrick, D.T. & Funder, D.C. (1988). Profiting from controversy. Lessons from the person-situation debate. *American Psychologist, 43*(1), 23–34.

Kersten, K. (2011). From Florida, school reform that works. In *Minneapolis Star Tribune* (February 13), p. OP3.

Kincaid, M. (1925). A study of individual differences in learning. *Psychological Review, 32,* 34–53.

King, B.J. (2001). *Roots of human behavior.* Chantilly, VA: The Great Courses.

Kirwan, B. (1992a). Human error identification in human reliability assessment. Part 1: Overview of approaches. *Applied Ergonomics, 23*(5), 299–318.

Kirwan, B. (1992b). Human error identification in human reliability assessment. Part 2: Detailed comparison of techniques. *Applied Ergonomics, 23*(6), 371–381.

Kirwan, B. (2005). Human reliability assessment. In J.R. Wilson & N. Corlett (Eds.), *Evaluation of human work* (3rd edition) (Chapter 32, pp. 833–875). Boca Raton, FL: CRC Press.

Klein, J. (2011). Scenes from the class struggle. In *The Atlantic* (June), pp. 66–77.

Kleiner, B.M. (2008). Macroergonomics: Work system analysis and design. *Human Factors, 50*(3), 461–467.

Knight, G. & Noyes, J. (1999). Children's behavior and the design of school furniture. *Ergonomics, 42,* 747–760.

Koenig, G., Tamres, M., & Mann, R.W. (1994). The biomechanics of the shoe-ground interaction in golf. In *Proceedings of the 1994 Scientific Congress of Golf* (pp. 40–50). London: E&FN Spon.

Koestler, A. (1967). *The ghost in the machine.* New York: Macmillan.

Kohler, I. (1951a). Über Aufbau und Wandlungen der Wahrnehmungswelt, insbesondere über 'bedingte Empfindungen.' *Oesterreichische Akademie Wissenschaften, 227*(1). (Translation available by G. Krauthamer).

Kohler, I. (1951b). Warum sehen wir aufrecht? *Die Pyramide, 2,* 30–33.

Kohler, I. (1953). Rehabituation in perception. *Die Pyramide, 5, 6, 7.* (Translation available by H. Gleitman, edited by J.J. Gibson).

Kohler, I. (1955). Experiments with prolonged optical distortion. *Acta Psychologie, 11,* 176–178.

Kohn, A. (1993). Why incentive plans cannot work. *Harvard Business Review, September–October*, 54–63.

Kohn, A. (2006). *The homework myth: Why our kids get too much of a bad thing*. Philadelphia: Da Capo Press.

Konz, S. (1995). *Work design. Industrial ergonomics* (4th edition). Scottsdale, AZ: Publishing Horizons.

Kors, A.C. (1998). *The birth of the modern mind: The intellectual history of the 17th and 18th centuries*. Chantilly, VA: The Great Courses.

Koskelo, R., Vuorikari, K., & Hänninen, O. (2007). Sitting and standing postures are corrected by adjustable furniture with lowered muscle tension in high-school students. *Ergonomics, 50*, 1643–1656.

Koumpilova, M. (2011). For teachers, home work. In *St. Paul Pioneer Press* (November 13), pp. 1B, 4B.

Koumpilova, M. (2012a). Schools see a tech revolution. Will students see results? In *St. Paul Pioneer Press* (August 12), pp. 1A, 10A.

Koumpilova, M. (2012b). A tablet in every backpack. In *St. Paul Pioneer Press* (November 2), p. 1A.

Kramer, S.N. (1963). *The Sumerians*. Chicago: University of Chicago Press.

Krause, T.R., Hidley, J.H., & Hodson, S.J. (1990). Broad-based changes in behavior key to improving safety culture. *Occupational Health and Safety, 59*(7), 31–37.

Krishen, K. (1988a). Robotic vision technology and algorithms for space applications. In *Proceedings of the 39th Congress of the International Astronautical Federation* (Paper Number IAF-88-028). Paris: International Astronautical Federation.

Krishen, K. (1988b). Space robotic vision system technology. In *Proceedings of ISA-88 International Conference* (Paper Number 88-1615, pp. 1527–1537). Research Triangle Park, NC: International Society of Automation.

Krznaric, R. (2013). Have we all been duped by the Myers-Briggs test? *Fortune* (May 15).

Kubicek, S. (2011). Epigenetics: A primer. *The Scientist, 25*(3), 32–33.

Kurzban, R. & Barrett, H.C. (2012). Origins of cumulative culture. *Science, 335*(6072), 1056–1057.

LaBar, G. (1990). What if your workers had the "right-to-act"? *Occupational Hazards, 52*(2), 49–53.

Lajoie, S.P. & Azevedo, R. (2006). Teaching and learning in technology rich environments. In P.A. Alexander & P.H. Winne (Eds.), *Handbook of educational psychology* (2nd edition) (Chapter 35, pp. 803–821). Mahwah, NJ: Lawrence Erlbaum Associates.

Landes, D.S. (1998). *The wealth and poverty of nations. Why some are so rich and some so poor*. New York: W.W. Norton.

Langolf, G.D., Chaffin, D.B., & Foulke, J.A. (1976). An investigation of Fitts' law using a wide range of movement amplitudes. *Journal of Motor Behavior, 8*(2), 113–128.

Larson, R.K., Deprez, V., & Yamakido, H. (2010). *The evolution of human language. Biolinguistic perspectives*. Cambridge: Cambridge University Press.

Lathan, C., Cleary, K., & Traynor, L. (2000). Human-centered design of a spine biopsy simulator and the effects of visual and force feedback on path-tracking performance. *Presence, 9*(4), 337–349.

Leaman, A. & Bordass, B. (1999). Productivity in buildings: The killer variables. *Building Research and Information, 27*, 4–19.

Leaman, A. & Bordass, B. (2006). Productivity in buildings: The "killer" variables. In D. Clements-Croome (Ed.), *Creating the productive workplace* (2nd edition) (Chapter 10, p. 153). London: E&FN Spon.

Lederman, J. (2012). Sorry, kids. Schools adding hours to calendar. In *Minneapolis Star Tribune* (December 3), p. A3.

LeDoux (1995). Emotion: Clues from the brain. *Annual Review of Psychology, 46*, 209–235.

Lee, B.S. (1950a). Some effects of side-tone delay. *Journal of the Acoustical Society of America, 22*, 639–640.

Lee, B.S. (1950b). Effects of delayed speech feedback. *Journal of the Acoustical Society of America, 22*, 824–826.

Lee, B.S. (1951). Artificial stutter. *Journal of Speech and Hearing Disorders, 16*, 53–55.

Lee, R.B. (2005). *Cambridge encyclopedia of hunters and gatherers*. Cambridge: Cambridge University Press.

Legg, S.J. (2007). Ergonomics in schools. Special issue of *Ergonomics*: Guest editorial. *Ergonomics, 50*(10), 1–6.

Legg, S.J. & Jacobs, K. (2008). Ergonomics for schools. *Work: A Journal of Prevention, Assessment & Rehabilitation, 31*, 489–493.

Leplat, J. & Rasmussen, J. (1984). Analysis of human errors in industrial incidents and accidents for improvement of work safety. *Accident Analysis and Prevention, 16*(2), 77–88.

Lepore, J. (2009). Not so fast. Scientific management started as a way to work. How did it become a way of life? In *The New Yorker* (October 12), pp. 114–122.

Levenson, R.W. & Gottman, J.M. (1983). Marital interaction: Physiological linkage and affective exchange. *Journal of Personality and Social Psychology, 45*(3), 587–597.

Levine, M. (1953). Tracking performance as a function of exponential delay between control and display. *U.S. Air Force Wright Air Development Center Technical Report Number 53–236*. Dayton, OH: Wright Air Force Base.

Lewis, P.H. (1988). Computers are getting mixed grades in school. In *Minneapolis Star Tribune* (August 10), pp. 1E, 14E.

Lexington (2009). The underworked American. In *The Economist* (June 13), p. 40.

Li, Q. (1998). *Effects of feedback control delays on team performance and coordination in a computer-mediated projective tracking task* (unpublished PhD thesis). Storrs, CT: University of Connecticut.

Lieberman, P. (2006). *Toward an evolutionary biology of language*. Cambridge, MA: Harvard University Press.

Lindsay, A.D. (Ed.) (1911). *A treatise of human nature by David Hume* (Vol. 1). London: J.M. Dent.

Livermore, D. (2013). *Customs of the world: Using cultural intelligence to adapt, wherever you are*. Chantilly, VA: The Great Courses.

Locke, J. (1632–1704) (1974). *An essay concerning human understanding* (abridged and edited by A.D. Woozley). New York: Meridian.

Loh, V., Andrews, S., Hasketh, B., & Griffin, B. (2013). The moderating effect of individual differences in error-management training: Who learns from mistakes? *Human Factors, 55*(2), 435–448.

Lotterman, E. (2012a). Culture plays role in wealth of nations. In *St. Paul Pioneer Press* (August 5), pp. 1D, 3D.

Lotterman, E. (2012b). Financial success is a lot like cabbage. In *St. Paul Pioneer Press* (October 14), p. 1D.

Lottridge, D., Chignell, M., & Jovicic, A. (2011). Affective interaction: Understanding, evaluating, and designing for human emotion. In P.R. DeLucia (Ed.), *Reviews of human factors and ergonomics, Volume 7* (Chapter 5, 197–237). Santa Monica, CA: Human Factors and Ergonomics Society.

Loveless, T. (2006). *The 2006 Brown Center report on American education: How well are American students learning? With special sections on the nation's achievement, the happiness factor in learning, and honesty in state test scores*. Washington, DC: The Brookings Institution. Downloaded from: http://www.brookings.edu/~/media/Files/rc/reports/2006/10education_loveless/10education_loveless.pdf.

Lowrey, A. (2012). Income gap could threaten growth. In *Minneapolis Star Tribune* (October 17), pp. D1, D3.

Lueder, R. (2010). Through the rearview mirror: Ergonomics for children. *Human Factors and Ergonomics Society Bulletin, 53*(10), 1–2.

Lueder, R. & Berg Rice, V.J. (Eds.) (2008). *Ergonomics for children. Designing products and places for toddlers to teens.* New York: Taylor and Francis.

Magan, C. (2013). Schools hope to enter "innovation zone." In *St. Paul Pioneer Press* (February 20), p. 3A.

Magan, C. & Webster, M. (2012). State's online math tests earn grade "A." Educators say the payoff has been dramatic hikes in student achievement. In *St. Paul Pioneer Press* (September 12), pp. 1A, 18A.

Magnus, G. (2011). What Karl Marx can tell us now. In *Minneapolis Star Tribune* (August 20), p. A11.

Malcolm, N. (2002). *Aspects of Hobbes.* Oxford: Clarendon Press.

Mandal, A.C. (1982). The correct height of school furniture. *Human Factors, 24*(3), 257–269.

Mangu-Ward, K. (2010). Let go of the little red schoolhouse. In *St. Paul Pioneer Press* (April 2), p. 8B.

Margulis, L. (2006). The phylogenetic tree topples. *American Scientist, 94*(3), 194.

Marler, P. (1970). Birdsong and speech development: Could there be parallels? *American Scientist, 58*(6), 669–673.

Marler, P. (1997). Three models of song learning: Evidence from behavior. *Journal of Neurobiology, 33*(5), 501–516.

Marohn, C. (2012). *Thoughts on building strong towns, Volume 1.* Brainerd, MN: Strong Towns.

Marshack, A. (1964). Lunar notation on Upper Paleolithic remains. *Science, 146,* 743–746.

Martin, G. & Williams, A. (2005). *Domesday book: A complete translation.* London: Penguin.

Martin, J. (2006). Social cultural perspectives in educational psychology. In P.A. Alexander & P.H. Winne (Eds.), *Handbook of educational psychology* (2nd edition) (Chapter 25, pp. 595–614). Mahwah, NJ: Lawrence Erlbaum Associates.

Martín-de-Castro, G., Delgado-Verde, M., López-Sáez, P., & Navas-López, J.E. (2011). Towards "an intellectual capital-based view of the firm": Origins and nature. *Journal of Business Ethics, 98,* 649–662.

Marx, K. (1867, 1990). *Capital* (Volume I) (translation by B. Fowkes & D. Fernbach of 1867 original, *Das Kapital, Kritik der politischen Ökonomie*). London: Penguin Books.

Marx, K. & Engels, F. (1848, 2013). *The communist manifesto (Das kommunistische Manifest).* Northbridge, MA: Swenson & Kemp.

Massimino, M.J. & Sheridan, T.B. (1994). Teleoperator performance with varying force and visual feedback. *Human Factors, 36*(1), 145–157.

Matheson, K. (2013). Is this the "future of education"? In *Minneapolis Star Tribune* (February 24), p. A10.

Matthews, R.A., Gallus, J.A., & Henning, R.A. (2011). Participatory ergonomics: Development of an employee assessment questionnaire. *Accident Analysis & Prevention, 43*(1), 360–369.

Maurino, D.E. (2000). Human factors and aviation safety: What the industry has, what the industry needs. *Ergonomics, 43*(7), 952–959.

Maxwell, L.E. & Evans, G.W. (2000). The effects of noise on pre-school children's pre-reading skills. *Journal of Environmental Psychology, 20,* 91–97.

May, R.M. (1976). Simple mathematical models with very complicated dynamics. *Nature, 261*(5560), 459–467.

Mayer, S.E. (1997). *What money can't buy: Family income and children's life chances.* Cambridge, MA: Harvard University Press.

Mayer-Schonberger, V. & Cukier, K. (2013). *Big data: A revolution that will transform how we live, work, and think.* New York: Houghton Mifflin Harcourt.

Mazzucato, M. (2013). *The entrepreneurial state: Debunking public vs. private sector myths.* London: Anthem Press.

McAfee, R.B. & Winn, A.R. (1989). The use of incentives/feedback to enhance work place safety: A critique of the literature. *Journal of Safety Research, 20*(1), 7–19.

McCaslin, M. & Good, T.L. (1996). The informal curriculum. In D.C. Berliner and R.C. Calfee (Eds.), *Handbook of educational psychology* (pp. 622–672). New York: Simon & Schuster Macmillan.

McCloskey, D.N. (2010). *Bourgeois dignity: Why economics can't explain the modern world.* Chicago: University of Chicago Press.

McCrae, R.R. & Costa, P.T. Jr. (1989). Reinterpreting the Myers-Briggs type indicator from the perspective of the five-factor model of personality. *Journal of Personality, 57*(1), 17–40.

McCullough, D. (1992). *Truman.* New York: Simon & Schuster.

McDermid, C. & Smith, K.U. (1964). Compensatory reaction to angularly displaced visual feedback in behavior. *Journal of Applied Psychology, 48*(1), 63–68.

McGregor, D. (1960). *The human side of enterprise.* New York: McGraw-Hill.

McKechnie, J.L. (Ed.) (1983). *Webster's new universal unabridged dictionary.* New York: Simon & Schuster.

McKibben, B. (1989). *The end of nature.* New York: Random House.

Meadows, D.H., Meadows, D.L., Randers, J., & Behrens, W.W. III (1972). *The limits to growth: A report for the Club of Rome's project on the predicament of mankind.* New York: Universe Books.

Meister, D. (1982). Reduction of human error. In G. Salvendy (Ed.), *Handbook of industrial engineering* (Chapter 6.2, pp. 6.2.1–6.2.9). New York: Wiley.

Meister, D. (1984). Human reliability. In F.A. Muckler (Ed.), *Human factors review: 1984* (Chapter 2, pp. 13–53). Santa Monica, CA: Human Factors Society.

Meister, D. (1989). *Conceptual aspects of human factors.* Baltimore: Johns Hopkins University Press.

Meister, D. (1994). Behavioral design of automated systems. In *Proceedings of the Human Factors and Ergonomics Society 38th Annual Meeting* (pp. 521–525). Santa Monica, CA: Human Factors and Ergonomics Society.

Meister, D. (1995). *Divergent viewpoints: Essays on human factors questions.* Self-published (personal communication).

Meltz, B.F. (1998). Studies show computer use may hinder development. In *Minneapolis Star Tribune* (October 11), pp. E1, E8.

Menand, L. (2012). Today's assignment. In *The New Yorker* (December 17, Volume LXXXVIII, Number 40), pp. 25–26.

Merken, R.S. (1986). Human factors and human nature: Is psychological theory really necessary? *Human Factors Society Bulletin, 29*(9), 1–2.

Mervis, J. (2009). Study questions value of school software for students. *Science, 323,* 79.

Miall, R.C. & Reckess, G.Z. (2002). The cerebellum and the timing of coordinated eye and hand tracking. *Brain and Cognition, 48,* 212–226.

Miall, R.C., Weir, D.J., Wolpert, D.M., & Stein, J.F. (1993). Is the cerebellum a Smith predictor? *Journal of Motor Behavior, 25*(3), 203–216.

Michaels, C. & Beek, P. (1995). The state of ecological psychology. *Ecological Psychology, 7*(4), 259–278.

Miles, W.R. (1936). *Psychological studies of human variability.* Princeton, NJ: Psychological Review Co.

Miller, D.P. & Swain, A.D. (1987). Human error and human reliability. In G. Salvendy (Ed.), *Handbook of human factors* (Chapter 2.8, pp. 219–250). New York: Wiley.

Miller, J.M. (1982). The management of occupational safety. In G. Salvendy (Ed.), *Handbook of industrial engineering* (Chapter 6.14, pp. 6.14.1–6.14.18). New York: Wiley.

Minnesota Department of Education (2006, 2009, 2012). *Minnesota 2006, 2009 and 2012 Grade 8 comprehensive math and reading assessment test results.* St. Paul, MN: Minnesota Department of Education. Downloaded from: http://education.state.mn.us/mde/index.html.

Miranda, M. (2009). Yoga bringing calm to class. In *St. Paul Pioneer Press* (March 13), pp. 1B, 10B.

Mitchell, C. (2011a). "L" is for language learners. In *Minneapolis Star Tribune* (February 12), pp. B1, B5.

Mitchell, C. (2011b). Mobility hurts students, test scores. In *Minneapolis Star Tribune* (April 27), pp. B1, B2.

Miyake, A., Kost-Smith, L.E., Finkelstein, N.D., Pollock, S.J., Cohen, G.L., & Ito, T.A. (2010). Reducing the gender achievement gap in college science: A classroom study of values affirmation. *Science, 330*, 1234–1237.

Montgomery, W.E. (1956). Machine guarding. *The Journal of the American Society of Safety Engineers* (November), 57–60.

Moore, D. (2003). *The dependent gene: The fallacy of "nature vs. nurture."* New York: Holt.

Moore, D. & Barnard, T. (2012). With eloquence and humanity? Human factors/ergonomics in sustainable human development. *Human Factors, 54*(6), 940–951.

Moore, K.S., Gomer, J.A., Butler, S.N., & Pagano, C.C. (2007). Perception of robot passability and aperture width during direct line of sight and teleoperation conditions. *Proceedings of the Human Factors and Ergonomics Society 51st Annual Meeting* (pp. 1076–1080). Santa Monica, CA: Human Factors and Ergonomics Society.

Moran, S. (2009). Acing the math test. In *Minneapolis Star Tribune* (November 30), pp. E1, E8.

Moray, N. (1994). "De maximis non curat lex"; or How context reduces science to art in the practice of human factors. In *Proceedings of the Human Factors and Ergonomics Society 38th Annual Meeting* (pp. 526–530). Santa Monica, CA: Human Factors and Ergonomics Society.

Moray, N. (2008). The good, the bad, and the future: On the archaeology of ergonomics. *Human Factors, 50*(3), 411–417.

Moretz, S. (1988). Incentives provide a competitive edge. *Occupational Hazards, 50*(3), 31–35.

Morris, K.V. (2012). Lamarck and the missing lnc. *The Scientist, 26*(10), 29–33.

Morrow, D., North, R., & Wickens, C.D. (2005). Reducing and mitigating human error in medicine. In P.R. DeLucia (Ed.), *Reviews of human factors and ergonomics, Volume 1* (Chapter 6, pp. 254–296). Santa Monica, CA: Human Factors and Ergonomics Society.

Mosher, R.S. (1964). Industrial manipulators. *Scientific American, 211*(4), 88–98.

Mosher, R.S. & Murphy, W. (1965). Human control factors in walking machines. In *Proceedings of the American Society of Mechanical Engineers Human Factors Conference*. New York: ASME.

Murphy, K.R. (Ed.) (1996a). *Individual differences and behavior in organizations*. San Francisco: Jossey-Bass.

Murphy, K.R. (1996b). Individual differences and behavior in organizations: Much more than g. In K.R. Murphy (Ed.), *Individual differences and behavior in organizations*. San Francisco: Jossey-Bass.

Myers, I.B. & Myers, P.B. (1980). *Gifts differing. Understanding personality type*. Palo Alto, CA: Consulting Psychologists Press.

Mykerezi, E. & Temple, J. (2009). *Food insecurity and school performance of U.S. children through eighth grade*. Minneapolis, MN: Presentation to University of Minnesota Humphrey School of Public Affairs, November 17.

Nair, P. & Fielding, R. (2005). *The language of school design: Design patterns for 21st century schools*. Minneapolis, MN: Design-Share.com—The International Forum for Innovative Schools. Downloaded from: http://DesignShare.com/Patterns.

NASA Advanced Technology Advisory Committee (1985). *Advancing automation and robotics technology for the U.S. economy, Volumes I and II* (NASA TM -87566). Washington, DC: National Aeronautics and Space Administration.

Nasar, S. (2011). *Grand pursuit: The story of economic genius*. New York: Simon & Schuster.

National Commission on Teaching and America's Future (1996). *What matters most: Teaching for America's future.* New York: National Commission on Teaching and America's Future. Downloaded from: http://www.tc.columbia.edu/~teachcomm.

Neumann, O. & Sanders, A.F. (1996). *Handbook of perception and action* (in 3 volumes). New York: Academic Press.

Newell, A. & Rosenbloom, P.S. (1981). Mechanisms of skill acquisition and the law of practice. In J.R. Anderson (Ed.), *Cognitive skills and their acquisition* (pp. 1–55). Hillsdale, NJ: Erlbaum.

Newell, K.M. (1986). Constraints on the development of coordination. In M.G. Wade & H.T.A. Whiting (Eds.), *Motor development in children: Aspects of coordination and control* (pp. 341–360). Dordrecht, Netherlands: Martinus Nijhoff.

Newell, K.M. (2003). Schema theory (1975). Retrospectives and prospectives. *Research Quarterly for Exercise and Sport, 74*(4), 383–388.

Newell, K.M. & Corcos, D.M. (Eds.) (1993a). *Variability and motor control.* Champaign, IL: Human Kinetics.

Newell, K.M. & Corcos, D.M. (1993b). Issues in variability and motor control. In K.M. Newell & D.M. Corcos (Eds.), *Variability and motor control* (pp. 1–12). Champaign, IL: Human Kinetics.

Newell, K.M. & Slifkin, A.B. (1998). The nature of movement variability. In J.P. Piek (Ed.), *Motor behavior and human skill: A multidisciplinary approach* (pp. 143–160). Champaign IL: Human Kinetics.

Newman, L. & Carayon, P. (1994). Community ergonomics: Data collection methods and analysis of human characteristics. In *Proceedings of the Human Factors and Ergonomics Society 38th Annual Meeting* (pp. 739–743). Santa Monica, CA: Human Factors and Ergonomics Society.

Newton, C. (2013). Apple turns a new page. Firm offers education market tools to inexpensively build digital textbooks for iPads. In *St. Paul Pioneer Press* (January 20), p. 6A.

Nikolic, M.I. & Sarter, N.B. (2003). Converging on error management: A review of current findings and future needs. In *Proceedings of the Human Factors Society 47th Annual Meeting* (pp. 513–517). Santa Monica, CA: Human Factors and Ergonomics Society.

Nikolic, M.I. & Sarter, N.B. (2004). Error management on modern flight decks: How pilots explain and recover from unintended actions and outcomes. In *Proceedings of the Human Factors Society 48th Annual Meeting* (pp. 330–334). Santa Monica, CA: Human Factors and Ergonomics Society.

Nof, S.Y. (1985). Robot ergonomics: Optimizing robot work. In S.Y. Nof (Ed.), *Handbook of industrial robotics* (pp. 549–604). New York: Wiley.

Norman, D.A. (1981). Categorization of action slips. *Psychological Review, 88*(1), 1–15.

Norman, D.A. & Spohrer, J.C. (1996). Learner-centered education. *Communications of the ACM, 39*, 24–27.

Norris, F. (2013). Of market milestones and index millstones. In *St. Paul Pioneer Press* (November 23), p. 8A.

O'Conner, D. (1996). Spending isn't key to success. In *St. Paul Pioneer Press* (June 13), pp. 1A, 10A.

O'Donnell, A.M. (2006). The role of peers and group learning. In P.A. Alexander & P.H. Winne (Eds.), *Handbook of educational psychology* (2nd edition) (Chapter 34, pp. 781–802). Mahwah, NJ: Lawrence Erlbaum Associates.

Oakley, K.P. (1957). *Man the tool-maker.* Chicago: University of Chicago Press.

Occupational Safety and Health Reporter (1976). Washington, DC: Bureau of National Affairs.

Ojeda-Zapata, J. (2009). That teaching touch. Apple's iPod technology is helping to put learning at the fingertips of students in classrooms across the Twin Cities and the country. In *St. Paul Pioneer Press* (December 22), pp. 1A, 8A.

Olson, J. (2011). Divorce also hard on test scores. In *Minneapolis Star Tribune* (June 2), pp. A1, A7.

Orfield, S.J., Brand, J.L., & Hakkarainen, P. (2006). *Better lighting and daylighting solutions. Improving visual quality in office environments*. Washington, DC: American Society of Interior Designers.

Organization for Economic Cooperation and Development (2010). *PISA 2009 results: Executive summary*. Paris, France: OECD. Downloaded from: http://www.oecd.org/dataoecd/ 34/60/46619703.pdf.

Oslund, J. (2013). What does it take to make a top workplace? In *Minneapolis Star Tribune* (June 16, Top Workplaces section), p. 3.

Ottensmeyer, M.P., Hu, J., Thompson, J.M., Ren, J., & Sheridan, T.B. (2000). Investigations into performance of minimally invasive telesurgery with feedback delays. *Presence, 9*(4), 369–382.

Otto, M.W. & Smits, J.A.J. (2011). *Exercise for mood and anxiety. Proven strategies for overcoming depression and enhancing well-being*. New York: Oxford University Press.

Owens, J.A., Belon, K., & Moss, P. (2010). Impact of delaying school start time on adolescent sleep, mood, and behavior. *Archives of Pediatric and Adolescent Medicine, 164*, 608–614.

Page, L.C., Schimmenti, J., Bernstein, L., & Horst, L. (2002). *National evaluation of smaller learning communities. Literature Review*. Cambridge, MA: Abt Associates Inc. Downloaded from: http://www.abtassociates.com/reports/SMALLER.pdf.

Paris, S.G. & Cunningham, A.E. (1996). Children becoming students. In D.C. Berliner & R.C. Calfee (Eds.), *Handbook of educational psychology* (pp. 117–147). New York: Simon & Schuster Macmillan.

Park, K.S. (1987). *Human reliability. Analysis, prediction, and prevention of human errors*. Amsterdam: Elsevier.

Parks, G. (2000). The High/Scope Perry preschool project. In *OJJDP Juvenile Justice Bulletin* (October), pp. 1–8. Washington, DC: Office of Juvenile Justice and Delinquency Prevention, Office of Justice Programs, U.S. Department of Justice. Downloaded from: http://www.ncjrs.gov/pdffiles1/ojjdp/181725.pdf.

Patterson, G.A. (2009a). Student PCs: Worth the cost? In *Minneapolis Star Tribune* (March 18), pp. B1, B8.

Patterson, G.A. (2009b). A longer school day, a smarter kid? In *Minneapolis Star Tribune* (May 11), pp. A1, A5.

Pazuchanics, S.L. (2006). The effects of camera perspective and field of view on performance in teleoperated navigation. In *Proceedings of the Human Factors and Ergonomics Society 50th Annual Meeting* (pp. 1528–1532). Santa Monica, CA: Human Factors and Ergonomics Society.

Peach, R.W. (Ed.) (1994). *The ISO 9000 Handbook* (2nd edition). Fairfax, VA: Irwin Professional.

Peacock, B. (2002). Laws and rules. Murphy's law: If it can happen, it will. *Ergonomics in Design, 10*(Winter), 4–5, 31.

Pearson, A.R., West, T.V., Dovidio, J.F., Powers, S.R., Buck, R., & Henning, R.A. (2008). The fragility of intergroup relations: Divergent effects of delayed audio-visual feedback in intergroup and intragroup interaction. *Psychological Science, 19*(12), 1272–1279.

Pentland, A. (2013). The data-driven society. *Scientific American, 309*(4), 78–83.

Perl, R.E. (1934). An application of Thurstone's method of factor analysis to practice series. *Journal of General Psychology, 11*, 209–212.

Perrow, C. (1999). *Normal accidents. Living with high-risk technologies*. Princeton, NJ: Princeton University Press.

Peters, G.A. & Peters, B.J. (2006). *Human error. Causes and control*. Boca Raton, FL: CRC Press.

Peters, G.A. & Peters, B.J. (2013). *Human error* (Volumes 1 & 2). Santa Monica, CA: Peters & Peters.

Peters, T.J. & Waterman, R.H. Jr. (2004). *In search of excellence. Lessons from America's best-run companies.* New York: Harper Business Essentials.

Petersen, D. (1971). *Techniques of safety management.* New York: McGraw-Hill.

Pfitzinger, J. (2010). On the move (and calm as a result). In *Minneapolis Star Tribune* (October 17), p. E2.

Pflüger, E. (1875). Beiträge zur Lehre von der Respiration. I. Über die physiologische Verbrennung in den lebendigen Organismen. *Archiv für die Gesamte Physiologie des Menschen und der Tiere, 10*, 251–367.

Phillipson, N. (2010). *Adam Smith: An enlightened life.* New Haven, CT: Yale University Press.

Picciano, A.G. & Seaman, J. (2007). *K-12 online learning: A survey of U.S. school district administrators.* Newburyport, MA: The Sloan Consortium.

Plutchik, R. (2001). The nature of emotions. *American Scientist, 89*(4), 344–350.

Poffenburger, A.T. (1915). The influence of improvement in one mental process upon other related processes. *Journal of Educational Psychology, 6*, 459–474.

Poincaré, H. (1900). The theory of Lorentz and the principle of reaction. *Archives néerlandaises des Sciences Exactes et Naturelles* (Series 2), *5*, 252–278.

Pope, W.C. (1990). *Managing for performance perfection. The changing emphasis.* Weaverville, NC: Bonnie Brae Publications.

Poulton, E.C. (1966). Engineering psychology. *Annual Review of Psychology, 17*, 177–200.

Powers, S.R., Rauh, C., Buck, R., Henning, R.A., & West, T.V. (2011). The effect of video feedback delay on frustration and emotion communication accuracy. *Computers in Human Behavior, 27*, 1651–1657.

Preatoni, E., Hamill, J., Harrison, A.J., Hayes, K., van Emmerik, R.E.A., Wilson, C., & Rodano, R. (2013). Movement variability and skills monitoring in sports. *Sports Biomechanics, 12*(2), 69–92.

Probasco, P.D. (1969). *Social feedback factors in vocal shadowing* (unpublished MSc thesis). Madison, WI: University of Wisconsin-Madison.

Pulat, B.M. (1997). *Fundamentals of industrial ergonomics* (2nd edition). Prospect Heights, IL: Waveland Press.

Rafferty, A. & Kirschner, J. (2013). Cars, drivers plunge into river after Washington I-5 bridge collapse. *NBC News* (May 24).

Rahimi, M. (1987). Human factors engineering and safety in robotics and automation. *Human Factors Society Bulletin, 30*(7), 3–5.

Raibert, M.H. (1987). Machines that run. In *Robotics and Motor Control: Commonalities and Differences* (June 3, Conference Paper Number 5). Burnaby, British Columbia, Canada: Simon Fraser University.

Raley, C., Stripling, R., Kruse, A., Schmorrow, D., & Patrey, J. (2004). Augmented cognition overview: Improving information intake under stress. In *Proceedings of the 48th Annual Meeting of the Human Factors & Ergonomics Society* (pp. 1150–1154). Santa Monica, CA: Human Factors and Ergonomics Society.

Ramaprasad, A. (1983). On the definition of feedback. *Behavioral Science, 28*(1), 4–13.

Ramirez, G. & Beilock, S.L. (2011). Writing about testing worries boosts exam performance in the classroom. *Science, 331*, 211–213.

Rasmussen, J. (1982). Human errors. A taxonomy for describing human malfunction in industrial installations. *Journal of Occupational Accidents, 4*, 311–333.

Rasmussen, J. (1990). The role of error in organizing behaviour. *Ergonomics, 33*(10/11), 1185–1199.

Rawnsley, A.I. & Harris, J.D. (1954). Comparative analysis of normal speech and speech with delayed side-tone by means of sound spectrograms. *U.S. Navy Medical Research Laboratory Report Number 248.* Washington, DC: U.S. Navy Medical Center.

Ray, P.S., Bishop, P.A., & Wang, M.Q. (1997). Efficacy of the components of a behavioral safety program. *International Journal of Industrial Ergonomics, 19*, 19–29.

Raynor, M. & Ahmed, M. (2013). *The three rules: How exceptional companies think.* New York: Portfolio/Penguin.

Reason, J. (1990). *Human error.* Cambridge: Cambridge University Press.

Reason, J. (1997). *Managing the risks of organizational accidents.* Brookfield, VT: Ashgate.

Reason, J. (2000). Human error: Models and management. *British Medical Journal, 320*, 768–770.

Reason, J. (2013). *A life in error. From little slips to big disasters.* Williston, VT: Ashgate.

Reason, J.T. & Hobbs, A. (2003). *Managing maintenance error: A practical guide.* Aldershot, UK: Ashgate.

Reid, R. (1987). Workers turn on to safety training. *Occupational Hazards, 49*(6), 43–47.

Rhee, M. (2010). What I've learned. In *Newsweek* (December 13), pp. 36–41.

Richardson, V.L. (1973). *Hazard surveys at select employers.* Madison, WI: Wisconsin Department of Industry, Labor and Human Relations.

Ridley, M. (2003). *Nature via nurture. Genes, experience, and what makes us human.* New York: HarperCollins.

Riley, J.M. & Kaber, D.B. (1999). The effects of visual display type and navigational aid on performance, presence, and workload in virtual reality training of telerover navigation. In *Proceedings of the Human Factors and Ergonomics Society 43rd Annual Meeting* (pp. 1251–1255). Santa Monica, CA: Human Factors and Ergonomics Society.

Riley, K.A. (2007). *Surviving and thriving as an urban leader. Reflective and analytical tools for leaders of our city schools.* London: Institute of Education, London Centre for Leadership in Learning. Downloaded from: http://www.leru.org.uk/publications_and_resources/Publication_LEADERSHIP_Surviving_and_Thriving_as_an_Urban_Leader_%28Riley_2007%29.

Riley, K.A. (2009). Taking your leadership pulse. *Developing School Leaders, 67*(2).

Riley, M.A. & Turvey, M.T. (2002).Variability and determinism in motor behavior. *Journal of Motor Behavior, 34*(2), 99–125.

Riley, M.W. & Bishu, R.R. (1997). Quality control principles and their relationship to ergonomics. In P. Seppälä, T. Luopajärvi, C.-H. Nygård, & M. Mattila (Eds.), *Proceedings of the 13th Triennial Congress of the International Ergonomics Association, Tampere, Finland. 1997* (Vol. 1, pp. 250–252). Helsinki, Finland: Finnish Institute of Occupational Health.

Ringen, S. (2013). *Nation of devils: Democratic leadership and the problem of obedience.* New Haven, CT: Yale University Press.

Ripley, A. (2010). A call to action for public schools. In *Time* (September 20), pp. 32–42.

Robertson, M., Henning, R.A., Warren, N., Nobrega, S, Dove-Steinkamp, M., Tibirica, L., Bizarro, A., & the CPH-NEW Research Team (2013). The intervention design and analysis scorecard: A planning tool for participatory design of integrated health and safety interventions in the workplace. *Journal of Occupational and Environmental Medicine, 55*, S86–S88.

Rodriguez, P. (2010). *Why economies rise or fall.* Chantilly, VA: The Great Courses.

Rogers, R., Sewell, K.W., Harrison, K.S., & Jordan, M.J. (2006). The MMPI-2 restructured clinical scales: A paradigmatic shift in scale development. *Journal of Personality Assessment, 87*(2), 139–147.

Roscoe, S.N., Hasler, S.G., & Dougherty, D.J. (1966). Flight by periscope: Making takeoffs and landings; the influence of magnification, practice, and various conditions of flight. *Human Factors, 8*, 13–40.

Rosenblueth, A., Wiener, N., & Bigelow, J. (1943). Behavior, purpose and teleology. *Philosophy of Science, 10*, 18–24.

Rosin, H. (2013). The touch-screen generation. In *The Atlantic* (April, Volume 311, Number 3), pp. 57–65.

Rothe, M. (1973). *Social tracking in children as a function of age* (unpublished MSc thesis). Madison, WI: University of Wisconsin-Madison.

Rothschild, E. (2001). *Economic sentiments: Adam Smith, Condorcet, and the enlightenment.* Cambridge, MA: President and Fellows of Harvard College.

Rubow, R. & Smith, K.U. (1971). Feedback parameters of electromyographic learning. *American Journal of Physical Medicine, 50*, 115–131.

Russell, J.A. (1980). A circumplex model of affect. *Journal of Personality and Social Psychology, 39*(6), 1161–1178.

Rutter, M. (2006). *Genes and behavior. Nature-nurture interplay explained.* Malden, MA: Blackwell.

Salas, E., Wilson, K.A., Burke, C.S., & Wightman, D.C. (2006). Does crew resource management training work? An update, an extension, and some critical needs. *Human Factors, 48*(2), 392–412.

Salminen, S. & Tallberg, T. (1996). Human errors in fatal and serious occupational accidents in Finland. *Ergonomics, 39*(7), 980–988.

Salvendy, G. (1985). Human factors in planning robotic systems. In S.Y. Nof (Ed.), *Handbook of industrial robots* (pp. 639–664). New York: Wiley.

Salvendy, G. (Ed.) (1982). *Handbook of industrial engineering.* New York: Wiley.

Salvendy, G. (Ed.) (1987, 2006, 2012). *Handbook of human factors and ergonomics* (1st edition, 3rd edition, 4th edition). New York: Wiley.

Samuelson, P. (1976). Illogic of neo-Marxian doctrine of unequal exchange. In D.A. Belsey, E.J. Kane, P.A. Samuelson, & R.M. Solow (Eds.), *Inflation, trade and taxes: Essays in honor of Alice Bourneuf* (pp. 96–107). Columbus, OH: Ohio State University Press.

Sanders, M.S. & McCormick, E.J. (1993). *Human factors in engineering and design* (7th edition). New York: McGraw-Hill.

Sandler, W. & Lillo-Martin, D. (2006). *Sign language and linguistic universals.* Cambridge: Cambridge University Press.

Sarter, N.B. & Woods, D.D. (1991). Situation awareness: A critical but ill-defined phenomenon. *International Journal of Aviation Psychology, 1*, 45–57.

Sarter, N.B. & Woods, D.D. (1995). How in the world did we ever get into that mode? Mode error and awareness in supervisory control. *Human Factors, 37*(1), 5–19.

Saulny, S. (2009). They stand when called upon, and when not. In *The New York Times* (February 25), pp. A1, A15.

Sauter, S. (1971). *Psychophysiological feedback components of social tracking* (unpublished MSc thesis). Madison, WI: University of Wisconsin-Madison.

Sauter, S. & Smith, K.U. (1971). Social feedback: Quantitative division of labor in social interactions. *Journal of Cybernetics, 1*(2), 80–93.

Schafer, L. (2014). What makes a region an incubator of prosperity? In *Minneapolis Star Tribune* (January 29), pp. D1, D3.

Schmidt, R.A. (1975). A schema theory of discrete motor learning. *Psychological Review, 82*, 225–260.

Schmidt, R.A. (1988). *Motor control and learning. A behavioral emphasis* (2nd edition). Champaign, IL: Human Kinetics.

Schmidt, R.A. (2003). Motor schema theory after 27 years: Reflections and implications for a new theory. *Research Quarterly for Exercise and Sport, 74*(4), 366–375.

Schmidt, R.A. (2011). Jack Adams, a giant of motor behavior, has died. *Journal of Motor Behavior, 43*(1), 83–84.

Schmidt, R.A. & Lee, T.D. (1999). *Motor control and learning. A behavioral emphasis* (3rd edition). Champaign, IL: Human Kinetics.

Schmidt, R.A. & Wrisberg, C.A. (2014). *Motor learning and performance: From principles to application* (5th edition). Champaign, IL: Human Kinetics.

Schneider, M. (2002). *Do school facilities affect academic outcomes?* Washington, DC: National Clearinghouse for Educational Facilities. Downloaded from: http://www.edfacilities.org/pubs/outcomes.pdf.

Schneider, W. (1985). Training high performance skills: Fallacies and guidelines. *Human Factors, 27*, 285–300.

Schumpeter (2009). The three habits of highly irritating management gurus. In *The Economist* (October 24), p. 78.

Schumpeter (2012). Simplify and repeat. The best way to deal with growing complexity may be to keep things simple. In *The Economist* (April 28), p. 76.

Schumpeter (2013a). Companies' moral compasses. Some ideas for restoring faith in firms. In *The Economist* (March 2), p. 66.

Schumpeter (2013b). The transience of power. In *The Economist* (March 16), p. 70.

Schumpeter (2013c). The age of smart machines. In *The Economist* (May 25), p. 70.

Schumpeter (2013d). Too much of a good thing. Leaders need to learn to beware of their strengths. In *The Economist* (June 8), p. 72.

Schumpeter (2013e). Cronies and capitols. In *The Economist* (August 10), p. 59.

Schumpeter (2013f). In praise of laziness. Businesspeople would be better off if they did less and thought more. In *The Economist* (August 17), p. 58.

Schumpeter (2013g). The entrepreneurial state. In *The Economist* (August 31), p. 59.

Schumpeter (2013h). It's complicated. In *The Economist* (November 23), p. 68.

Schumpeter (2014). Measuring management. In *The Economist* (January 18), p. 69.

Schunk, D.H. & Zimmerman, B.J. (2006). Competence and control beliefs: Distinguishing the means and the ends. In P.A. Alexander & P.H. Winne (Eds.), *Handbook of educational psychology* (2nd edition) (Chapter 16, pp. 349–367). Mahwah, NJ: Lawrence Erlbaum Associates.

Schuster, D.H. & Guilford, J.P. (1964). The psychometric prediction of problem drivers. *Human Factors, 6*(4), 393–421.

Seielstad, G. (2013). *Dawn of the Anthropocene-humanity's defining moment.* Alexandria, VA: American Geosciences Institute.

Semuels, A. (2013). As employers push efficiency, the daily grind wears down workers. In *St. Paul Pioneer Press* (April 14), p. 5D.

Shane, S. (2010). *Born entrepreneurs, born leaders. How your genes affect your work life.* Oxford: Oxford University Press.

Shappell, S.A. (2011). Managing human error in complex systems. *Human Factors and Ergonomics Society webinar* (October 20). Santa Monica, CA: Human Factors and Ergonomics Society.

Sharit, J. (1999). Human and system reliability analysis. In W. Karwowski & W.S. Marras (Eds.), *The occupational ergonomics handbook* (pp. 601–642). Boca Raton, FL: CRC Press.

Sharit, J. (2006). Human error. In G. Salvendy (Ed.), *Handbook of human factors and ergonomics* (3rd edition) (Chapter 27, pp. 708–760). New York: Wiley.

Sheridan, T.B. (1989). Telerobotics. *Automatica, 25*(4), 487–507.

Sheridan, T.B. (1992a). Musings on telepresence and virtual presence. *Presence: Teleoperators and Virtual Environments, 1*, 120–125.

Sheridan, T.B. (1992b). *Telerobotics, automation, and human supervisory control.* Cambridge, MA: The MIT Press.

Sheridan, T.B. (1999). Descartes, Heidegger, Gibson, and God: Toward an eclectic ontology of presence. *Presence: Teleoperators and Virtual Environments, 8*(5), 551–559.

Sheridan, T.B. (2004). Driver distraction from a control theory perspective. *Human Factors, 46*(4), 587–599.

Sheridan, T.B. (2008). Risk, human error, and system resilience: Fundamental ideas. *Human Factors, 50*(3), 418–426.

Shermer, M. (2013). Is God dying? *Scientific American, 309*(6), 82.

Siegele, L. (2010). It's a smart world. In *The Economist* (November 6, special report on smart systems), pp. 1–18.

Siipola, E.M. (1935). Studies in mirror drawing. *Psychological Monographs, 46*(6), 66–77.

Silberman, C.E. (1970). *Crisis in the classroom. The remaking of American education.* New York: Random House.

Silver, N. (2012). *The signal and the noise. Why so many predictions fail—But some don't.* New York: Penguin Press.

Simon, C.W. (1987). Will egg-sucking ever become a science? *Human Factors Society Bulletin, 3*(6), 1–4.

Skinner, B.F. (1953). *Science and human behavior.* New York: Free Press.

Smetanka, M.K. & Hotakainen, R. (1996). Thousands fail math, reading. In *Minneapolis Star Tribune* (May 29), pp. A1, A8.

Smith, A. (1776, 1977). *An inquiry into the nature and causes of the wealth of nations.* Chicago: University of Chicago Press.

Smith, A. (1759, 2010). *The theory of moral sentiments.* London: Penguin.

Smith, D. (2009). The economist and the kids. In *Minneapolis Star Tribune* (December 18), p. OP1.

Smith, J.H. & Smith, M.J. (1994). Community ergonomics: An emerging theory and engineering practice. In *Proceedings of the Human Factors and Ergonomics Society 38th Annual Meeting* (pp. 729–733). Santa Monica, CA: Human Factors and Ergonomics Society.

Smith, J.H., Cohen, W J., Conway, F.T., Carayon, P., Bayeh, A.D., & Smith, M.J. (2003). Community ergonomics. In H.W. Hendrick & B.M. Kleiner (Eds.), *Macroergonomics: Theory, methods and applications* (pp. 67–96). Mahwah, NJ: Lawrence Erlbaum.

Smith, K.U. (1945). *Behavioral systems analysis of aircraft gun systems. Special report.* Washington, DC: U.S. Air Force Air Materials Command.

Smith, K.U. (1962a). *Delayed sensory feedback and behavior.* Philadelphia: Saunders.

Smith, K.U. (1962b). *Work theory and economic behavior* (Indiana Business Paper No. 5). Bloomington, IN: Indiana University Foundation for Economic and Business Studies.

Smith, K.U. (1965a). *Behavior organization and work: A new approach to industrial behavioral science* (revised). Madison, WI: College Printing & Typing.

Smith, K.U. (1965b). *Human factors analysis of the mechanism of walking with special reference to design of a pedipulator vehicle* (Technical Report). Madison, WI: University of Wisconsin Behavioral Cybernetics Laboratory.

Smith, K.U. (1965c). *Sensory-feedback analysis of manlike machine systems: A new behavioral theory for human engineering* (Technical Report). Madison, WI: University of Wisconsin Behavioral Cybernetics Laboratory.

Smith, K.U. (1966). Cybernetic theory of time perception and its evolution. In *Proceedings of the 18th International Congress of Psychology. Symposium on Perception of Space and Time* (pp. 152–159). Moscow: University of Moscow.

Smith, K.U. (1970). Inversion and delay of the retinal image: Feedback systems analysis of a classical problem of space perception. *American Journal of Optometry and Physiological optics, 47,* 175–204.

Smith, K.U. (1972). Cybernetic psychology. In R.N. Singer (Ed.), *The psychomotor domain* (pp. 285–348). New York: Lea and Febiger.

Smith, K.U. (1973). Physiological and sensory feedback of the motor system: Neural metabolic integration for energy regulation in behavior. In J. Maser (Ed.), *Efferent organization and integration of behavior* (pp. 19–66), New York: Academic Press.

Smith, K.U. (1974). *Industrial social cybernetics.* Madison, WI: University of Wisconsin Behavioral Cybernetics Laboratory.

Smith, K.U. (1975a). *Hazard management: Behavioral practices in risk management, industrial safety and workers' compensation.* Madison, WI: Occupational Safety and Health Research Unit, Wisconsin Department of Industry, Labor and Human Relations.

Smith, K.U. (1975b). *Social and human factors design of safety and health programming of risk management of workers' compensation.* Madison, WI: Occupational Safety and Health Research Unit, Wisconsin Department of Industry, Labor and Human Relations.

Smith, K.U. (1979). *Human-factors and systems principles for occupational safety and health.* Cincinnati, OH: NIOSH, Division of Training and Manpower Development.

Smith, K.U. (1987). Origins of human factors science. *Human Factors Society Bulletin, 30*(4), 1–3.

Smith, K.U. (1988). Human factors in hazard control. In P. Rentos (Ed.), *Evaluation and control of the occupational environment* (pp. 1–7). Cincinnati, OH: NIOSH, Division of Training and Manpower Development.

Smith, K.U. (1990). Hazard management: Principles, applications and evaluation. In *Proceedings of the Human Factors and Ergonomics Society 34th Annual Meeting* (pp. 1020–1024). Santa Monica, CA: Human Factors and Ergonomics Society.

Smith, K.U. & Arndt, R. (1970). Self-generated control mechanisms in posture. *American Journal of Physical Medicine, 49*(4), 241–252.

Smith, K.U. & Greene, P. (1963). A critical period in maturation of performance with space-displaced vision. *Perceptual and Motor Skills, 17*, 627–639.

Smith, K.U. & Kao, H. (1971). Social feedback: Determination of social learning. *Journal of Nervous and Mental Disease, 152*, 289–297.

Smith, K.U. & Molitor, K. (1969). Adaptation to combined reversal and delay of eyemovement-retinal feedback. *Journal of Motor Behavior, 1*(4), 296–306.

Smith, K.U. & Murphy, T.J. (1963). Sensory feedback mechanisms of handwriting motions and their neurogeometric bases. In M. Herrick (Ed.), *New horizons for research in handwriting* (pp. 111–157). Madison, WI: University of Wisconsin Press.

Smith, K.U. & Putz, V. (1970). Steering feedback delay: Effects on binocular coordination and hand-eye synchronization. *American Journal of Optometry, 47*, 234–238.

Smith, K.U. & Ramana, D. (1969). Characteristics of head motion with variable visual feedback delay, velocity and displacement. *American Journal of Physical Medicine, 48*, 289–300.

Smith, K.U. & Schappe, R. (1970). Feedback analysis of the movement mechanisms of handwriting. *Journal of Experimental Education, 38*(4), 61–68.

Smith, K.U. & Smith, M.F. (1966). *Cybernetic principles of learning and educational design.* New York: Holt, Rinehart and Winston.

Smith, K.U. & Smith M.F. (1973). *Psychology. An introduction to behavior science.* Boston: Little, Brown & Co.

Smith, K.U. & Smith, T.J. (1970). Feedback mechanisms of athletic skill and learning. In L. Smith (Ed.), *Psychology of motor learning* (pp. 83–195). Chicago: Athletic Institute.

Smith, K.U. & Smith, T.J. (1988). Analysis of the human factors in automation. In K.M. Blache (Ed.), *Success factors for implementing change—A manufacturing viewpoint* (pp. 259–338). Dearborn, MI: Society of Manufacturing Engineers.

Smith, K.U. & Smith, T.J. (1991). A study of handwriting and its implications for cognitive considerations in human-computer interactions. *International Journal of Human-Computer Interaction, 3*(1), 1–30.

Smith, K.U. & Smith, W.M. (1962). *Perception and motion: An analysis of space-structured behavior.* Philadelphia: Saunders.

Smith, K.U. & Sussman, H. (1969). Cybernetic theory and analysis of motor learning and memory. In E.A. Bilodeau & I.M. Bilodeau (Eds.), *Principles of skill acquisition* (pp. 103–139). New York: Academic Press.

Smith, K.U., Ansell, S., & Smith, W.M. (1963a). Feedback analysis in medical research. I. Delayed sensory feedback in behavior and neural function. *American Journal of Physical Medicine, 42*(6), 228–262.

Smith, K.U., Cambria, R., & Steffan, J. (1964). Sensory-feedback analysis of reading. *Journal of Applied Psychology, 48*(5), 275–286.

Smith, K.U., Kaplan, R., & Kao, H. (1970). Experimental systems analysis of simulated vehicle steering and safety training. *Journal of Applied Psychology, 54*, 364–376.

Smith, K.U., Myziewski, M., Mergen, J., & Koehler, J. (1963b). Computer systems control of delayed auditory feedback. *Perceptual and Motor Skills, 17*, 343–354.

Smith, K.U., Putz, V., & Molitor, K. (1969). Eyemovement-retina delayed feedback. *Science, 166*, 1542–1544.

Smith, K.U., Wargo, L., Jones, R., & Smith, W.M. (1963c). Delayed and space-displaced sensory feedback and learning. *Perceptual and Motor Skills, 16*, 781–796.

Smith, M.J. (1973). *Social tracking of heart rate* (unpublished PhD thesis). Madison, WI: University of Wisconsin-Madison.

Smith, M.J. (1994). Employee participation and preventing occupational diseases caused by new technologies. In G.E. Bradley & H.W. Hendrick (Eds.), *Human factors in organizational design and management–IV* (pp. 719–724). Amsterdam: North-Holland.

Smith, M.J. & Beringer, D.B. (1987). Human factors in occupational injury evaluation and control. In G. Salvendy (Ed.), *Handbook of human factors* (pp. 767–789). New York: Wiley.

Smith, M.J. & Carayon-Sainfort, P. (1989). A balance theory of job design for stress reduction. *International Journal for Industrial Ergonomics, 4*, 67–79.

Smith, M.J., Bauman, R.D., Kaplan, R.P., Cleveland, R., Derks, S., Sydow, M., & Coleman, P.J. (1971). *Inspection effectiveness.* Washington, DC: Occupational Safety and Health Administration.

Smith, M.J., Carayon, P., Smith, J., Cohen, W., & Upton, J. (1994). Community ergonomics: A theoretical model for rebuilding the inner city. In *Proceedings of the Human Factors and Ergonomics Society 38th Annual Meeting* (pp. 724–728). Santa Monica, CA: Human Factors and Ergonomics Society.

Smith, M.J., Cohen, H.H., Cohen, A., & Cleveland, R.J. (1978). Characteristics of successful safety programs. *Journal of Safety Research, 10*(1), 5–15.

Smith, R.L. (1988). Human visual requirements for control and monitoring of a space telerobot. In W. Karwowski, H.R. Parsaei, & M.R. Wilhelm (Eds.), *Ergonomics of hybrid automated systems I* (pp. 233–240). Amsterdam: Elsevier.

Smith, R.L. & Stuart, M.A. (1989). The effects of spatially displaced visual feedback on remote manipulator performance. In *Proceedings of the Human Factors Society 33rd Annual Meeting* (pp. 1430–1434). Santa Monica, CA: Human Factors Society.

Smith, T.J. (1980). *Systems principles of occupational science.* Presentation to International Congress on Applied Systems Research and Cybernetics (December 13). Acapulco, Mexico.

Smith, T.J. (1993). The scientific basis of human factors—A behavioral cybernetic perspective. *Proceedings of the Human Factors and Ergonomics Society 37th Annual Meeting* (pp. 534–538). Santa Monica, CA: Human Factors and Ergonomics Society.

Smith, T.J. (1994a). Core principles of human factors science. *Proceedings of the Human Factors and Ergonomics Society 38th Annual Meeting* (pp. 536–540). Santa Monica, CA: Human Factors and Ergonomics Society.

Smith, T.J. (1994b). Human factors and design factors: Two sides of the same coin? Symposium and panel summaries. *Proceedings of the Human Factors and Ergonomics Society 38th Annual Meeting* (pp. 520, 541–543). Santa Monica, CA: Human Factors and Ergonomics Society.

Smith, T.J. (1998). Context specificity in performance—The defining problem for human factors/ergonomics. *Proceedings of the Human Factors/Ergonomics Society 42nd Annual Meeting* (pp. 692–696). Santa Monica, CA: Human Factors and Ergonomics Society.

Smith, T.J. (1999). Synergism of ergonomics, safety, and quality—A behavioral cybernetic analysis. *International Journal of Occupational Safety and Ergonomics, 5*(2), 247–278.

Smith, T.J. (2001). Work. In W. Karwowski (Ed.), *International encyclopedia of ergonomics and human factors*. New York: Taylor & Francis.

Smith, T.J. (2002). Macroergonomics of hazard management. In H.W. Hendrick & B. Kleiner (Eds.), *Macroergonomics: Theory, methods, and applications* (pp. 199–221). Mahwah, NJ: Lawrence Erlbaum.

Smith, T.J. (2007). The ergonomics of learning: Educational design and learning performance. *Ergonomics, 50*(10), 1530–1546.

Smith, T.J. (2009). The nature of learning and the design of the learning environment— A behavioral cybernetic perspective (Invited comments for panel session entitled "Macroergonomics in education: On your mark, set, GO!"). In *Proceedings of the Human Factors and Ergonomics Society 53rd Annual Meeting* (pp. 1042–1046). Santa Monica, CA: Human Factors and Ergonomics Society.

Smith, T.J. (2011). Ergonomics of learning environments—Designs with strong, equivocal or poor returns on educational investment. In *Proceedings of the Human Factors and Ergonomics Society 55th Annual Meeting* (pp. 560–564). Santa Monica, CA: Human Factors and Ergonomics Society.

Smith, T.J. (2012). Integrating community ergonomics with educational ergonomics— Designing community systems to support classroom learning. *Work: A Journal of Prevention, Assessment & Rehabilitation, 41*, 3676–3684.

Smith, T.J. (2013). Designing learning environments to promote student learning—Ergonomics in all but name. *Work: A Journal of Prevention, Assessment & Rehabilitation, 44*, S39–S60.

Smith, T.J., Bridger, B., Fostervold, K.I., Jacobs, K., Lueder, R., & Straker, L. (2009). The future of ergonomics in education—Panel session. In *IEA 2009. Proceedings of the International Ergonomics Association 17th World Congress on Ergonomics*. London: Taylor & Francis.

Smith, T.J. & Henning, R.A. (2005). Cybernetics of augmented cognition as an alternative to information processing. In D.D. Schmorrow (Ed.), *Foundations of augmented cognition* (pp. 641–650). New York: CRC Press.

Smith, T.J. & Henning, R.A. (2006). Social cybernetics of augmented cognition—A control systems analysis. In D.D. Schmorrow, K.M. Stanney, & L.M. Reeves (Eds.), *Foundations of augmented cognition* (2nd edition) (pp. 45–54). Arlington, VA: Strategic Analysis, Inc. & Augmented Cognition International (ACI) Society.

Smith, T.J. & Larson, T.L. (1991). Integrating quality management and hazard management: A behavioral cybernetic perspective. *Proceedings of the Human Factors and Ergonomics Society 35th Annual Meeting* (pp. 903–907). Santa Monica, CA: Human Factors and Ergonomics Society.

Smith, T.J. & Mehri, O. (2006). Ergonomic performance standards and regulations—Their scientific and operational basis. In W. Karwowski (Ed.), *Handbook of standards and guidelines in ergonomics and human factors* (Chapter 3, pp. 79–108). Mahwah, NJ: Lawrence Erlbaum.

Smith, T.J. & Orfield, S.J. (2007). Occupancy quality predictors of office worker perceptions of job productivity. In *Proceedings of the Human Factors and Ergonomics Society 51st Annual Meeting* (pp. 539–543). Santa Monica, CA: Human Factors and Ergonomics Society.

Smith, T.J. & Smith, K.U. (1974). Systems theory of human evolution and its creative selective basis. In *Proceedings of the Wisconsin Academy of Sciences, Arts and Letters*. Green Bay, WI: WASAL.

Smith, T.J. & Smith, K.U. (1978). Cybernetic theory of natural selection and evolution. In *Proceedings of the 144th National Meeting of the American Association for the Advancement of Science* (pp. 130–131). Washington, DC: AAAS.

Smith, T.J. & Smith, K.U. (1985). Cybernetic factors in motor performance and development. In D. Goodman, R.B. Wilberg, & I.M. Franks (Eds.), *Differing perspectives in motor learning, memory, and control* (pp. 239–283). Amsterdam: Elsevier.

Smith, T.J. & Smith, K.U. (1987a). Feedback-control mechanisms of human behavior. In G. Salvendy (Ed.), *Handbook of human factors* (Chapter 2.9, pp. 251–293). New York: Wiley.

Smith, T.J. & Smith, K.U. (1987b). Motor feedback control of human cognition—Implications for the cognitive interface. In G. Salvendy, S.L. Sauter, & J.J. Hurrell, Jr. (Eds.), *Social, ergonomic and stress aspects of work with computers* (pp. 239–254). Amsterdam: Elsevier.

Smith, T.J. & Smith, K.U. (1988a). The cybernetic basis of human behavior and performance. In G. Williams and P. Williams (Eds.), *Continuing the conversation. A newsletter of ideas in cybernetics. Special issue on behavioral cybernetics* (Number 15, pp. 1–28, Winter). Gravel Switch, KY: HortIdeas.

Smith, T.J. & Smith, K.U. (1988b). The social cybernetics of human interaction with automated systems. In W. Karwowski, H.R. Parsaei, & M.R. Wilhelm (Eds.), *Ergonomics of hybrid automated systems I* (pp. 691–711). Amsterdam: Elsevier.

Smith, T.J. & Smith, K.U. (1990). The human factors of workstation telepresence. In S. Griffin (Ed.), *Third annual workshop on space operations automation and robotics* (SOAR'89) (pp. 235–250). Houston Johnson Space Center, TX: National Aeronautics and Space Administration.

Smith, T.J. & Stuart, M.A. (1990). Human factors of teleoperation in space. In *Proceedings of the Human Factors Society 34th Annual Meeting* (pp. 116–120). Santa Monica, CA: Human Factors Society.

Smith, T.J., Henning, R.A., & Li, Q. (1998). *Teleoperation in space—Modeling effects of displaced feedback and microgravity on tracking performance* (SAE Technical Report Number 981701). Warrendale, PA: Society of Automotive Engineers International.

Smith, T.J., Henning, R.H., & Smith, K.U. (1994a). Sources of performance variability. In G. Salvendy & W. Karwowski (Eds.), *Design of work and development of personnel in advanced manufacturing* (Chapter 11, pp. 273–330). New York: Wiley.

Smith, T.J., Henning, R.A., & Smith, K.U. (1995). Performance of hybrid automated systems— A social cybernetic analysis. *International Journal of Human Factors in Manufacturing, 5*(1), 29–51.

Smith, T.J., Keran, C.M., Mathison, P.K., & Koehler, E.J. (1996). Control of tracking behavior under displaced visual feedback: Implications for teleoperation. In N.A. Duffie, R.R. Teeter, & A.J. Butts (Eds.). *Advanced Developments in Space Robotics. Proceedings of the AIAA Robotics Technology Forum* (pp. 4-1–4-10). Madison, WI: AIAA Space Automation and Robotics Technical Committee and University of Wisconsin-Madison.

Smith, T.J., Koehler, E.J., Keran, C.M., & Mathison, P.K. (1994b). *The behavioral cybernetics of telescience—An analysis of performance impairment during remote work* (SAE Technical Paper Number 94138). Warrendale, PA: Society of Automotive Engineers International.

Smith, T.J., Lockhart, R.W., & Smith, K.U. (1983). Safety cybernetics: Theory and practice of involving workers in hazard management programs. In Xth World Congress on the Prevention of Occupational Accidents and Diseases Specialist Day Papers: *Analysis of the risk of accidents at work, methods and applications* (pp. 43–60). Ottawa-Hull, Canada: International Social Security Association for Research and Prevention of Occupational Risks.

Smith, T.J., Orfield, S.J., & Brand, J. (2004). *Occupancy quality—A new agenda for sustainable design*. Presentation to EnvironDesign 8 Conference (April 22, Workshop 3A). Minneapolis, MN: Orfield Laboratories.

Smith, T.J., Smith, R.L., Stuart, M.A., Smith, S.T., & Smith, K.U. (1989). Interactive performance in space—The role of perturbed sensory feedback. In M.J. Smith & G. Salvendy (Eds.)., *Work with computers: Organizational, management, stress and health aspects* (pp. 484–495). Amsterdam: Elsevier.

Smith, T.J., Stuart, M.A., Smith, R.L., & Smith, K.U. (1990). Interactive performance variability in telerobot operation: Nature and causes. In W. Karwowski & M. Rahimi (Eds.), *Ergonomics of hybrid automated systems II* (pp. 857–870). Amsterdam: Elsevier.

Smith, W.M. & Bowen, K.F. (1980). The effects of delayed and displaced visual feedback on motor control. *Journal of Motor Behavior, 12*(2), 91–101.

Snoddy, G.S. (1926). Learning and stability. A psychophysiological analysis of a case of motor learning with clinical applications. *Journal of Applied Psychology, 10*(1), 1–36.

Snow, R.E., Corno, L., & Jackson, D. III (1996). Individual differences in affective and cognative functions. In D.C. Berliner and R.C. Calfee (Eds.), *Handbook of educational psychology* (pp. 243–310). New York: Simon & Schuster Macmillan.

Snyder, F.W. & Pronko, N.H. (1952). *Vision with spatial inversion*. Wichita, KS: University of Wichita Press.

Solomon, D. (2010). The school of hard drives. The secretary of education talks about the need for computers in the classroom. In *The New York Times Magazine* (September 19), p. 26.

Sorkin, A.R., Anderson, J., de la Merced, M., & White, B. (2008). Bank of America in talks to acquire Merrill Lynch. In *New York Times* (September 14), dealbook.nytimes.com/2008/09/14/bank-of-america-in-talks-to-buy-merrill-lynch/ (retrieved December 2, 2013).

South Metro School Briefs (2012). Rotarians buy iPads for schools. In *Minneapolis Star Tribune* (September 19), p. AA2.

Spector, T. (2012). *Identically different: Why you can change your genes*. London: Weidenfeld & Nicolson.

Spong, M. & Hokayem, P. (2006). Bilateral teleoperation: An historical survey. *Automatica, 42*(12), 2035–2057.

St. Anthony, N. (2013). Low worker engagement is pulling everyone down. In *Minneapolis Star Tribune* (June 17), p. D1.

Stafford, D. (2013). Study: CEO pay doesn't guarantee performance. In *St. Paul Pioneer Press* (August 29), p. 12A.

Standard and Poor's (2005). *SchoolMatters*. New York: McGraw-Hill. Downloaded from: http://www.SchoolMatters.com.

Steele, C.M. (2010). *Whistling Vivaldi, and other clues to how stereotypes affect us*. New York: Norton.

Stergiou, N. & Decker, L.M. (2011). Human movement variability, nonlinear dynamics, and pathology: Is there a connection? *Human Movement Science, 30*(5), 869–888.

Stergiou, N., Harbourne, R.T., & Cavanaugh, P.T. (2006). Optimal movement variability: A new theoretical perspective for neurologic physical therapy. *Journal of Neurologic Physical Therapy, 30*(3), 120–129.

Stergiou, N., Yawen, Y., & Kyvelidou, A. (2013). A perspective on human variability with application to infant motor development. *Kinesiology Review, 2*, 93–102.

Stetson, R.H. (1951). *Motor phonetics: A study of speech movements in action*. Amsterdam: North Holland.

Stone, N.J. (2008). Human factors and education: Evolution and contributions. *Human Factors, 50*, 534–539.

Straker, L., Jones, K.J., & Miller, J. (1997). A comparison of the postures assumed when using laptop computers and desktop computers. *Applied Ergonomics, 28*(4), 263–268.

Straker, L., Maslen, B., Burgess-Limerick, R., Johnson, P., & Dennerlein, J. (2010). Evidence-based guidelines for the wise use of computers by children: Physical development guidelines. *Ergonomics, 53*, 458–477.

Straker, L., Pollock, C., & Maslen, B. (2009). Principles for the wise use of computers by children. *Ergonomics, 52*, 1386–1401.

Stratton, G.M. (1896). Some preliminary experiments in vision without inversion of the retinal image. *Psychological Review, 3*, 611–617.

Stratton, G.M. (1897). Vision without inversion of the retinal image. *Psychological Review, 4*, 341–360, 463–481.

Stratton, G.M. (1899). The spatial harmony of touch and sight. *Mind, 8*, 463–505.

Strauch, B. (2004). *Investigating human error: Incidents, accidents, and complex systems.* Burlington, VT: Ashgate.

Strickler, J. (2013). Don't be the office dinosaur. In *Minneapolis Star Tribune* (October 27), pp. E1, E6.

Struebing, L. (1996). 9000 standards. *Quality Progress* (January), 23–28.

Stuart, M.A. & Smith, R.L. (1989). Spatially displaced visual feedback and performance of Fitts'-like tasks (NASA Technical Report JSC-23547). Houston, TX: National Aeronautics and Space Administration.

Stuart, M.A., Manahan, M.K., Bierschwale, J.M., Sampaio, C.E., & Legendre, A.J. (1991). Adaptive strategies of remote systems operators exposed to perturbed camera-viewing conditions. In *Proceedings of the Human Factors Society 35th Annual Meeting* (pp. 151–155). Santa Monica, CA: Human Factors Society.

Subramaniam, M. & Youndt, M.A. (2005). The influence of intellectual capital on the types of innovative capabilities. *Academy of Management Journal, 48*, 450–463.

Sussman, H. & Smith, K.U. (1969). Analysis of memory as a feedforward control mechanism. *Journal of Motor Behavior, 1*(2), 101–117.

Sussman, H. & Smith, K.U. (1971). Jaw movements under delayed auditory feedback. *Journal of the Acoustical Society of America, 49*(6), 1874–1880.

Swain, A.D. (1973). An error-cause removal program for industry. *Human Factors, 15*(3), 207–221.

Swain, A.D. (1974). *The human element in systems safety: A guide for modern management.* London: Industrial and Commercial Techniques.

Swanson, C.B. (2009). *Closing the education gap. Educational and economic conditions in America's largest cities.* Bethesda, MD: Editorial Projects in Education, Inc. Downloaded from: http://www.edweek.org/media/cities_in_crisis_2009.pdf.

Swingle, C.A. (1997). *The relationship between the health of school age children and learning: Implications for schools.* Lansing, MI: Michigan Department of Community Health.

Taylor, F.V. & Birmingham, H.F. (1959). That confounded system performance measure—A demonstration. *Psychology Review, 66*, 178–182.

Taylor, F.W. (1911). *The principles of scientific management.* New York: Harper and Brothers.

Tessier, C., Zhang, L., & Cao, C.G.L. (2012). Ergonomic considerations in natural orifice translumenal endoscopic surgery (Notes): A case study. *Work, 41*, 4683–4688.

Thaler, R.H. & Sunstein, C.R. (2009). *Nudge: Improving decisions about health, wealth, and happiness.* New York: Penguin.

Thomas, D. (2010). Gene-environment-wide association studies: Emerging approaches. *Nature Reviews-Genetics, 11*, 259–272.

Thomsen, L., Frankenhuis, W.E., Ingold-Smith, M., & Carey, S. (2011). Big and mighty: Preverbal infants mentally represent social dominance. *Science, 331*, 477–480.

Thorndike, E.L. & Woodworth, R.S. (1901a). The influence of improvement in one mental function upon the efficiency of other functions. I. *The Psychological Review, 8*(3), 247–261.

Thorndike, E.L. & Woodworth, R.S. (1901b). The influence of improvement in one mental function upon the efficiency of other functions. II. The estimation of magnitudes. *The Psychological Review, 8*(4), 384–395.

Thorndike, E.L. & Woodworth, R.S. (1901c). The influence of improvement in one mental function upon the efficiency of other functions. III. Functions involving attention, observation and discrimination. *The Psychological Review, 8*(6), 553–564.

Ting, T., Smith, M., & Smith, K.U. (1972). Social feedback factors in rehabilitative processes and learning. *American Journal of Physical Medicine, 51*, 86–101.

Toet, A., Jansen, S.E.M., & Delleman, N.J. (2008). Effects of field-of-view restriction on manoeuvring in a 3-D environment. *Ergonomics, 51*(3), 385–394.

Tononi, G., Edelman, G.M., & Sporns, O. (1998). Complexity and coherency: Integrating information in the brain. *Trends in Cognitive Science, 2*, 474–484.

Topf, M. & Preston, R. (1991). Behavior modification can heighten safety awareness, curtail accidents. *Occupational Health & Safety, 60*(2), 43–49.

Tough, P. (2011). The poverty clinic. In *The New Yorker* (March 21), pp. 25–32.

Tough, P. (2012). *How children succeed. Grit, curiosity, and the hidden power of character.* Boston: Houghton Mifflin.

Trist, E.L. (1981). The sociotechnical perspective—The evolution of sociotechnical systems as a conceptual framework and as an action research program. In A.H. Van de Ven & W.F. Joyce (Eds.), *Perspectives on organization design and behavior* (Chapter 2). New York: Wiley; 1981.

Turner, B.A. (1978). *Man-made disasters.* London: Wykeham Publishing.

Turner, G.M. (2008). A comparison of *The Limits to Growth* with 30 years of reality. *Global Environmental Change, 18*, 397–411.

Turvey, M.T., Shaw, R.E., & Mace, W. (1978). Issues in the theory of action: Degrees of freedom, coordinative structures and coalitions. In J. Requin (Ed.), *Attention and performance VII* (pp. 557–595). Hillsdale, NJ: Lawrence Erlbaum.

Tuttle, C.C., Gill, B., Gleason, P., Knechtel, V., Nichols-Barrer, I., & Resch, A. (2013). *KIPP middle schools: Impacts on achievement and other outcomes. Final report* (February 27). Washington, DC: Mathematica Policy Research. Downloaded from: http://www.mathematica-mpr.com/publications/SearchList2.aspx?jumpsrch=yes&txtSearch=KIPP.

U.S. Department of Education (1983). *A nation at risk.* Washington, DC: U.S. Government Printing Office.

United Nations Population Information Network (2013), http://www.un.org/popin/ (retrieved December 8, 2013).

United States Department of Labor (1989). *The role of labor-management committees in safeguarding worker safety and health.* Washington, DC: U.S. Department of Labor, Bureau of Labor-Management Relations and Cooperative Programs.

Vaillancourt, D.E. & Newell, K.M. (2002). Changing complexity in human behavior and physiology through aging and disease. *Neurobiology of Aging, 23*(1), 1–11.

Van Cott, H.P. (1984). From control systems to knowledge systems. *Human Factors, 26*(1), 115–122.

van der Waal, E., Borgeaud, C., & Whiten, A. (2013). Potent social learning and conformity shape a wild primate's foraging decisions. *Science, 340*(6131), 483–485.

van Emmerik, R.E.A. (2007). The functional role of variability in movement coordination and disability. In W.E. Davis, & G.D. Broadhead (Eds.), *Ecological task analysis and movement.* Champaign, IL: Human Kinetics.

van Emmerik, R.E.A. & van Wegen, E.E.H. (2000). On variability and stability in human movement. *Journal of Applied Biomechanics, 16*, 394–406.

van Emmerik, R.E.A., Rosenstein, M.T., McDermott, W.J., & Hamill, J. (2004). A nonlinear dynamics approach to human movement. *Journal of Applied Biomechanics, 20*, 396–420.

Van Rossum, J.H.A. (1990). Schmidt's schema theory: The empirical base of the variability of practice hypothesis. A critical analysis. *Human Movement Science, 9*, 387–435.

VanDoren, V.J. (1996). The Smith predictor: A process engineer's crystal ball. *Control Engineering, 43*(5), 1–4. Downloaded from: http://www.controleng.com/single-article/the-smith-predictor-a-process-engineer-s-crystal-ball/1e02746896161e157f0a7175f67d37db.html.

Vercher, J.-L., Volle, M., & Gauthier, G.M. (1993). Dynamic analysis of human visuo-oculo-manual coordination control in target tracking tasks. *Aviation, Space, and Environmental Medicine, 64*, 500–506.

Vertut, J. & Coiffet, P. (1985). *Teleoperation and robotics. Evolution and development.* London: Kogan Page.

Vincente, K.J. & Rasmussen, J. (1992). Ecological interface design: Theoretical foundations. *IEEE Transactions on Systems, Man, and Cybernetics, 22*, 589–605.

Vischer, J.C. (1989). *Environmental quality in offices.* New York: Van Nostrand Reinhold.

Waber, D.P., De Moor, C., Forbes, P.W., Almli, C.L., Botteron, K.N., Leonard, G., Milovan, D., Paus, T., Rumsey, J., & The Brain Development Cooperative Group (2007). The NIH MRI study of normal brain development: Performance of a population based sample of healthy children aged 6 to 18 years on a neuropsychological battery. *Journal of the International Neuropsychological Society, 13*, 1–18.

Wahlstrom, K. (2002). Changing times: Findings from the first longitudinal study of later high school start times. *NASSP Bulletin, 86*, 3–21. Downloaded from: http://www.cehd.umn.edu/carei/reports/docs/SST-2002Bulletin.pdf.

Wainer, H. (2007). The most dangerous equation. *American Scientist, 95*, 249–256.

Wallace, B. & Ross, A. (2006). *Beyond human error. Taxonomies and safety science.* Boca Raton, FL: CRC Press.

Warrick, M.J. (1949). Effect of transmission-type control lags on tracking accuracy. *U.S. Air Force Air Material Command Technical Report Number 5916.* Washington, DC: U.S. Air Force Air Materials Command.

Watson, P. (2005). *Ideas: A history of thought and invention, from fire to Freud.* New York: HarperCollins.

Welch, R.B. (1978). *Perceptual modification. Adapting to altered sensory environments.* New York: Academic Press.

Werner, E. (2010). A second call made for more class time. In *Minneapolis Star Tribune* (September 28), p. A4.

West, B.J. (2006). *Where medicine went wrong: Rediscovering the path to complexity.* Hackensack, NJ: World Scientific.

West, G. (2013). Wisdom in numbers. *Scientific American, 308*(5), 14.

West, G. & Bettencourt, L. (2013). *Cities, scaling and sustainability.* Santa Fe, NM: Santa Fe Institute.

Westgaard, R.H. & Winkel, J. (1997). Review article. Ergonomic intervention research for improved musculoskeletal health: A critical review. *International Journal of Industrial Ergonomics, 20*, 463–500.

Whittingham, R.B. (2004). *The blame machine: Why human error causes accidents.* Oxford, UK: Elsevier.

Wickens, C. (2002). Multiple resources and performance prediction. *Theoretical Issues in Ergonomics Science, 3*(2), 159–177.

Wiegmann, D.A. (2010). Analysis of error management strategies during cardiac surgery: Theoretical and practical implications. In *Proceedings of the Human Factors Society 54th Annual Meeting* (pp. 872–875). Santa Monica, CA: Human Factors and Ergonomics Society.

Wiener, E.L. (1993). Life in the second decade of the glass cockpit. In R.S. Jensen and D. Neumeister (Eds.). *Proceedings of the Seventh International Symposium on Aviation Psychology.* Columbus, OH: Ohio State University Department of Aviation.

Wiener, N. (1948). *Cybernetics: Or control and communication in the animal and the machine.* Cambridge, MA: MIT Press.

Wilde, O. (1890). *The picture of Dorian Gray.* London: Lippincott's.

Wilson, K.G. & Daviss, B. (1995). *Redesigning education.* New York: Holt.

Wilson, W. (2010). The learning machines. Classroom technology, from the writing slate to the electronic tablet. In *The New York Times Magazine*, pp. 52–53.

Wilson, W. & Kuhn, C. (2010). Build a better brain—With sleep, protein, exercise … and belief. In *St. Paul Pioneer Press* (September 2), p. 9B.

Wingert, P. (2010). Blackboard jungle. Freshly minted teachers have passed every test but one: How to control their classrooms. In *Newsweek* (March 15), p. 33.

Wolf, W. (1954). *Die Welt die Aegypter.* Stuttgart: Gustav Kilpper Verlag.

Woodrow, H. (1938). The effects of practice on groups of different initial ability. *Educational Psychology, 29*, 268–278.

Woodrow, H. (1939). Factors in improvement with practice. *Journal of Psychology, 7*, 55–70.

Woodrow, H. (1946). The ability to learn. *Psychological Review, 53*, 147–158.

Woods, D.D., Dekker, S., Cook, R., Johannesen, L., & Sarter, N. (2010). *Behind human error* (2nd edition). Burlington, VT: Ashgate.

Woodworth, R.S. (1938). *Experimental psychology.* New York: Holt.

Woolley, L. (1946). *Ur: The first phases.* London: Penguin.

Wuethrich, B. (2000). Learning the world's languages—Before they vanish. *Science, 288*(5469), 1156–1159.

Yates, A. (1963). Delayed auditory feedback. *Psychological Bulletin, 60*, 213–251.

Young, S.B. (2012). "Nanny state." In *Minneapolis Star Tribune* (July 29), pp. OP1, OP4.

Young, S.B. (2013). The middle class is flying apart. In *Minneapolis Star Tribune* (November 10), pp. OP1, OP3.

Youngblood, D. (1991). In its quest for quality, Tennant shows talent. In *Minneapolis Star Tribune* (March 27), pp. D1–D2.

Zaidel, D.M. (1991). *Specification of a methodology for investigating the human factors of advanced driver information systems.* Ottawa: Transport Canada Publication No. TP 11199.

Index

Page numbers followed by f and t indicate figures and tables, respectively.

Printed and bound by CPI Group (UK) Ltd, Croydon, CR0 4YY

18/10/2024

01776269-0007